Advances in
Nutritional
Research
Volume 1

Advances in
Nutritional Research

A Continuation Order Plan is available for this series. A continuation order will bring delivery of each new volume immediately upon publication. Volumes are billed only upon actual shipment. For further information please contact the publisher.

Advances in
Nutritional
Research
Volume 1

Edited by Harold H. Draper
University of Guelph
Guelph, Ontario, Canada

Plenum Press · New York and London

ISBN 0-306-34321-5

© 1977 Plenum Press, New York
A Division of Plenum Publishing Corporation
227 West 17th Street, New York, N.Y. 10011

Printed in the United States of America

Contributors

Gerald Harvey Anderson, Department of Nutrition and Food Science, University of Toronto, Toronto, Ontario, Canada

David H. Baker, Department of Animal Science, University of Illinois, Urbana, Illinois

Charles H. Barrows, Laboratory of Cellular and Comparative Physiology, Gerontology Research Center, National Institute on Aging, National Institutes of Health, Department of Health, Education and Welfare, Bethesda, Maryland, and Baltimore City Hospital, Baltimore, Maryland

William R. Beisel, Scientific Advisor, U.S. Army Medical Research Institute of Infectious Diseases, Fort Detrick, Frederick, Maryland

Neville Colman, State University of New York—Downstate Medical Center, and Veterans Administration Hospital, Hematology and Nutrition Laboratory, Bronx, New York

Gary W. Evans, Agricultural Research Service, Human Nutrition Laboratory, U.S. Department of Agriculture, Grand Forks, North Dakota

Cihad T. Gürson, Department of Pediatrics, University of Istanbul, Istanbul Medical Faculty, Çapa, Istanbul, Turkey

Leslie M. Klevay, Agricultural Research Service, Human Nutrition Laboratory, U.S. Department of Agriculture, Grand Forks, North Dakota

Gertrude C. Kokkonen, Laboratory of Cellular and Comparative Physiology, Gerontology Research Center, National Institute on Aging, National Institutes of Health, Department of Health, Education and Welfare, Bethesda, Maryland, and Baltimore City Hospital, Baltimore, Maryland

Mitchell Rubinoff, Division of Medical Oncology, Cancer Chemotherapy Foundation Laboratory, Department of Medicine, The Mount Sinai School of Medicine of The City University of New York, New York, New York

Carol Schreiber, Division of Medical Oncology, Cancer Chemotherapy Foundation Laboratory, Department of Medicine, The Mount Sinai School of Medicine of The City University of New York, New York, New York

J. W. Suttie, Department of Biochemistry, College of Agricultural and Life Sciences, University of Wisconsin–Madison, Madison, Wisconsin

Samuel Waxman, Division of Medical Oncology, Cancer Chemotherapy Foundation Laboratory, Department of Medicine, The Mount Sinai School of Medicine of The City University of New York, New York, New York

Lloyd A. Witting, Supelco, Inc., Supelco Park, Bellefonte, Pennsylvania

Foreword

Advances in Nutritional Research was conceived by the Public and Professional Information Committee of the American Institute of Nutrition as a service to the membership of the Institute and to others engaged in research and teaching in the nutritional sciences.

This publication, which will consist of a series of annual volumes, is intended to serve primarily as a source of authoritative information on the status of research on topics of active current investigation. In addition, it will contain reviews on subjects of research that have been characterized by progressive, if unspectacular, developments over a period of years. The coverage will extend to basic research in human, animal, and microbial nutrition, nutritional biochemistry, clinical nutrition, nutritional toxicology and pathology, and related specialties.

The editors hope that this series will assist investigators to keep abreast of new developments in the broad field of nutritional research. It should be of particular value to graduate students and to investigators who wish to acquire a grasp of the current state of knowledge in fields other than their own. If the response of authors of the chapters in this first volume and the quality of their contributions are indicative, this hope will be realized.

H. H. Draper
Chairman, Board of Editors

Contents

Chapter 5   Metabolic and Nutritional Consequences of Infection ... 125

William R. Beisel

Chapter 1

Role of Vitamin K in the Synthesis of Clotting Factors

J. W. Suttie

1. Historical Background

The discovery of vitamin K in the mid-1930s grew out of experiments by Henrik Dam on the possible essentiality of cholesterol in the diet of the chick. Dam was investigating reports that chicks did not thrive on diets that had been extracted with nonpolar solvents to remove the sterols; he noted that in such animals subdural or muscular hemorrhages developed and that blood taken from these animals clotted slowly. It was shown in these initial experiments that the addition of cholesterol, lemon juice, yeast, or cod liver oil to the diet did not decrease the incidence of this lesion. The disease was subsequently observed by others, and a period followed during which attempts were made by various groups to identify the active component of the diet. Dam continued to study the distribution and lipid solubility of the active component in vegetable and animal sources, and proposed that the antihemorrhagic vitamin of the chick was a new fat-soluble vitamin, which he called vitamin K (Dam, 1935). K was the first letter of the German word *koagulation* and was also the first letter of the alphabet which had not been used to describe an existing or postulated vitamin activity at that time. Dam's report of the discovery of a new vitamin was followed by an independent report by Almquist and Stokstad (1935), which described their success in curing the disease with ether extracts of alfalfa and clearly pointed out that microbial action in fish meal and bran preparations could lead to the development of antihemorrhagic activity.

J.W. Suttie • Department of Biochemistry, College of Agricultural and Life Sciences, University of Wisconsin–Madison, Madison, Wisconsin 53706.

Indirect evidence indicated that the clotting defect might be related to a lowered concentration of plasma prothrombin. Dam *et al.* (1936) succeeded in preparing a crude plasma prothrombin fraction and in demonstrating that its activity was decreased when obtained from vitamin K-deficient chick plasma. Early workers did not recognize that the synthesis of proteins other than prothrombin might have been influenced, and it was widely believed that the defect in the plasma of animals fed vitamin K-deficient diets was solely attributable to a lack of prothrombin. A real understanding of the factors involved in the generation of thrombin from prothrombin did not begin until the mid-1950s; during the next 10 years, Factors VII, X, and IX were discovered and subsequently shown to be dependent on vitamin K for their synthesis.

Following Dam's clear demonstration of the need for a dietary antihemorrhagic factor, attempts were made to isolate and characterize this new vitamin. Dam *et al.* (1939) succeeded in isolating the vitamin as a yellow oil from alfalfa, and, although they described some of its properties, they did not recognize it as a quinone derivative. It was soon recognized that the active preparations were quinones, and Almquist and Klose (1939) demonstrated that phthiocol (2-methyl-3-hydroxy-1,4-naphthoquinone), which had previously been isolated from *Mycobacterium tuberculosis,* had biological activity. Vitamin K_1 was then isolated from alfalfa, characterized as 2-methyl-3-phytyl-1,4-naphthoquinone, and synthesized by Doisey's group (MacCorquodale *et al.,* 1939). Their identification was independently confirmed by other groups, and subsequent studies demonstrated that the compound obtained from putrified fish meal and called vitamin K_2 contained an unsaturated polyprenyl side chain. The elucidation of the structure of vitamin K was an extremely competitive research area involving a number of large groups; it is therefore difficult at this time to assign priority and credit with fairness and accuracy. A more detailed discussion of the historical aspects of the field can be found in earlier reviews (Almquist, 1941; Dam, 1942, 1948; and Isler and Wiss, 1959) and in a recent review of these historical developments (Almquist, 1975).

2. Chemical and Nutritional Aspects

2.1. Chemistry and Nomenclature

Nomenclature of those compounds that possess vitamin K activity has been subject to a number of modifications since the discovery of the vitamin. The nomenclature in general use at the present time is based on the tentative rules adopted by a IUPAC–IUB subcommittee on nomenclature of quinones (1966). The term "vitamin K" is used as a generic descriptor for 2-methyl-1,4-naphthoquinone (I) and for all derivatives of this compound that exhibit an antihemorrhagic activity in animals fed a vitamin K-deficient diet. The compound

2-methyl-3-phytyl-1,4-naphthoquinone (II) is generally called vitamin K_1 or phyl-loquinone. The compound first isolated from putrified fish meal and called vitamin K_2 is one of a series of vitamin K's with unsaturated side chains called multiprenylmenaquinones, which are found in animal tissues and bacteria. This particular form of the vitamin had 7 isoprenoid units or 35 carbons in the side chain and has been called vitamin $K_{2(35)}$ or menaquinone-7 (III). Vitamins of the menaquinone series with up to 13 prenyl groups have been identified, as well as several partially saturated members of this series. The parent com-pound of this series, 2-methyl-1,4-naphthoquinone has often been called vitamin K_3, but it is more commonly designated as menadione. The nomenclature for these vitamins is summarized in Fig. 1. Nomenclature in this field has recently been complicated by an attempt of the International Union of Nutritional Sciences to alter the IUPAC nomenclautre. This revision suggests that phylloquinone be called phytylmenaquinone (PMQ), and that a vitamin of the K_2 series, such as $K_{2(20)}$, be called prenylmenaquinone-4 (MQ-4) rather than menaquinone-4 (MK-4). These differences have not been resolved, and at the present time most workers in the field see little advantage in a change from the IUPAC nomen-clature.

2.2. Biological Activity of Various Forms of Vitamin K

Following the discovery of vitamin K in the early 1930s, a large number of related compounds were synthesized and their biological activity compared. Data

Comparison of Vitamin K Nomenclature

Chemical name	Old	IUPAC (abbr.)	AIN (abbr.)
2-Methyl-1,4-naphthoquinone (I)	K_3	Menadione	Menaquinone
2-Methyl-3-phytyl-1, 4-naphthoquinone (II)	K_1	Phylloquinone (K)	Phytylmenaquinone (PMQ)
2-Methyl-3-multiprenyl-1, 4-naphthoquinone (class)	$K_{2(n)}$	Menaquinone-n (MK-n)	Prenylmenaquinone-n (MQ)
2-Methyl-3-farnesylgeranylgeranyl- 1,4- naphthoquinone (III)	$K_{2(35)}$	Menaquinone-7 (MK-7)	Prenylmenaquinone-7 (MQ-7)

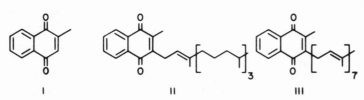

Fig. 1. Structures and nomenclature of compounds with vitamin K activity.

from different studies are somewhat difficult to compare because of variations in assay methods, but a number of generalities can be made. Although there were early suggestions that menadione itself might be functioning as a vitamin, it is usually assumed that it is alkylated to a biologically active menaquinone either by intestinal organisms or by tissue alkylating enzymes. The 2-methyl group is usually considered essential for activity, and alterations at this position, such as the 2-ethyl derivative, result in inactive compounds. This is not because of the inability of this compound to be aklylated, as 2-ethyl-3-phytyl-1,4-naphthoquinone is also inactive. The hydroquinone form of the substituted 1,4-naphthoquinones will spontaneously oxidize back to the quinone, but it can be stabilized by esterification, and compounds such as the diphosphate, the disulfate, and the diacetate have been prepared and shown to be active forms. Studies with substituted 2-methyl-1,4-naphthoquinones have demonstrated that polyisoprenoid side chains are the most effective substituents at the 3-position, and those compounds that contain more isoprenoid units are more active. The biological activity of phylloquinone (2-methyl-3-phytyl-1,4-naphthoquinone) is reduced by saturation of the double bond, but this compound is more active than 2-methyl-3-octadecyl-1,4-naphthoquinone, which has an unbranched alkyl side chain of similar size. Natural phylloquinone is the *trans* isomer, and the *cis* isomer is essentially inactive. Comparisons of the biological activity of many of the numerous forms of the vitamin that have been synthesized are available (Griminger, 1966; Weber and Wiss, 1971).

2.3. Nutritional Requirements

The establishment of the dietary level of vitamin K required for various species has been difficult. Although all species of animals need vitamin K to synthesize the essential clotting factors, the varying degrees to which various species can utilize the large amount of vitamin K synthesized by intestinal bacteria, and the degree to which various species practice coprophagy influence the development of a deficiency. A spontaneous deficiency of vitamin K was first noted in chicks, and symptoms of a dietary deficiency are much more likely to develop in poultry than in any other species. There appears to be little evidence that ruminants, because of vitamin K production by the rumen microflora, need a vitamin source in the diet. However, it has been possible to produce a deficiency in most monogastric species.

Values from different studies are difficult to compare because different forms of the vitamin have been used, and both curative and preventive assays have been used to establish the requirement. Although phylloquinone has generally been used as the vitamin source in experimental nutrition studies, other forms have often been used for consideration in more practical diets. The activity of menadione depends a great deal on the type of assay used. It is rather ineffective in a curative assay where the rate of alkylation is probably the important factor, but it often shows activity nearly equal to phylloquinone in a long-term preventive assay. Practical nutritionists have often preferred to add a water-soluble form of menadione, such as the menadione sodium bisulfite complex (MSBC), to rations; this source is widely used in poultry rations. Detailed discussions of the vitamin K requirements of various species are available in articles by Scott (1966), Doisy and Matschiner (1970), and Griminger (1971). The data indicate that the daily requirement for most species falls in a range of 2–200 μg vitamin K/kg body weight. The data in Table I, which have been adapted from a table presented by Griminger (1971), indicate the magnitude of the requirement for various species. It should be remembered that this requirement can be altered by variations in age, sex, or strain, and that any condition that influences lipid absorption or conditions altering intestinal flora will have a concomitant influence on these values.

The requirement of the adult human for vitamin K is extremely low, and there is no possibility of a simple dietary deficiency developing in the absence of complicating factors. Frick et al. (1967) studied the vitamin K requirement of starved, debilitated, intravenously fed patients given antibiotics to decrease intestinal vitamin K synthesis. Their data would indicate that the requirement is on the order of 1 μg kg^{-1} day^{-1}. Doisy (1971) fed a chemically defined diet containing less than 10 μg vitamin K/day to two normal subjects and was able to deplete prothrombin concentrations to less than 50% by about 20 weeks. Mineral oil and antibiotics were administered during a portion of this period to decrease vitamin absorption and synthesis. These patients responded to the administration of about 0.5 μg vitamin K_1/kg by rapidly restoring clotting activity to normal. It was concluded from this study that about 1 μg vitamin K kg^{-1} day^{-1} was sufficient to maintain normal clotting factor synthesis in the normal adult human. O'Reilly (1971) fed four normal volunteers a diet containing about 25 μg vitamin K/day, and his data would suggest that prothrombin concentrations can be maintained near the low end of the normal range on a diet containing about 0.5 μg vitamin K kg^{-1} day^{-1}. The limited studies available would therefore suggest that the

Table I. Vitamin K Requirement of Various Species[a]

Species	Body wt. basis (μg kg^{-1} day^{-1})	Dietary basis (μg/kg diet)
Dog	1.25	60
Pig	5	50
Rhesus monkey	2	60
Rat, male	11–16	100–150
Chicken	80–120	530
Turkey poult	180–270	1200

[a] Data taken from a more extensive table (Griminger, 1971). Requirement based on the amount of vitamin needed to prevent the development of a deficiency, with no correction for any difference in potency of different forms of the vitamin on a weight basis.

vitamin K requirement of the human is in the range of 0.5–1.0 μg kg^{-1} day^{-1}. This amount of vitamin is exceeded in almost any diet that is in other respects nutritionally adequate, and a simple dietary deficiency appears to be of little concern.

2.4. Antagonists of Vitamin K Action

Studies of the mechanism of action of vitamin K have been aided by the availability of a number of compounds that antagonize the action of the vitamin. These can be divided into two general classes—the "indirect" antagonists (coumarin or indandione derivatives) and "direct" antagonists (e.g., the 2-halo analogues of the vitamin), which appear to act as true competitive inhibitors of the action of the vitamin. The indirect anticoagulants have been widely used, both in clinical medicine and as rodenticides. The relative potency of some of the many antagonists synthesized has been reviewed (Green, 1966; Weber and Wiss, 1971), some representative compounds of which are shown in Fig. 2.

2.4.1. Coumarin Derivatives

The history of the discovery of the anticoagulant action of coumarin derivatives was documented by Link (1959). A hemorrhagic disease of cattle was described in Canada and in the Midwest of the United States in the 1920s and was

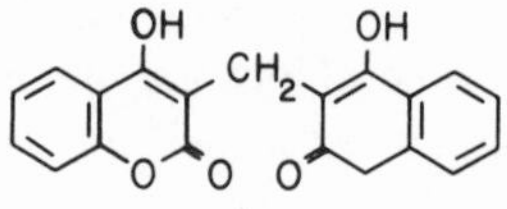

Dicoumarol

[3,3- methylene-bis-(4-hydroxycoumarin)]

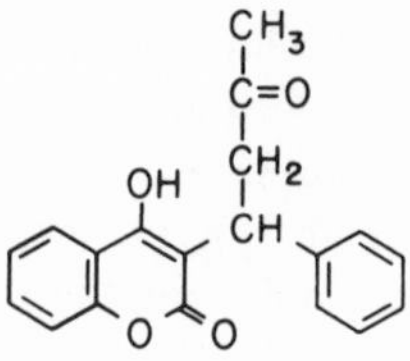

Warfarin

[3-(α-acetonylbenzyl)-4-hydroxycoumarin]

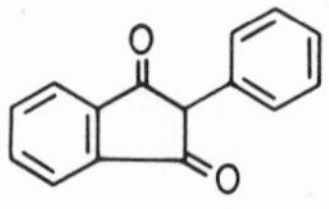

Phenindione

(2- phenyl-1,3- indandine)

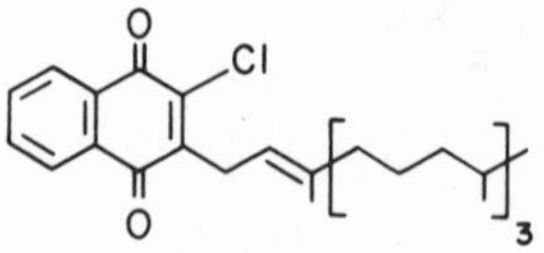

Chloro – K

(2-chloro -3- phytyl-1,4 -naphthoquinone)

Fig. 2. Structures of some compounds which antagonize the action of vitamin K.

soon traced to the consumption of improperly cured sweet clover hay. By the early 1930s, it was established that the cause of the hemorrhagic symptom was a decrease in the prothrombin activity of blood. The compound present in spoiled sweet clover that was responsible for this disease was isolated and characterized by Link's group (Campbell and Link, 1941; Stahmann et al., 1941) as 3,3'-methylene-bis (4-hydroxycoumarin) and called dicoumarol. A large number of substituted 4-hydroxycoumarins have been sythesized in Link's laboratory and elsewhere. The most successful of these, both clinically for long-term suppression of the vitamin K-dependent clotting factors and as a rodenticide, have been warfarin [3-(α-acetonylbenzyl)-4-hydroxycoumarin] and Tromexan [3,3'-carboxymethylene-bis(4-hydroxycoumarin) ethyl ester]. The various drugs used differ in the degree to which they are absorbed from the intestine and in their plasma half-life and presumably in their effectiveness as a vitamin K antagonist at the active site. Much of the information on the structure–activity relationships of the 4-hydroxycoumarins has been reviewed by Renk and Stoll (1968); the clinical use and phramacodynamic interactions of these compounds have recently been reviewed by O'Reilly (1975).

Warfarin is a widely used rodenticide. Concern has been expressed in recent years because of the identification of anticoagulant-resistant rat populations first identified in northern Europe (Boyle, 1960; Lund, 1964) and subsequently in the United States (Jackson and Kaukeinen, 1972). These rats have both an increased resistance to warfarin and an increased requirement for vitamin K (Hermodson et al., 1969; Thierry et al., 1970) and appear to have the same genetic alteration described by O'Reilly (1971) in human patients. Concern over the spread of the anticoagulant-resistant rat population led to renewed interest in synthesizing coumarin derivatives; hence, Hadler and Shadbolt (1975) synthesized a number of coumarins that not only are effective in the warfarin-resistant rat, but are much more active in normal animals than were most compounds synthesized in the past.

Although coumarin anticoagulants have been available for 35 years, there is no general consensus as to the mechanism of action. Because of their lack of structural similarity to the vitamin and the nature of the dose-response curves observed when they are used, it has usually been assumed that they do not directly compete with the vitamin for a protein-binding site. It has been suggested that the coumarin anticoagulants may interfere with some cellular transport site for the vitamin, or that they interfere with the metabolic interconversion of the vitamin to its 2,3-epoxide and back to the vitamin (discussed below). Current data are insufficient to describe definitively the antagonistic action of these compounds.

2.4.2. Indandione Derivatives

A second class of chemical compounds with anticoagulant activity that can be reversed by vitamin K administration was first identified by Kabat et al.

(1944) as 2-substituted-1,3-indandiones. A large number of these compounds have also been synthesized, of which two of the more commonly used members of the series are 2-phenyl-1,3-indandione (phenindione) and 2-pivalyl-1,3-indandione (pival). These compounds have had commercial use as rodenticides, but they are not used as clinical anticoagulants. Observations that warfarin-resistant rats are also resistant to the action of indandiones (Ren *et al.*, 1973) would suggest that the mechanism of action of these compounds is similar to that of the 4-hydroxycoumarins.

2.4.3. 2-Halo-3-Phytyl-1,4-Naphthoquinones

During the course of a series of investigations into the structural requirements for vitamin K activity, Lowenthal *et al.* (1960) found that the replacement of the 2-methyl group of phylloquinones by a chlorine or bromine atom resulted in a compound that was a potent antagonist of vitamin K. The most active of these two compounds is the 2-chloro derivative (commonly known as chloro-K); Lowenthal (1970) has shown that in contrast to the coumarin and indandione derivatives, chloro-K acts as if it were a true competitive inhibitor of the vitamin at its active site(s). Further evidence for its different mechanism of action is its efficacy in anticoagulant resistant rats (Suttie, 1973*b*). Chloro-K may also interfere with those enzymes responsible for metabolizing the vitamin, as it has been shown to cause significant alterations in the tissue distribution of radioactive vitamin K (Thierry and Suttie, 1971).

3.   Mechanism of Action of Vitamin K

Knowledge of the cellular events responsible for the production of prothrombin and the regulation of these metabolic events has come largely from investigations spanning only the last 10–15 years. During the 25-year period following Dam's discovery of the vitamin a great deal was learned about the biological activity of various forms of both the vitamin and its antagonists and about the significance of the vitamin in animal nutrition and human medicine. Although Anderson and Barnhart (1964) demonstrated conclusively that prothrombin was produced in the liver, the lack of a general understanding of the mechanism of protein biosynthesis prevented serious experimental approaches to the cellular or molecular mechanisms involved until the last decade. Progress has been rapid, however, since the mid-1960s, and a number of recent reviews of the mechanism of action of the vitamin are now available (Woolf and Babior, 1974; Olson, 1974; Suttie, 1974, 1975; Suttie *et al.*, 1974; Olson *et al.*, 1974). Investigations during the last few years have tended to rule out other possible mechanisms of synthesis and have indicated that prothrombin is formed in the liver by a vitamin K-dependent carboxylation of a liver precursor protein.

3.1. Development of the Prothrombin Precursor Concept

Based on their observations of coagulation in plasma from anticoagulant treated patients, Hemker *et al.* (1963) clearly stated the possibility that a precursor protein was involved in the formation of prothrombin. The nature of the response observed (Pyörälä, 1965; Hill *et al.*, 1968; Bell and Matschiner, 1969; Suttie, 1970) when vitamin K was administered to severely hypoprothrombinemic rats was consistent with the presence of a significant pool of a precursor that could be converted to prothrombin following vitamin administration. Both Pyörälä (1965) and Bell and Matschiner (1969) clearly pointed out that the rate of prothrombin synthesis observed during this initial period exceeded the theoretical induction curve based on the experimentally determined half-life of prothrombin. The possibility of a precursor was strengthened when it was shown Shah and Suttie, 1972) that the appearance of plasma prothrombin was preceded by a transient increase of prothrombin in liver microsomal preparations. This response suggested that a pool of precursor could be converted to prothrombin in a vitamin-dependent step; following depletion of this pool, the rate of synthesis would decrease and become dependent on the rate of synthesis of this precursor. These early studies in vitamin K-deficient rats also demonstrated that the vitamin K-stimulated initial burst of prothrombin was decreased only slightly by prior administration of the protein synthesis inhibitor cycloheximide, but that the increase in plasma prothrombin seen 1–2 hr after vitamin administration was blocked by cycloheximide treatment.

These studies, which utilized inhibitors of protein biosynthesis in intact animals, strongly suggested that protein synthesis was not involved in the vitamin K-dependent step of prothrombin synthesis but did not offer final proof. More conclusive and direct evidence of the presence of a precursor was obtained when Shah and Suttie (1971) had demonstrated that the prothrombin produced when hypoprothrombinemic rats were given vitamin K and cycloheximide did not contain radioactive amino acids if they were administered at the same time as the vitamin. These data strongly suggested that plasma prothrombin must have been derived from an existing precursor pool. This study further indicated that when radioactive amino acids were administered to the hypoprothrombinemic rats prior to cycloheximide and vitamin K administration, the prothrombin formed subsequently did contain radioactive amino acids. This observation was consistent with the presence of a precursor protein pool that was rapidly being synthesized and which could be converted to prothrombin in a step that did not require protein synthesis.

The hypothesis that there was a liver precursor to prothrombin was strengthened by observations that the plasma of humans or animals treated with coumarin anticoagulants contains a protein that in many ways is similar to prothrombin. The existence of such a protein was first postulated to be present in the plasma of human patients receiving anticoagulant therapy by Hemker *et al.*

(1963) by indirect means, and later a protein antigenetically similar to prothrombin, but lacking biological activity was demonstrated in such plasma by a number of workers (Ganrot and Nilehn, 1968; Josso *et al.*, 1968; Hemker *et al.*, 1970; Denson, 1971; Brozovic and Gurd, 1973; Cesbron *et al.*, 1973). A similar protein was first demonstrated in bovine plasma by Stenflo (1970), and its existence has been confirmed by others (Malhotra and Carter, 1971; Nelsestuen and Suttie, 1972*a*; Reekers *et al.*, 1973; Wallin and Prydz, 1975).

The existence of such a protein in the plasma of other species given oral anticoagulants has been somewhat controversial; its presence has been sought both by immunochemical methods and by thrombin generation using nonphysiological activators. It has been claimed (Johnson *et al.*, 1972; Pereira and Couri, 1972) that plasma from rats administered anticoagulants does contain appreciable amounts of an abnormal form of prothrombin; others (Olson, 1974; Carlisle *et al.*, 1975) have found very little evidence of such a protein in this species. Both Olson (1974) and Carlisle *et al.* (1975) found some abnormal prothrombin in the plasma of anticoagulant-treated chicks, but the latter group also reported that this protein appeared to be missing in plasma from anticoagulant-treated mice, hamsters, guinea pigs, rabbits, and dogs. In addition to the reports of immunochemically similar, but biologically inactive, forms of prothrombin in plasma following anticoagulant treatment, there is evidence of similar proteins corresponding to the other vitamin K-dependent clotting factors.

3.2. Isolation and Characterization of the Abnormal Prothrombin

The identification of these new proteins in the plasma of patients on anticoagulant therapy provided the stimulus for a series of investigations that culminated in an understanding of the chemical nature of the postribosomal modification of prothrombin. The protein from human plasma has been purified but has not been subjected to extensive chemical characterization, whereas the protein (abnormal prothrombin) from bovine plasma was purified and extensively studied both by Stenflo and Suttie.

The initial studies of this protein (Stenflo, 1972; Nelsestuen and Suttie 1972*a*; Stenflo and Ganrot, 1972) indicated that it had the same molecular weigh and amino acid composition as normal prothrombin, but that it did not adsorb to insoluble barium salts as did normal prothrombin. The lack of barium salt adsorption and the calcium-dependent electrophoretic and immunochemical propertie suggested a difference in calcium-binding properties of these two proteins, which was directly demonstrated by Nelsestuen and Suttie (1972*b*) and confirmed by Stenflo and Ganrot (1973). The difference in calcium binding was shown by Stenflo (1973) to be a property of an amino-terminal peptide that could be derived from the two proteins. The observation (Nelsestuen and Suttie, 1972*a* that the abnormal prothrombin could yield thrombin when treated with trypsin o snake venoms indicated that this portion of the molecule was normal and that the critical difference in the two proteins was the inability of the abnormal protein to bind to calcium ions and therefore to interact with a phospholipid surface. It was later shown (Esmon *et al.*, 1975) that the abnormal prothrombin will not bind to a phospholipid surface in the presence of calcium ions, and that phospholipid addition drastically stimulates the Factor X_a and Ca^{2+} activation of prothrombin, but not the abnormal prothrombin. A comparison of some of the properties of normal bovine prothrombin and the abnormal prothrombin is shown in Table II. The initial studies of the abnormal prothrombin clearly implicated the calcium-binding region of prothrombin as the vitamin K-dependent region but provided no evidence of the chemical nature of this region. Nelsestuen and Suttie (1973) isolated an acidic peptide from a tryptic digest of normal bovine prothrombin that would both adsorb to insoluble barium salts and bind calcium ions in solution. This peptide came from the amino terminal region of prothrombin, contained a high proportion of acidic amino acid residues, and had an anomalously high apparent molecular weight on molecular sieve columns. Stenflo (1974) later isolated two acidic peptides from prothrombin by different methods, and both groups postulated the existence of some unknown acidic, nonpeptide, prosthetic group attached to this portion of the molecule. These peptides could not be obtained when similar isolation procedures were applied to preparations of abnormal prothrombin.

3.3.   Characterization of γ-Carboxyglutamic Acid

Stenflo (Stenflo *et al.*, 1974; Fernlund *et al.*, 1975) succeeded in isolating a tetrapeptide, residues 6–9 of prothrombin, which had an apparent sequence of

Table II. Properties of Abnormal Bovine Prothrombin

Property[a]	Comparison to prothrombin
Molecular weight	Indistinguishable
Amino acid composition	Apparently identical
End-terminal residues	Apparently identical
Carbohydrate composition	Apparently identical
Immunochemical determinants	Similar or identical
Electrophoretic mobility	Similar without Ca^{2+}
Hydrodynamic properties	Indistinguishable
Circular dichroism spectra	Indistinguishable
Adsorption to Ba salts	Very low
Ca^{2+} binding	Very low
Ca^{2+}-dependent phospholipid binding	Lacking
Biological activity	Lacking or very low
Activation by trypsin	Apparently identical
Activation by *E. carinatus* venom	Apparently identical

[a] Properties determined from various studies, see text for references.

Leu-Glu-Glu-Val, and demonstrated by a combination of mass fragmentation, NMR spectra, and chemical synthesis that the glutamic acid residues of this peptide were modified so that they were present as γ-carboxyglutamic acid (3-amino-1,1,3-propanetricarboxylic acid) residues (Fig. 3). Independently, Nelsestuen *et al.* (1974), by rather similar methods, characterized γ-carboxyglutamic acid from a dipeptide (residues 33 and 34 of prothrombin), which originally appeared to be Glu-Ser. These characterizations of the modified glutamic acid residues have been confirmed by Magnusson *et al.* (1974) who showed that all 10 of the first 33 Glu residues in prothrombin have been modified in this fashion (Fig. 4). Factor X is also a calcium-binding vitamin K-dependent clotting factor, which has been shown (Howard and Nelsestuen, 1975) to contain γ-carboxyglutamic acid residues. Although not directly demonstrated, presumably Factors IX and VII also contain these vitamin K-dependent modifications.

3.4. Isolation of Liver Precursor Proteins

Studies indicated that although the abnormal bovine prothrombin was activated at a very slow rate by physiological activators, it was possible to generate a rapid thrombin-like activity from it by treatment with *Echis carinatus* venom, thereby implying that if the concentration of the hypothesized liver precursor built up in the liver of hypoprothrombinemic animals, it might be detected by the release of thrombin following incubation with these snake venoms. Suttie then demonstrated (Shah *et al.*, 1973; Suttie, 1973a) that when microsomes were isolated from vitamin K-deficient or anticoagulant-treated rats and then solubilized with detergent, and the extract was treated with *Echis carinatus* venom, thrombin activity was generated. The amount of this microsomal precursor decreased rapidly when vitamin K was injected, and as its level fell, the amount of microsomal prothrombin rose and then fell as it moved out of the liver into the plasma. A similar protein has also been observed by Morrissey *et al.* (1973). A protein has now been isolated (Esmon *et al.*, 1975a) from the livers of warfarin-

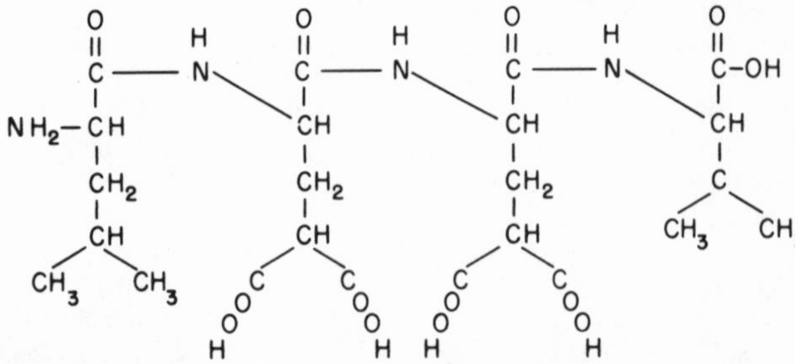

Fig. 3. Structure of the peptide (residues 6–9) from bovine prothrombin first shown to contain γ-carboxyglutamic acid.

```
 1    2    3    4    5    6    7    8    9   10   11   12   13   14   15
Ala-Asn-Lys-Gly-Phe-Leu-GLA-GLA-Val-Arg-Lys-Gly-Asn-Leu-GLA-

16   17   18 , 19   20   21   22   23   24   25   26   27   28   29   30
Arg-GLA-Cys-Leu-GLA-GLA-Pro-Cys-Ser-Arg-GLA-GLA-Ala-Phe-GLA-

31   32   33   34
Ala-Leu-GLA-Ser---
```

Fig. 4. Amino acid sequence of the first 34 residues of bovine prothrombin GLA = γ-carboxyglutamic acid (Magnusson *et al.*, 1974).

treated rats that has the properties predicted for this precursor. It is a glycoprotein immunochemically similar to prothrombin the molecular weight of which is indistinguishable from rat prothrombin. Electrophoretic and isofocusing analyses indicate that the precursor is less negatively charged than prothrombin, and specific proteolysis of the precursor yields fragments indistinguishable from those formed by similar proteolysis of prothrombin. This protein does not adsorb to $BaSO_4$, and its rate of activation to thrombin by Factor X_a and Ca^{2+} was not stimulated by the addition of phospholipid. Therefore, the isolated protein has most of the properties postulated for the prothrombin precursor and apparently differs from rat prothrombin only in that it lacks the γ-carboxyglutamic acid residues and the sialic acid residues present in plasma prothrombin. More recently, a second protein has been isolated (Grant and Suttie, 1976) from the same source, which has similar properties but is much more basic.

3.5. Vitamin K-Dependent Carboxylase

The first vitamin K-dependent *in vitro* system to produce prothrombin was that described by Shah and Suttie (1974). Postmitochondrial supernates from vitamin K-deficient rats were shown to respond to the addition of vitamin K by producing a significant amount of prothrombin as assayed by the standard two-stage assay. Prothrombin production was dependent on O_2 and an energy supply; it was inhibited by antagonists of vitamin K but not by cycloheximide. After the vitamin K-dependent step in prothrombin synthesis was shown to be the formation of γ-carboxyglutamic acid residues, Esmon *et al.* (1975b) demonstrated that the same postmitochondrial supernatant would catalyze a vitamin K-dependent carboxylation of the endogenous microsomal precursor (Fig. 5). It was possible to isolate radioactive prothrombin from this system following incubation and to show that essentially all the radioactivity was present as γ-carboxyglutamic acid residues in the amino-terminal region of prothrombin. These observations would appear to offer final proof of the role of vitamin K in the biosynthesis of prothrombin. An *in vivo* demonstration of $H^{14}CO_3^-$ incorporation into prothrombin has also been claimed (Girardot *et al.*, 1974). The vitamin K-dependent carboxylase has been studied by Sadowski *et al.* (1976) in washed microsomes where the activity requires the presence of microsomal precursor, O_2, vitamin K,

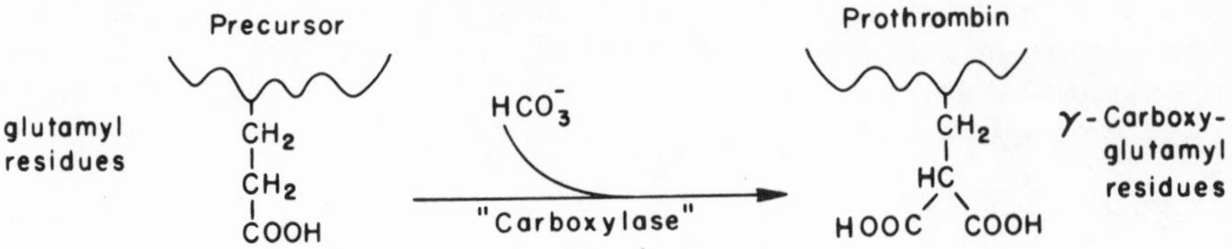

Fig. 5. The vitamin K-dependent reaction of liver microsomes.

and HCO_3^-, and is stimulated by an energy source, and factor(s) present in the postmicrosomal supernatant. A major factor in the supernatant is protein(s) acting as a NAD^+ ($NADP^+$) reductase. This requirement can be replaced with the addition of NAD(P)H to the system, and the requirement for reducing equivalents from pyridine nucleotides in the systems has now been shown (Friedman and Shia, 1976; Sadowski et al., 1976) to be largely a requirement for the reduced form of vitamin K. Friedman and Shia (1976) have also reported that dithiothreitol (DTT) can be used as the source for reducing equivalents for this reaction, and that this reducing agent might also be functioning to protect an essential sulfhydryl group in the enzyme system. They have also shown that vitamins with a geranyl or farnesyl group at the 3-position of the vitamin are considerably more active than the phytyl derivative (phylloquinone). The carboxylase activity in this microsomal preparation can be inhibited by warfarin, and this inhibition can be overcome by high concentrations of the vitamin.

The carboxylase activity has now been solubilized in various detergents (Esmon and Suttie, 1976; Girardot et al., 1976; Mack et al., 1976) and the solubilized preparation retains many of the properties of the membrane associated system. The solubilized system is still stimulated by DTT and inhibited by mercuricals (Mack et al., 1976). It has been reported that the solubilized system is inhibited by the spin-trapping agent 5,5-dimethyl-1-pyroline-N-oxide and that O_2 is not required in the solubilized system when the reduced form of the vitamin is used (Girardot et al., 1976). The latter finding, however, is not supported by other data (Esmon and Suttie, 1976) that show the requirement for O_2 in such a system. The solubilized system is not inhibited by warfarin, but it is still sensitive to a direct vitamin K antagonist, such as 2-chloro-3-phytyl-1,4-naphthoquinone. The solubilized system has been particularly useful in clarifying the need for ATP in the system. Incubation in the absence of ATP and the presence of an ATP inhibitor AMPP(NH)P does inhibit the membrane-bound carboxylase activity, but not the solubilized system (Esmon and Suttie, 1976). These data suggest that the energy to drive the carboxylation comes from the reoxidation of the reduced vitamin in the system, but the mechanism remains unclear. There is no evidence to indicate that either biotin or the coenzyme A esters of the glutamyl residues are involved in this carboxylase, and it is usually assumed that the vitamin must function either to "activate" the residue.

Studies of the mechanism should be facilitated by the observation of Suttie *et al.* (1976) that the pentapeptide, Phe-Leu-Glu-Glu-Val, will serve as a substrate for the carboxylase. The properties of this carboxylase as they are presently understood are summarized in Table III.

4. Vitamin K Epoxide

Any theory of the mechanism of action of vitamin K must take into consideration the possibility that the formation of the 2,3-epoxide of the vitamin (K-oxide) (Matschiner *et al.*, 1970) is involved in the reaction (Fig. 6). Bell and Matschiner (1970, 1972) demonstrated that warfarin administration blocks the action of a liver enzyme that reduces K-oxide to the vitamin so that during anticoagulation treatment there is a high ratio of K-oxide to vitamin K in the liver. They postulated that K-oxide is a competitive inhibitor of the action of the vitamin and that warfarin exerts its anticoagulant effect through a buildup of this metabolite. Although this theory was supported (Bell *et al.*, 1972; Bell and Caldwell, 1973; Ren *et al.*, 1973; Zimmerman and Matschiner, 1974) by a considerable number of indirect data and observations that the epoxide reductase was less sensitive in warfarin-resistant animals, the theory has now been shown (Goodman *et al.*, 1974; Sadowski and Suttie, 1974) to be untenable.

More recently, Willingham and Matschiner (1974) have postulated that the formation of the epoxide ("epoxidase" activity) is an obligatory step in vitamin action promoting prothrombin biosynthesis. This hypothesis was originally based on observations that "epoxidase" activity increased in liver under various treatments in much the same manner as concentrations of the prothrombin precursor. The theory is also supported by observations on the effects of various anticoagu-

Table III. Properties of the Vitamin K-Dependent Carboxylase[a]

A.	Absolute requirements	B.	Known inhibitors
	Vitamin K and NAD(P)H		Chloro-K
	or Vitamin KH_2		Warfarin[b]
	O_2		p-Hydroxymercuribenzoate
	HCO_3^- (CO_2)		Spin-trapping agents (high concn.)
	Presence of precursor		N_2
C.	Noninhibitory conditions	D.	Stimulatory conditions
	ATP analog AMPP(NH)P[c]		Dithiothreitol
	Avidin		Additional substrate (peptide)
	Cyt P_{450} inhibitors		
	EDTA		

[a] There is not complete agreement in the published literature on all points; the properties assigned represent the authors evaluation of the consensus of the published literature.
[b] Only when intact microsomes are present.
[c] Some inhibition when intact microsomes are present.

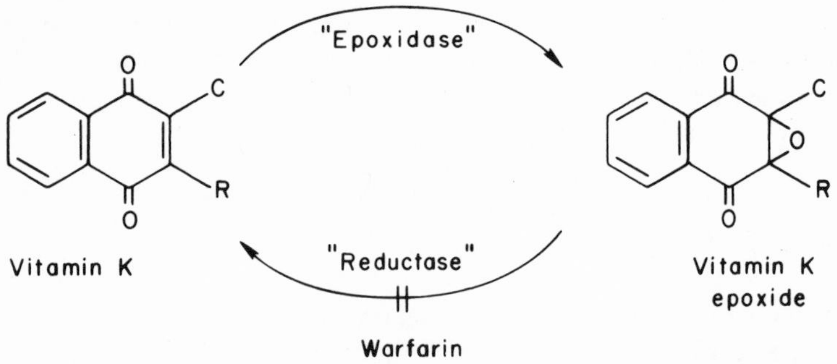

Fig. 6. Metabolism of vitamin K in microsomal preparations. The conversion of the epoxide to the vitamin "epoxidase activity" is inhibited by warfarin.

lants in normal and warfarin-resistant rats (Bell *et al.*, 1976; Willingham *et al.*, 1976) and observations (Sadowski, 1976) that many of the requirements for *in vitro* epoxidation and vitamin K-dependent carboxylation are similar. The available evidence is far from conclusive, but it does suggest that both reactions involve some components of a microsomal redox system, and that they might somehow be coupled. The mechanism whereby the coumarin anticoagulants inhibit the action of vitamin K is also not yet clarified, and the vitamin K-dependent carboxylase system should also serve as a tool for investigating these interactions.

5. Other Vitamin K-Dependent Proteins

Although for many years it was thought that the only defect in a vitamin K deficiency was the lack of synthesis of plasma clotting Factors II, VII, IX, and X, it is now clear that such is not the case. Stenflo (1976) described the isolation of a fifth plasma protein, the concentration of which decreased upon anticoagulant treatment. This protein is composed of two polypeptide chains, binds calcium ions, and contains γ-carboxyglutamic acid residues. The amino acid sequence in the amino terminal region of this protein is homologous to the other vitamin K-dependent clotting factors. Subsequent studies (Esmon *et al.*, 1976) indicated that this protein cannot be a modified form of clotting Factors II, VII, IX, or X, and that it is a serine protease that can be activated by limited trypsin digestion. It was also shown that this protein will bind to phospholipid surfaces both in the presence and absence of calcium ions. These studies have not provided any evidence to indicate what the physiological role of this protein might be, but it has been suggested (Seegers *et al.*, 1976) that it is an inhibitor of coagulation and involved in epinephrine-induced platelet aggregation.

Not only has an additional vitamin K-dependent protein been found in plasma, but it has been shown (Hauschka *et al.*, 1975; Price *et al.*, 1976) that there is a γ-carboxyglutamic acid-containing protein present in high concentrations in skeletal tissue. The amino acid sequence of this protein does not show homology to the vitamin K-dependent plasma clotting factors. The biological function of this protein is as of yet also unknown, and it may function in some direct way in the calcification process, or it may function as a regulator or inhibitor of calcification. A significant concentration of protein-bound γ-carboxyglutamic acid has also been found in kidney tissues, and it has been shown (Hauschka *et al.*, 1976) that kidney microsomes will carry out a vitamin K-dependent *in vitro* synthesis of γ-carboxyglutamic acid. These findings suggest that the function of vitamin K might be far more widespread than was once realized and may make available other systems with which to probe the role of this fat-soluble vitamin at the molecular level.

6. Summary

The last 10–15 years have seen rapid advances in our understanding of the function of vitamin K in metabolism. It has been demonstrated that the vitamin functions as an essential cofactor for a microsomal carboxylase that converts peptide-bound glutamyl residues to γ-carboxyglutamyl residues. This carboxylase appears to be unique in that it requires oxygen and has no demonstrable requirement for ATP. Present evidence would indicate that the energy to drive this carboxylation event comes from the reoxidation of the reduced form of the vitamin, but how this is accomplished has not been determined. Recent discoveries of γ-carboxyglutamic acid residues in proteins other than the long-recognized vitamin K-dependent clotting factors have opened new areas of investigation that promise to yield further insight into the physiological and biochemical role of this vitamin.

References

Almquist, H. J., 1941, Vitamin K, *Physiol. Rev.* **21**:194–216.
Almquist, H. J., 1975, The early history of vitamin K, *Am. J. Clin. Nutr.* **28**:656–659.
Almquist, H. J., and Klose, A. A., 1939, Color reactions in vitamin K concentrates, *J. Am. Chem. Soc.* **61**:1610–1611.
Almquist, H. J., and Stokstad, E. L. R., 1935, Dietary haemorrhagic disease in chicks, *Nature (London)* **136**:31.
Anderson, G. F., and Barnhart, M. I., 1964, Intracellular localization of prothrombin, *Proc. Soc. Exp. Biol. Med.* **116**:1–6.
Bell, R. G., and Caldwell, P. T., 1973, The mechanism of Warfarin resistance. Warfarin and the metabolism of vitamin K_1, *Biochemistry* **12**:1759–1762.
Bell, R. G., and Matschiner, J. T., 1969, Synthesis and destruction of prothrombin in the rat, *Arch. Biochem. Biophys.* **135**:152–159.
Bell, R. G., and Matschiner, J. T., 1970, Vitamin K activity of phylloquinone oxide, *Arch. Biochem. Biophys.* **141**:473–476.

Bell, R. G., and Matschiner, J. T., 1972, Warfarin and the inhibition of vitamin K activity by an oxide metabolite, *Nature (London)* **237**:32–33.

Bell, R. G., Caldwell, P. T., and Holm, E. E. T., 1976, Coumarins and the vitamin K–K epoxide cycle. Lack of resistance to coumatetralyl in warfarin-resistant rats, *Biochem. Pharmacol.* **25**:1067–1070.

Bell, R. G., Sadowski, J. A., and Matschiner, J. T., 1972, Mechanism of action of Warfarin. Warfarin and metabolism of vitamin K_1, *Biochemistry* **11**:1959–1961.

Boyle, C. M., 1960, Case of apparent resistance of *Rattus norvegicus* barkenhout to anticoagulant poisons, *Nature (London)* **188**:517.

Brozovic, M., and Gurd, L. J., 1973, Prothrombin during Warfarin treatment. *Br. J. Haematol.* **24**:579–588.

Campbell, H. A., and Link, K. P., 1941, Studies on the hemorrhagic sweet clover disease. IV. The isolation and crystallization of the hemorrhagic agent, *J. Biol. Chem.* **138**:21–33.

Carlisle, T. L., Shah, D. V., Schlegel, R., and Suttie, J. W., 1975, Plasma abnormal prothrombin and microsomal prothrombin precursor in various species, *Proc. Soc. Exp. Biol. Med.* **148**:140–144.

Cesbron, N., Boyer, C., Guillin, M.-C., and Menache, D., 1973, Human coumarin prothrombin. Chromatographic, coagulation and immunologic studies, *Thrombos. Diathesth. Haemorrh. (Stuttg.)* **30**:437–450.

Dam, H., 1935, The antihaemorrhagic vitamin of the chick, *Biochem. J.* **29**:1273–1285.

Dam, H., 1942, Vitamin K, its chemistry and physiology, *Advan. Enzymol.* **2**:285–324.

Dam, H., 1948, Vitamin K, *Vitamins Hormones* **VI**:27–53.

Dam, H., Geiger, A., Glavind, J., Karrer, P., Karrer, W., Rothschild, E., and Solomon, H., 1939, Isolierung des vitamins K in hochgereinigter form, *Helv. Chim. Acta* **22**:310–313.

Dam, H., Schønheyder, F., and Tage-Hansen, E., 1936, Studies on the mode of action of vitamin K, *Biochem. J.* **30**:1075–1079.

Denson, K. W. E., 1971, The levels of factors II, VII, IX and X by antibody neutralization techniques in the plasma of patients receiving phenindione therapy, *Br. J. Haematol.* **20**:643–648.

Doisy, E. A., 1971, Vitamin K in human nutrition, in *Symposium Proceedings on the Biochemistry, Assay and Nutritional Value of Vitamin K and Related Compounds,* pp. 79–92, Association of Vitamin Chemistry, Chicago.

Doisy, E. A., and J. T. Matschiner, 1970, Biochemistry of vitamin K, in *Fat-Soluble Vitamins* (R. A. Morton, ed.), pp. 293–331, Pergamon Press, Oxford.

Esmon, C. T., and Suttie, J. W., 1976, Vitamin K-dependent carboxylase: Solubilization and properties, *J. Biol. Chem.* **251**:6238–6243.

Esmon, C. T., Grant, G. A., and Suttie, J. W., 1975a, Purification of an apparent rat liver prothrombin precursor: Characterization and comparison to normal rat prothrombin, *Biochemistry* **14**:1595–1600.

Esmon, C. T., Jackson, C. M., and Suttie, J. W., 1975b, The functional significance of vitamin K action. Difference in phospholipid binding between normal and abnormal prothrombin, *J. Biol. Chem.* **250**:4095–4099.

Esmon, C. T., Sadowski, J. A., and Suttie, J. W., 1975c, A new carboxylation reaction. The vitamin K-dependent incorporation of $H^{14}CO_3^-$ into prothrombin, *J. Biol. Chem.* **250**:4744–4748.

Esmon, C. T., Suttie, J. W., and Jackson, C. M., 1975d, The functional significance of vitamin K action. Difference in phospholipid binding between normal and abnormal prothrombin, *J. Biol. Chem.* **250**:4095–4099.

Esmon, C. T., Stenflo, J., Jackson, C. M., and Suttie, J. W., 1976, A new vitamin K-dependent protein. A phospholipid-binding zymogen of a serine esterase, *J. Biol. Chem.* **251**:3052–3056.

Fernlund, P., Stenflo, J., Roepstorff, R., and Thomsen, J., 1975, Vitamin K and the biosynthesis of prothrombin. V. γ-Carboxyglutamic acids, the vitamin K-dependent structures in prothrombin. *J. Biol. Chem.* **250**:6125–6133.

Frick, P. G., Riedler, G., and Brogli, H., 1967, Dose response and minimal daily requirement for vitamin K in man, *J. Appl. Physiol.* **23**:387–389.

Friedman, P. A., and Shia, M., 1976, Some characteristics of a vitamin K-dependent carboxylating system from rat liver microsomes, *Biochem. Biophys. Res. Commun.* **70**:647–654.

Ganrot, P. O., and Nilehn, J. E., 1968, Plasma prothrombin during treatment with Dicumarol. II. Demonstration of an abnormal prothrombin fraction, *Scand. J. Clin. Lab. Invest.* **22**:23–28.

Girardot, J.-M., Delaney, R., and Johnson, B. C., 1974, Carboxylation, the completion step in prothrombin biosynthesis, *Biochem. Biophys. Res. Commun.* **59**: 1197–1203.

Girardot, J.-M., Mack, D. O., Floyd, R. A., and Johnson, B. C., 1976, Evidence for vitamin K semiquinone as the functional form of vitamin K in the liver vitamin K-dependent protein carboxylation reaction, *Biochem. Biphys. Res. Commun.* **70**:655–662.

Goodman, S. R., Houser, R. M., and Olson, R. E., 1974, Ineffectiveness of phylloquinone epoxide as an inhibitor of prothrombin synthesis in the rat, *Biochem. Biophys. Res. Commun.* **61**:250–257.

Grant, G. A., and Suttie, J. W., 1976, Rat liver prothrombin precursors: Purification of a second, more basic form, *Biochemistry* **15**:5387–5393.

Green, J., 1966, Antagonists of vitamin K, *Vit. Horm.* **24**:619.

Griminger, P., 1966, Biological activity of the various vitamin K forms, *Vit. Horm.* **24**:605–618.

Griminger, P., 1971, Nutritional requirements for vitamin K-animal studies, in *Symposium Proceedings on the Biochemistry, Assay, and Nutritional Value of Vitamin K and Related Compounds*, pp. 39–59, Association of Vitamin Chemistry, Chicago.

Hadler, M. R., and Shadbolt, R. S., 1975, Novel 4-hydroxycoumarin anticoagulants active against resistant rats, *Nature (London)* **253**:275–277.

Hauschka, P. V., Lian, J. B., and Gallop, P. M., 1975, Direct identification of the calcium-binding amino acid γ-carboxyglutamate in mineralized tissue, *Proc. Natl. Acad. Sci. USA* **72**:3925–3929.

Hauschka, P. V., Friedman, P. A., Traverso, H. P., and Gallop, P. M., 1976, Vitamin K-dependent γ-carboxyglutamic acid formation by kidney microsomes *in vitro, Biochem. Biophys. Res. Commun.* (in press).

Hemker, H. C., Veltkamp, J. J., Hensen, A., and Loeliger, E. A., 1963, Nature of prothrombin biosynthesis: Preprothrombinaemia in vitamin K-deficiency, *Nature (London)* **200**:589–590.

Hemker, H. C., Muller, A. D., and Loeliger, E. A., 1970, Two types of prothrombin in vitamin K deficiency, *Thrombos. Diath. Haemorrh.* **23**:633–637.

Hermodson, M. A., Suttie, J. W., and Link, K. P., 1969, Warfarin metabolism and vitamin K requirement in the Warfarin-resistant rat, *Am. J. Physiol.* **217**:1316–1319.

Hill, R. B., Gaetani, S., Paolucci, A. M., RamaRao, P. B., Alden, R., Ranhotra, G. S., Shah, D. V., Shah, V. K., and Johnson, B. C., 1968, Vitamin K and biosynthesis of protein and prothrombin, *J. Biol. Chem.* **243**:3930–3939.

Howard, J. B., and Nelsestuen, G. L., 1975, Isolation and characterization of vitamin K-dependent region of bovine blood clotting Factor X, *Proc. Natl. Acad. Sci. USA* **72**:1281–1285.

Isler, O., and Wiss, O., 1959, Chemistry and biochemistry of the K vitamins, *Vit. Horm.* **17**:53–90.

IUPAC–IUB—Tentative Rules, 1966, Nomenclature of quinones with isoprenoid side chains, *J. Biol. Chem.* **241**:2989–2991.

Jackson, W. B., and Kaukeinen, D., 1972, Resistance of wild Norway rats in North Carolina to Warfarin rodenticide, *Science* **176**:1343–1344.

Johnson, H. V., Boyd, C., Martinovic, J., Valkovich, G., and Johnson, B. C., 1972, A new blood protein which increases with vitamin K deficiency, *Arch. Biochem. Biophys.* **148**:431–442.

Josso, F., Lavergne, J. M., Gouault, M., Prou-Wartelle, O., and Soulier, J. P., 1968, Différents états moléculaires du facteur II (prothrombine). Leur étude à l'aide de la staphylocoagulase et d'anticorps anti-facteur II, *Thrombos. Diath. Haemorrh.* **20**:88–98.

Kabat, H., Stohlman, E. F., and Smith, M. J., 1944, Hypoprothrombinemia induced by administration of indandione derivatives, *J. Pharmacol. Exp. Ther.* **80**:160–170.

Link, K. P., 1959, The discovery of Dicumarol and its sequels, *Circulation* **19**:97–107.

Lowenthal, J., 1970, Vitamin K analogs and mechanisms of action of vitamin K, in *The Fat-Soluble Vitamins* (H. F. DeLuca and J. W. Suttie, eds.), pp. 431–446, Univ. of Wisconsin Press, Madison.

Lowenthal, J., MacFarlane, J. A., and McDonald, K. M., 1960, The inhibition of the antidotal activity of vitamin K_1 against coumarin anticoagulant drugs by its chloro analogue, *Experientia* **16**:428–429.

Lund, M., 1964, Resistance to Warfarin in the common rat, *Nature (London)* **203**:778.

MacCorquodale, D. W., Cheney, L. C., Binkley, S. B., Holcomb, W. F., McKee, R. W., Thayer, S. A., and Doisy, E. A. 1939, The constitution and synthesis of vitamin K_1, *J. Biol. Chem.* **131**:357–370.

Mack, D. O., Suen, E. T., Girardot, J.-M., Miller, J. A., Delaney, R., and Johnson, B. C., 1976, Soluble enzyme system for vitamin K-dependent carboxylation, *J. Biol. Chem.* **251**:3269–3276.

Magnusson, S., Sottrup-Jensen, L., Petersen, T. E., Morris, H. R., and Dell, A., 1974, Primary structure of the vitamin K-dependent part of prothrombin, *FEBS Lett.* **44**:189–193.

Malhotra, O. P., and Carter, J. R., 1971, Isolation and purification of prothrombin from dicumarolized steers, *J. Biol. Chem.* **246**:2665–2671.

Matschiner, J. T., Bell, R. G., Amelotti, J. M., and Knauer, T. E., 1970, Isolation and characterization of a new metabolite of phylloquinone in the rat, *Biochim. Biophys. Acta* **201**:309–315.

Morrissey, J. J., Jones, J. P., and Olson, R. E., 1973, Isolation and characterization of isoprothrombin in the rat, *Biochem. Biophys. Res. Commun.* **54**:1075–1082.

Nelsestuen, G. L., and Suttie, J. W., 1972a, The purification and properties of an abnormal prothrombin protein produced by Dicumarol-treated cows. A comparison to normal prothrombin, *J. Biol. Chem.* **247**:8176–8182.

Nelsestuen, G. L., and Suttie, J. W., 1972b, Mode of action of vitamin K. Calcium binding properties of bovine prothrombin, *Biochemistry* **11**:4961–4964.

Nelsestuen, G. L., and Suttie, J. W., 1973, The mode of action of vitamin K. Isolation of a peptide containing the vitamin K-dependent portion of prothrombin, *Proc. Natl. Acad. Sci. USA* **70**:3366–3370.

Nelsestuen, G. L., Zytkovicz, T. H., and Howard, J. B., 1974, The mode of action of vitamin K. Identification of γ-carboxyglutamic acid as a component of prothrombin, *J. Biol. Chem.* **249**:6347–6350.

Olson, R. E., 1974, New concepts relating to the mode of action of vitamin K, *Vit. Horm.* **32**:483–511.

Olson, R. E., Kipfer, R. K., Morrissey, J. J., and Goodman, S. R., 1974, Function of vitamin K in prothrombin synthesis, *Thrombos. Diath. Haemorrh.* [*Suppl.*] **57**:31–44.

O'Reilly, R. A., 1971, Vitamin K in hereditary resistance to oral anticoagulant drugs, *Am. J. Physiol.* **221**:1327–1330.

O'Reilly, R. A., 1975, Vitamin K and the oral anticoagulant drugs, *Annu. Rev. Med.* **27**:245–261.

Pereira, M. A., and Couri, D., 1972, Site of inhibition of dicoumarol of prothrombin biosynthesis: Carbohydrate content of prothrombin from dicoumarol-treated rats, *Biochim. Biophys. Acta* **261**:375–378.

Price, P. A., Otsuka, A. S., Poser, J. W., Kristaponis, J., and Raman, N., 1976, Characterization of a γ-carboxyglutamic acid-containing protein from bone, *Proc. Natl. Acad. Sci. USA* **73**:1447–1451.

Pyörälä, K., 1965, Determinants of the clotting factor response to Warfarin in the rat, *Ann. Med. Exp. Biol. Fenn.* **43**:Suppl. 3, p. 99.

Reekers, P. P. M., Lindhout, M. J., Kop-Klaassen, B. H. M., and Hemker, H. C., 1973, Demonstration of three anomalous plasma proteins induced by vitamin K antagonist, *Biochim. Biophys. Acta* **317**:559–562.

Ren, P., Laliberte, R. E., and Bell, R. G., 1973, Effects of Warfarin, phenylindanedione, tetrachloropyridinol, and chloro-vitamin K_1 on prothrombin synthesis and vitamin K metabolism in normal and warfarin-resistant rats, *Mol. Pharmacol.* **10**:373–380.

Renk, E., and Stoll, W. G., 1968, Orale Antikoagulantien, *Prog. Drug Res.* **11**:226–355.

Sadowski, J. A., 1976, The *in vivo* and *in vitro* metabolism of vitamin K_1 during the synthesis of prothrombin: The significance of the cyclic interconversion of vitamin K_1 and its 2,3-epoxide, Ph.D. thesis, Univ. of Wisconsin, Madison.

Sadowski, J. A., and Suttie, J. W., 1974, Mechanism of action of coumarins. Significance of vitamin K epoxide, *Biochemistry* **13**:3696–3699.

Sadowski, J. A., Esmon, C. T., and Suttie, J. W., 1976, Vitamin K-dependent carboxylase. Requirements of the rat liver microsomal enzyme system, *J. Biol. Chem.* **251**:2770–2775.

Scott, M. L., 1966, Vitamin K in animal nutrition, *Vit. Horm.* **24**:633–647.

Seegers, W. H., Novoa, E., Henry, R. L., and Hassovna, H. I., 1976, Relationship of "new" vitamin K dependent protein C and "old" autoprothrombin II-A, *Thromb. Res.* **8**:543–552.

Shah, D. V., and Suttie, J. W., 1971, Mechanism of action of vitamin K: Evidence for the conversion of a precursor protein to prothrombin in the rat, *Proc. Natl. Acad. Sci. USA* **68**:1653–1657.

Shah, D. V., and Suttie, J. W., 1972, The effect of vitamin K and warfarin on rat liver prothrombin concentrations, *Arch. Biochem. Biophys.* **150**:91–15.

Shah, D. V., and Suttie, J. W., 1974, The vitamin K dependent, *in vitro* production of prothrombin, *Biochem. Biophys. Res. Commun.* **60**:1397–1402.

Shah, D. V., Suttie, J. W., and Grant, G. A., 1973, A rat liver protein with potential thrombin activity: Properties and partial purification, *Arch. Biochem. Biophys.* **159**:483–491.

Stahmann, M. A., Huebner, C. F., and Link, K. P., 1941, Studies on the hemorrhagic sweet clover disease. V. Identification and synthesis of the hemorrhagic agent, *J. Biol. Chem.* **138**:513–527.

Stenflo, J., 1970, Dicumarol-induced prothrombin in bovine plasma, *Acta Chem. Scand.* **24**:3762–3763.

Stenflo, J., 1972, Vitamin K and the biosynthesis of prothrombin II. Structural comparison of normal and dicoumarol-induced bovine prothrombin, *J. Biol. Chem.* **247**:8167–8175.

Stenflo, J., 1973, Vitamin K and the biosynthesis of prothrombin III. Structural comparison of an NH_2 terminal fragment from normal and from dicoumarol-induced bovine prothrombin, *J. Biol. Chem.* **248**:6325–6332.

Stenflo, J., 1974, Vitamin K and the biosynthesis of prothrombin IV. Isolation of peptides containing prosthetic groups from normal prothrombin and the corresponding peptides from dicoumarol-induced prothrombin, *J. Biol. Chem.* **249**:5527–5535.

Stanflo, J., 1976, A new vitamin K-dependent protein, purification from bovine plasma and preliminary characterization, *J. Biol. Chem.* **251**:355–363.

Stenflo, J., and Ganrot, P. O., 1972, Vitamin K and the biosynthesis of prothrombin I. Identification and purification of a dicoumarol-induced abnormal prothrombin from bovine plasma, *J. Biol. Chem.* **247**:8160–8166.

Stenflo, J., and Ganrot, P. O., 1973, Binding of Ca^{2+} to normal and dicoumarol-induced prothrombin, *Biochem. Biophys. Res. Commun.* **50**:98–104.

Stenflo, J., Fernlund, P., Egan, W., and Roepstorff, P., 1974, Vitamin K dependent modifications of glutamic acid residues in prothrombin, *Proc. Natl. Acad. Sci. USA* **71**:2730–2733.

Suttie, J. W., 1970, The effect of cycloheximide administration on Vitamin K-stimulated prothrombin formation, *Arch. Biochem. Biophys.* **141**:571–578.

Suttie, J. W., 1973a, Mechanism of action of vitamin K: Demonstration of a liver precursor of prothrombin, *Science* **179**:192–194.

Suttie, J. W., 1973b, Anticoagulant-resistant rats: Possible control by the use of the chloro analog of vitamin K, *Science* **180**:741–743.

Suttie, J. W., 1974, Metabolism and properties of a liver precursor to prothrombin, *Vit. Horm.* **32**:463–481.

Suttie, J. W., 1975, Postribosomal synthesis of prothrombin under the influence of vitamin K, in

Prothrombin and Related Coagulation Factors (H. C. Hemker and J. J. Veltkamp, eds.), pp. 134–151, Leiden Univ. Press, Leiden.

Suttie, J. W., Grant, G. A., Esmon, C. T., and Shah, D. V., 1974, Postribosomal function of vitamin K in prothrombin synthesis, *Mayo Clin. Proc.* **49**:933–940.

Suttie, J. W., Hageman, J. M., Lehrman, S. R., and Rich, D. H., 1976, Vitamin K dependent carboxylase: Development of a peptide substrate, *J. Biol. Chem.* **251**:5827–5830.

Thierry, M. J., and Suttie, J. W., 1971, Effect of Warfarin and the chloro analog of vitamin K on phylloquinone metabolism, *Arch. Biochem. Biophys.* **147**:430–435.

Thierry, M. J., Hermodson, M. A., and Suttie, J. W., 1970, Vitamin K and Warfarin distribution and metabolism in the Warfarin-resistant rat, *Am. J. Physiol.* **219**:854–859.

Wallin, R., and Prydz, H., 1975, Purification from bovine blood of the warfarin-induced precursor of prothrombin, *Biochem. Biophys. Res. Commun.* **62**:398–406.

Weber, F., and Wiss, O., 1971, Vitamin K group: Active compounds and antagonists, in *The Vitamins* (W. H. Sebrell and R. S. Harris, eds.), Vol. III, 2nd ed., pp. 457–466, Academic Press, New York.

Willingham, A. K., and Matschiner, J. T., 1974, Changes in phylloquinone epoxidase activity related to prothrombin synthesis and microsomal clotting activity in the rat, *Biochem. J.* **140**:435–441.

Willingham, A. K., Laliberte, R. E., Bell, R. G., and Matschiner, J. T., 1976, Inhibition of vitamin K epoxidase by two non-coumarin anticoagulants, *Biochem. Pharmacol.* **25**:1063–1066.

Woolf, I. L., and Babior, B. M., 1972, Vitamin K and Warfarin. Metabolism, function and interaction, *Am. J. Med.* **53**:261–267.

Zimmermann, A., and Matschiner, J. T., 1974, Biochemical basis of hereditary resistance to Warfarin the rat, *Biochem. Pharmacol.* **23**:1033–1040.

Chapter 2

The Metabolic Significance of Dietary Chromium

Cihad T. Gürson

1. Introduction

The recognition of the role of chromium as a trace element in nutrition during the course of the past 15 to 20 years (Schwarz and Mertz, 1959), along with a realization of its importance as an occupational health hazard when used in its 6-valence state in the galvanotechnical field, has led to intensive research on chromium. The incidence of diseases of the respiratory system is high among workers who deal with chromium (Baetjer, 1950). The frequent occurrence of bronchial and lung cancer in individuals working in atmospheres highly polluted with chromium prompted efforts toward prevention of, as well as the development of methods of measurement of, excessive exposure to chromium. As a result of investigations in this area, it was established that chromium excreted in the urine could be used as a criterion for the assessment of degree of exposure to chromium (Franzen et al., 1970). Urine is also the biological sample used in studying chromium nutrition in humans.

Our information on chromium as a trace element dates back to experimental data on rats. This series of experiments is also considered the first to draw attention to the metabolic significance of chromium in the diet (Mertz and Schwarz, 1955, 1959). Sprague-Dawley rats fed a 30% Torula yeast diet in the postweaning period, were shown to have impaired glucose metabolism by means of glucose tolerance tests. It was also observed that when small amounts of brewer's yeast or remnants of a hospital diet were added to Torula yeast diet, the glucose tolerance test (GTT) remained normal. These observations pointed to a

Cihad T. Gürson • Department of Pediatrics, University of Istanbul, Istanbul Medical Faculty, Çapa, Istanbul, Turkey.

nutritional factor, the absence of which resulted in impaired glucose metabolism. Because the metabolic significance of this factor appeared to be limited to glucose tolerance, it was named the "glucose tolerance factor" (GTF). Subsequent studies showed that 3-valence chromium was the element responsible for the metabolic effect exerted by brewer's yeast and pig kidney extract (Schwarz and Mertz, 1959).

2.   Chromium Metabolism—Glucose Tolerance Factor

2.1.   Chemical Structure of Glucose Tolerance Factor

The chemical structure of GTF has not yet been described in full. Mertz *et al.* (1974) have compared the metabolic effects of synthetic organic and inorganic chromium salts with those of brewer's yeast extracts and foodstuffs known to contain GTF. Only quantitative differences were noted between the two groups of substances in (1) enhancing the effect of insulin *in vitro,* (2) correcting impaired GTT, and (3) influencing intestinal absorption. However, salts of GTF character, unlike the synthetic chromium salts, entered the chromium pool of the body in addition and to crossing the placenta. But because these conclusions are based on animal studies, it is not known to what extent they apply to human subjects. The molecular weight of GTF is between 300 and 500. It is water-soluble and resistant to most chemical treatments (Mertz, 1974a). Experimental data support the hypothesis that chromium exists in the form of a tetra-aqua-dinicotinate complex in the factor that potentiates the insulin effect. The binding of one amino acid (glycine, glutamic acid, and cysteine) to each of the four coplanar coordination sites provides for the stability of chromium during its transport in the body (Fig. 1.).

2.2.   Sources of Chromium

2.2.1.   Chromium Content in Food

Assessment of the chromium content of foodstuffs poses many technical problems. Variations among the reports of different workers result mainly from differences in methodology. It is reported that with the existing methods for chromium determination, some chromium is lost in the form of volatile organic compounds (Mertz *et al.,* 1974; Wolf, 1975). Furthermore, measurement of total chromium content in food substances may not be indicative of the amount that is biologically active. It is not as yet possible to differentiate chemically those chromium compounds that are biologically active from those that are biologically inert. The biological activity of chromium in the tested material is defined as the degree of *in vitro* oxidation of ^{14}C-glucose in the presence of insulin found

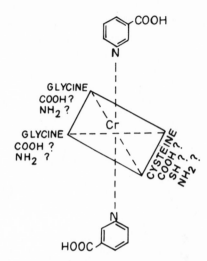

Fig. 1. Possible structure of dinicotinato–amino acid–Cr complex (reproduced by permission of Federation Proceedings, from Mertz et al., 1974).

in the adipose tissue of chromium-deficient rats. Although no correlation exists between biological activity and the chromium content of food as determined by extraction with acids, a significant correlation was found when the extractions were carried out using 50% ethanol (Toepfer et al., 1973).

Jones and co-workers (1975) cultivated brewer's yeast cells with ^{51}Cr(III), and performed an extraction with 50% ethanol, thereby obtaining samples known to contain high amounts of GTF. They then subjected these samples to oven-drying, freeze-drying, wet digestion, and dry-ashing. No losses attributable to volatilization occurred during the course of these procedures. Determinations repeated in platinum crucibles yielded the same results. These results led the authors to infer that these same procedures might be applicable to other biological materials as well. In contrast, in experiments done with brewer's yeast and beef liver, Wolf (1975) established lower concentrations of chromium in these substances when the assessment was done by direct- and low-temperature ashing, as opposed to a procedure whereby the samples are oven-dried first. In this author's experience, congruent results were obtained by dry-ashing and by digestion with HNO_3/H_2SO_4 in a closed system. These results have led the author to speculate that the methods of digestion or ashing of the food samples either cause loss of chromium or are inefficient in transforming all the available chromium in the sample into a state analyzable by flameless atomic absorption spectrometry (AAS).

Behne et al. (1976) also used brewer's yeast as their test material. The chromium content of brewer's yeast was determined directly by AAS, as well as subsequent to ashing the sample in a variety of ways, such as in open vessels with different temperatures and in closed pressure vessels, to determine how the various methods affected the volatile organic chromium salts in the samples. These authors also determined chromium in brewer's yeast by neutron activation analysis. Similar to the previous experiments with AAS, chromium was irradiated directly, as well as subsequent to heating, or after oxygen plasma ashing. Behne *et al.* (1976) point out that in the course of drying and ashing, heat may cause a loss of volatile organic salts present in brewer's yeast.

Following the work of Schroeder and associates (1962) the chromium content of foodstuffs has been investigated by several other workers (Maxia *et al.*, 1972; Toepfer *et al.*, 1973). The substance with the highest known concentration of chromium was brewer's yeast. Spices such as black pepper, thyme, and cloves came next. Fruits and vegetables are low in chromium content. On the other hand, vegetable oils and unrefined sugars have relatively high chromium concentrations. With the exception of kidney and liver, meat and fish are generally poor in chromium. It has been established that human milk has a higher chromium content than cow's milk.

2.2.2.   Placental Transport of Chromium

Experimental evidence indicates that chromium salts, administered orally or intravenously to pregnant animals, do not cross the placenta unless they are of GTF composition (Mertz and Roginski, 1971). Consequently, in order to reach the fetus, chromium compounds must have a structure different from that of simple chromium compounds. When extracts from brewer's yeast grown in labeled chromium containing culture medium were administered to pregnant rats, ^{51}Cr was accumulated at a higher rate by fetal tissue as compared to the corresponding maternal tissue. Postmortem chromium determinations have also shown that fetal tissue is richer in chromium than is adult tissue.

2.2.3.   Chromium Content of Drinking Water

The investigation on the chromium content of drinking waters has been limited to a few countries. Studies by Durfor and Becker (1962) brought to light the very low concentrations of chromium in drinking water in the United States. Other authors have suggested a relationship between chromium deficiency in malnutrition and chromium content in food and drinking water (Majaj and Hopkins, 1966; Carter *et al.*, 1968).

2.3.   Intestinal Absorption of Chromium

The intestinal absorption rate of 3-valence chromium salts is low. The experimental work of Donaldson and Barreras (1966) in humans and animals has shown that when ^{51}Cr and $Na_2{}^{51}CrO_4$ are administered orally in trace quantities, they are almost completely recovered in the stools. When $Na_2{}^{51}CrO_4$ was directly administered to the intestines by jejunal perfusion, significant absorption occurred. Gastric acid secretion was found to inhibit the intestinal absorption of this substance *in vitro* and *in vivo*. The authors suggest that a conversion of 6-valence chromium to poorly absorbed 3-valence chromium compounds by gastric acidity may explain these findings. Experiments with $^{51}CrCl_3$ showed that this substance had an intestinal absorption rate of 0.5–3.0% and chromic acetate was found to be absorbed at a rate of 1.5% (Mertz and Roginski, 1971). Therefore, considering that only 0.5–3.9% of ingested chromium is absorbed, it becomes obvious that the average daily intake of 60 μg (5–115 μg) of chromium will not compensate for the chromium lost in the urine and by other routes. It may therefore be assumed that chromium compounds in foodstuffs have a higher intestinal absorption rate than that determined for inorganic chromium salts.

Inorganic chromium salts have to be transformed into more easily absorbable compounds, probably into GTF, in the intestines before they can be absorbed. The rate of chromium absorption is not age-dependent, absorption rate of inorganic chromium compounds in aged subjects being essentially the same as that in younger subjects (Doisy *et al.,* 1971). The intestinal absorption of chromium is enhanced in a number of conditions. In insulin-dependent diabetics it is reported to be two to four times greater than absorption in normal individuals (Doisy *et al.,* 1971).

In malnourished infants and children a single dose will suffice to revert the GTT to normal (Majaj and Hopkins, 1966; Gürson and Saner, 1971). This difference is probably attributable to the fact that inorganic chromium salts are more easily converted to absorbable chromium compounds in the pediatric age group. However, reports of an increased rate of chromium absorption in disease states make it difficult to decide whether enhanced chromium absorption in malnourished infants is a function of age or is the reflection of a pathological state. We observed an increase in glucose removal rate (GRR) from an initial value of 2.13%/min to 5.18%/min following the administration of a single dose of chromium in an 11-year-old patient suffering from malnutrition. A mechanism similar to that encountered in insulin-dependent diabetics might be at work in this situation, a probability that requires further exploration. A variety of substances may impair or facilitate intestinal chromium absorption, i.e., absorption could be impaired in the presence of Fe, Mn, and Ti, while being enhanced with chromium chelates. Cu and Zn are known to have no effect on chromium absorption.

Part of the absorbed chromium is bound to the transferrin fraction of β-globulin, the remainder becoming GTF-bound chromium. It is in these two states that this trace element is transported to the sites involved in the metabolism of chromium.

2.4. Deposition and Content of Chromium in the Tissues

Radioactive chromium is taken up by all organs and tissues but concentrates mainly in the kidneys, spleen, epididymis, testes, and bones. Organic chromium salts extracted from brewer's yeast (i.e., GTF) show a predilection for liver tissue when administered to animals. This selective deposition of GTF chromium is possibly evidence for the existence of a chromium-specific homeostatic system (Mertz, 1969). Chromium determinations on postmortem material from man in several parts of the world and in several animal species, wild and domestic, have demonstrated that chromium is always present in liver and kidney tissue, regardless of species (Tipton and Cook, 1963; Tipton *et al.*, 1965; Schroeder, 1968). Chromium content showed a decrease with increasing age in humans; this correlation with age was most striking in studies in the United States. Differences in chromium content related to age were not observed in laboratory animals. These findings are taken as evidence that, in countries in which the diet contains refined sugar, marginal chromium deficiency occurs.

2.5. Excretion of Chromium

The major route of excretion for chromium is the urine. Chromium is also excreted in stools in smaller amounts. Laboratory experiments with animals have shown 80% of administered chromium can be recovered in the urine. The contamination of animal feces with urine makes quantitative assessment of fecal chromium difficult. In several studies in which chromium was determined in the stools collected for periods of 8 hr to 8 days following an intravenously administered chromium load, stool chromium was found to vary from 0 to 20% of the administered dose (Mertz, 1969).

Collins and associates (1961) studied urinary chromium excretion in dogs. According to these authors, urinary excretion is regulated by glomerular filtration rate and tubular reabsorption. Chromium excreted by the renal route is dialyzable. At least 63% of filtered chromium is reabsorbed. A recent study performed on dogs has shown a reabsorption rate of more than 97% (Davidson and Parker, 1974).

2.6. The Mode of Action of Chromium

2.6.1. Chromium in Carbohydrate Metabolism

Chromium plays an active role in carbohydrate metabolism. Experimental data suggest that chromium acts as a cofactor in initiating insulin activity. Figure 2 depicts the hypothesis advanced for the mechanism of chromium effect (Mertz *et al.*, 1974). It will be noted that for a chromium complex with 6-coordination sites, two of these sites become attached to the sulfhydryl groups on the cell membrane and two join with disulfides of the A chain of the insulin molecule. It is postulated that the differences in metabolic effect that exist between GTF extracted from brewer's yeast and the synthetic chromium compounds are related to the remaining two coordination sites. By combining with

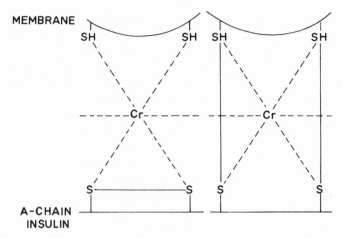

Fig. 2. Hypothetical ternary complex of chromium at site of action (reproduced by permission of Federation Proceedings, from Mertz *et al.*, 1974).

the basic groups of insulin and those of tissue receptors, two acid ligands bring insulin closer to its site of activity. It is speculated that the four water molecules of this complex are exchanged with the four sulfur atoms involved in the reactions of disulfide exchange. Whereas the amino acids increase the solubility of the complex, they also play a role in the formation of more stable compounds for transport of chromium in the body. The addition of active chromium salts to chromium-deficient tissues, such as the adipose tissue of rat epididymis, enhances the glucose uptake in response to insulin, with a concomitant increase in the galactose uptake (Mertz and Roginski, 1963). This increase in intracellular concentration of galactose is not followed by any metabolic activity related to insulin. These observations indicate that chromium exerts its effect on the insulin-sensitive cell membrane.

Experiments aiming to establish the relationship between blood chromium and blood glucose were initiated by Glinsmann and Mertz (1966) who investigated the variations in blood chromium level following GTT. According to these authors, a glucose load leads to increased blood chromium levels in healthy normal subjects and in diabetics in whom GTT has reverted to normal. Furthermore, blood chromium shows a similar increase following glucose intake in chromium-supplemented diabetic subjects in whom GTT remains abnormal, whereas no rise is observed in diabetics with abnormal GTT who do not receive chromium supplementation.

Levine *et al.* (1968) demonstrated a significant increase in serum chromium levels following a glucose load in young as well as in aged subjects. However, the increase was of smaller magnitude in aged subjects whose GTT did not respond to chromium supplementation. The data of Behne and Diehl (1972), obtained on four normal healthy subjects, show that the maximal increase in blood chromium occurred between 60 and 120 min of glucose load administration. Hambidge (1971), along with a number of other workers, considered the change in blood chromium level in response to an oral glucose load an index for assessing the chromium nutrition state. Hambidge reported that a concomitant increase occurred in immunoreactive insulin in nearly 70% of the cases in which chromium levels rose. It has recently been reported that following insulin administration, an acute increase in blood chromium level occurs in experimental animals and in human subjects. The increase observed in plasma chromium with glucose is probably an indirect effect mediated through insulin. Both glucose and insulin can exert their plasma chromium-increasing effect only in the presence of an adequate chromium pool (Mertz *et al.*, 1974). Doisy and associates (1976), reviewing other workers' experience with oral GTT and incorporating their own results with these, have concluded that the magnitude of the insulin increase observed following glucose loading is inversely proportional to tissue chromium content, i.e., with body chromium pool. According to these authors, in old-age subjects, tissue chromium content is low, whereas insulin response to GTT is increased. In contrast to these findings, Davidson and Burt (1973) reported a decrease in plasma chromium level following oral or intravenous GTT in normal individuals.

2.6.2. Effect of Chromium on Enzymes

In addition to its role in glucose metabolism, chromium has been reported to have an activating effect on such enzymes as succinic cytochrome c dehydrogenase, phosphoglucomutase, and trypsin *in vitro*. This activating effect appears to be related to the concentration of chromium in the medium (Mertz, 1969). Chromium may also have an effect on ribosomes (Wacker and Vallee, 1959). It is possible that chromium facilitates the incorporation of amino acids into insulin-activated proteins. Mitochondrial membrane changes may be considered another example of the effect of chromium at the intracellular level; the swelling of mitochondria in chromium-deficient tissue increases on the addition of chromium (Campbell and Mertz, 1963). Experiments on rats under conditions of increased protein synthesis demonstrate a marked intracellular shift in chromium and a shift of ^{51}Cr from the nucleus to the cytoplasm (Mathur and Doisy, 1972).

2.6.3. Effect of Chromium on Lipid Metabolism

Chromium also exerts an effect on lipid metabolism. Using labeled acetate, Curran (1954) was the first to demonstrate that chromium enhances the synthesis of cholesterol and of fatty acids in rat liver. Later, Schroeder and associates (1962) observed that chromium supplementation caused a decrease in the blood cholesterol of rats kept on a low-chromium diet. It has also been demonstrated that when 5 ppm chromium are added daily to their drinking water, rats fed a chromium-free diet in the postweaning period have lower blood cholesterol levels than do control animals. Another relevant observation was than an increase in blood cholesterol levels occurred with increasing age in male rats fed a low-chromium diet, whereas no such increase occurred in animals receiving an adequate supply of chromium.

In a group of rats left to die a natural death, the lipid content of the aorta in those animals receiving chromium supplementation did not differ from that of the control group; however, a smaller number of lipid plaques were observed (Schroeder and Balassa, 1965). In rats kept under standard laboratory conditions on a hypercholesterolemic diet the administration of Cr(III) caused a fall in the serum cholesterol level (Staub *et al.*, 1969).

Schroeder *et al.* (1970) established a difference in chromium content in aortic tissue removed during autopsy from individuals who died as a result of atherosclerotic heart disease and from those who died in traffic accidents. Aortic tissue contained no chromium in the former group, whereas it could be found in traffic death group. Basing their hypothesis on these findings and on the role of chromium in the catabolism and excretion of cholesterol, these authors propose that chromium deficiency may have a basic role in atherosclerosis. Doisy and associates (1976) established a decrease in cholesterol levels in individuals on chromium supplementation and showed that this decrease was greater if the initial cholesterol level was higher than 240 mg/dl, a value taken as the upper limit of normal. Whereas normal levels of triglycerides were not affected by chromium supplementation, a fall was observed when the initial values were high.

3. Chromium Deficiency

Chromium deficiency is defined as the state of an organism in which the intake of biologically active chromium fails to meet the necessary requirements. Because the existing methods of analysis are still inadequate for accurate measurement of biologically active chromium compounds, various direct and indirect methods have been used to evaluate chromium nutrition.

3.1. Experiments on Animals

Under experimental conditions in an environment totally free of chromium contamination, alterations such as growth failure, glycosuria, fasting hyperglycemia, hypercholesterolemia, and sclerotic aortic plaques have been observed in rats fed a chromium-free diet (Schroeder, 1968). Results of experiments performed on monkeys have confirmed these findings (Davidson and Blackwell, 1968). Stress tolerance was significantly lower in chromium-deficient animals as compared to those receiving chromium (Mertz and Roginski, 1969).

3.2. Chromium Deficiency in Man and Assessment of Chromium Nutrition

Findings related to chromium deficiency in humans have been reported in older subjects, diabetics, pregnant women, and malnourished children. Presumably, future studies will link the established effect of chromium on lipid metabolism to the role of this trace element in the pathogenesis of atherosclerotic diseases.

In a unique case of isolated chromium deficiency, a subject kept on total parenteral nutrition for 3.5 years, Jeejeebhoy *et al.* (1975) observed impaired glucose tolerance and the glucose utilization. At the same time, insulin levels remained normal and caused no alterations either in the effect of insulin on amino acid uptake or on free fatty acid release. Results obtained in cases of malnutrition in the current experiment are indicated in Table I (Gürson and Saner, 1972, unpublished data). In agreement with Jeejeebhoy *et al.* (1975) this table shows that in the different stages of malnutrition, chromium, which did correct GTT, had no effect on free fatty acids.

Table I. Mean Values for FFA, Fasting, after I.V. GTT, and before and after Chromium [a]

	Blood FFA, mEq/liter					
	Before Cr (min after i.v. glucose)				After Cr (min after i.v. glucose)	
Fasting	30	60	Fasting	30	60
	Admission					
0.479 ±0.138 (10)	0.353 ±0.125 (10)	0.353 ±0.147 (10)	0.476 ±0.122 (9)	0.342 ±0.105 (9)	0.386 ±0.151 (9)
	After treatment					
0.493 ±0.163 (8)	0.325 ±0.134 (8)	0.343 ±0.110 (8)	0.438 ±0.153 (5)	0.282 ±0.079 (5)	0.337 ±0.076 (5)

[a]FFA: *At admission:*
  Before chromium, fasting vs. 30 min $p < 0.02$;
  after chromium, fasting vs. 30 min $p < 0.01$.
  After treatment:
  Before chromium, fasting vs. 30 min $p < 0.05$;
  after chromium, fasting vs. 30 min $p < 0.05$.
  Results are expressed as mean ±SD. Figures in parentheses are numbers of cases.

Methods used in diagnosing chromium deficiency in humans can be considered under two headings, i.e., direct and indirect methods. The direct method relies on the determination of chromium in blood, urine, hair, or tissues, whereas the indirect method uses the correction of abnormality in glucose metabolism by administration of chromium as the criterion for its assessment.

3.2.1. Blood Chromium

Considering recent developments in the field of chromium research the reliability of methods for determining blood chromium levels has been reevaluated. Values for blood chromium ranging from 1 to 4.7 ng/ml have been reported in studies from the United States (Doisy *et al.*, 1976). A series of experiments based on plasma chromium determinations and using ^{51}Cr, have led to the conclusion that fasting blood chromium levels do not adequately reflect, and cannot be used as the sole criterion for, the chromium nutritional state. On the other hand, as discussed earlier, fluctuations in the blood chromium level following GTT are significant indicators of chromium nutrition.

3.2.2. Urinary Chromium

Assessment of chromium excreted in the urine under various experimental conditions may be used as an indicator of the chromium nutritional state. The volatility of organic chromium salts leads to a number of difficulties in methodology.

It has been established that measurements of urinary chromium in AAS show higher values in urine samples subjected to low-temperature ashing prior to treatment in the graphite oven, as compared to samples for which this procedure was not used. These results point to the presence of chromium in two different fractions in the urine, one that can be recovered by direct analysis, in contrast to the assessment of the second fraction, which requires low-temperature ashing (Wolf, 1975). It has recently been suggested that chromium is probably excreted as a mixture of different compounds, with independent assessment of total chromium and the nonvolatile fraction being possible. The difference between these two determinations would give the value for volatile chromium salts (Canfield and Doisy, 1976).

We have limited information on the urinary chromium in normal children. Hambidge (1971) has reported a daily excretion of 5.5 ± 2.9 μg in 18 children of a mean age of 8 (4–15 years of age). A study performed in Turkey on newborns yielded daily excretion values of 119 ± 13.9 ng for the first day of life, 154 ± 20.2 ng for the second day, 183 ± 25 ng for the third day and 141 ± 30.5 ng for the fourth day (Saner, 1975b). Urinary chromium excretion showed a significant increase on day 3, which suggests that the initiation of chromium homeostasis in the human infant occurs on the third day of life. On day 1, urinary chromium excretion remained unaltered following an intravenous glucose load. This finding may be taken as another piece of evidence supporting the view that chromium homeostasis is not yet functioning in the early newborn (Saner, 1975a).

In normal Turkish children of older age groups, daily excretion of chromium was reported to be 0.75 ± 0.42 μg for infants, 1.70 ± 0.85 μg for children of preschool age, and 1.8 ± 0.9 μg for schoolchildren (Gürson *et al.,* 1975). The mean value for the total series of 31 children was 1.50 ± 0.89 μg/day. These values are low compared to those reported by Hambidge on American children. The concentration of chromium in urine in the Turkish series was 2.1 ± 1.25 ng/ml for infants, 4.82 ± 1.59 ng/ml for preschool age, and 2.75 ±1.43 ng/ml for schoolchildren. The overall mean value was 3.40 ± 1.83 ng/ml (Table II). These figures indicate that the highest urinary chromium concentrations are in the preschool age group. In contrast, daily chromium excretion showed a stepwise increase with age. In two independent studies in the United States, urinary chromium excretion was investigated in young adult subjects. Reported mean values were 8.4 μg (range 2 to 21 μg) and 8.18 ± 0.43 μg (Hambidge, 1971; Mitman *et al.,* 1974). In a previous study on five healthy normal individuals whose ages ranged between 20 to 35, the daily chromium excretion and concentration in urine were 3.08 ± 1.90 μg/day and 2.69 ± 1.54 ng/ml, respectively.

Recently we repeated these determinations in another series of eight normal subjects. Urinary chromium was measured on 24-hr samples and on 4-hr samples in the same subjects. The 4-hr samples were collected following a 10-hr fast. As noted in Table III, chromium excretion, expressed in terms of excretion per minute, chromium/creatinine ratios, and absolute and calculated daily excretion, was essentially the same. Higher values for 24-hr urinary excretion, as compared to our older data, were observed in this recent study. Since identical techniques were employed in both studies, the difference may be attributed to the small

Table II. Pediatric Age Group (Normally Developed Children)[a]

	Urine vol. (ml/day)	Urinary Cr concn. (ng/ml)	Cr excretion (ng/day)
Infants (<2 years)	361 ± 137 (7)	2.1 ± 1.25 (7)	748 ± 422 (7)
Preschool children (2–7 years)	341 ± 114 (12)	4.82 ± 1.59 (12)	1692 ± 905 (12)
School children (>7 years)	688 ± 271 (12)	2.75 ± 1.43 (12)	1808 ± 851 (12)
Total group	480 ± 252 (31)	3.40 ± 1.83 (31)	1524 ± 886 (31)

[a]Results are expressed as mean ± SD. Figures in parentheses are numbers of cases.

Table III. Urinary Chromium Excretion[a]

	Urine collected for 24 hr			Urine collected for 4 hr		
No.	μg/day	ng/min	Cr(ng)/creatinine (mg)	Expected Cr for 24 hr (μg)	ng/min	Cr(ng)/creatinine(mg)
1	9.404	6.53	7.89	6.960	4.83	6.59
2	6.522	4.53	5.23	7.776	5.40	5.59
3	17.214	11.95	8.85	17.112	11.88	9.38
4	11.638	8.08	9.20	11.232	7.80	9.18
5	9.967	6.92	6.64	11.952	8.30	9.40
6	9.098	6.32	5.79	11.347	7.88	6.96
7	13.780	9.57	8.41	12.082	8.39	7.21
8	12.858	8.93	6.69	8.842	6.14	5.10
Mean	11.310	7.85	7.34	10.913	7.58	7.43
±SD	3.311	2.30	1.46	3.183	2.21	1.74

[a]There is no significant difference between these values.

sample used in the first study, or it may point to the possible occurrence of isolated chromium deficiency unaccompanied by symptoms of other deficiencies in young adult subjects. This latter assumption is based on the important observation on isolated chromium deficiency reported by Jeejeebhoy and associates (1975).

The assessment of chromium nutrition by application of GTT in humans has two drawbacks—the necessity of making punctures for collecting the samples and the need for laboratory facilities. In an earlier study it was possible to show a significant negative correlation between glucose removal rate and urinary chromium concentrations and daily excretion of chromium in a series of 15 normal children ($r = -0.8425, p < 0.001; r = -0.5494, p < 0.05$, respectively) (Saner et al., 1974). The same correlation was found in infants with different degrees of malnutrition as well as in pregnant subjects. These findings are discussed in greater detail later.

The inadequacy of fasting blood chromium levels for the assessment of chromium nutrition has been reported by several independent workers who also suggested that estimation of changes in plasma chromium concentrations following GTT could be a reliable index of chromium nutrition. An increase in chromium following GTT occurs in normal individuals. In contrast to this latter finding, Davidson and Burt (1973) reported a fall in plasma chromium level following oral or intravenous GTT in normal individuals. An increase in urinary chromium excretion is an expected finding when plasma chromium is elevated. Schroeder (1968) observed a twofold increase in urinary chromium from the initial values, 2 hr after GTT, in diabetic subjects. In contrast, Davidson and Parker (1974) reported a significant fall in urinary chromium excretion and in the chromium/creatinine ratio after a standard oral GTT.

To assess the effect of oral GTT on urinary chromium excretion, 4-hr urine samples were collected following a 10- to 12-hr fasting state before and after GTT in 10 normal, apparently healthy adults. Plasma glucose values, chromium excretion per minute, and chromium/creatinine values for these subjects are given in Table IV.

Prior to oral GTT, urinary chromium excretion was 6.06 ± 1.45 ng/min and the chromium/creatinine ratio 7.17 ± 2.81. Following GTT, these values increased to 9.21 ± 3.16 ng/min and 10.70 ± 4.91, respectively. This increase was found significant for both parameters ($t = 2.9831$, $p < 0.02$ for chromium excretion per minute; $t = 2.8131$, $p < 0.05$ for chromium/creatinine). When the results for each case were analyzed individually, these increases were observed in all but two of the subjects.

3.3.3. Chromium in Hair

Determination of chromium in hair samples is another suggested direct method for the assessment of chromium nutrition. Postmortem tissue analyses in the newborn have shown a higher tissue chromium content for this period of life than for older age groups (Tipton *et al.*, 1965). Subsequent studies on hair samples have reported that newborns have higher concentrations of chromium in hair compared to their mothers (Hambidge, 1971). In this group of 15 newborns and their mothers, Hambidge observed only a single case in which the mother's hair-chromium concentration was higher than that of her newborn baby.

Table IV. Urinary Chromium Excretion in Normal Subjects before and after Oral GTT [a]

No.	Fasting	Blood glucose mg/dl min after oral GTT					Urinary chromium before oral GTT		Excretion after oral GTT	
		30	60	90	120	180	ng/min	Cr/creat.	ng/min	Cr/creat.
1	72	134	100	85	80	75	5.28	4.30	7.12	7.31
1	84	144	130	119	100	81	5.88	10.23	10.20	15.41
3	73	109	83	89	91	—	8.51	7.51	13.40	17.97
4	73	97	94	100	91	—	6.14	13.60	12.19	16.85
5	86	117	110	83	117	104	8.02	7.19	7.53	6.23
6	88	123	91	116	83	78	4.91	6.60	6.26	7.09
7	86	109	78	89	81	—	6.79	6.06	7.03	7.32
8	86	187	134	119	71	67	3.56	4.44	11.52	12.12
9	75	116	103	86	116	61	5.88	5.99	4.23	4.18
10	80	100	94	100	86	61	5.65	5.81	12.60	12.54
Mean	80	124	102	99	92	75	6.06	7.17	9.21	10.70
±SD	6	26	18	15	15	15	1.45	2.81	3.16	4.91

[a] Urinary chromium excretion: ng/min, before vs. after OGTT $p < 0.02$; chromium/creatinine ratio, before vs. after OGTT $p < 0.05$.

In a study done on 50 pregnant Turkish women and their newborn babies, mean hair-chromium concentrations for the mothers and for the newborns were 203 ± 107 ng/g and 119 ± 75 ng/g, respectively (Saner and Gürson, 1976a). However, when each mother–infant pair was considered individually, 12 newborns were found to have higher hair-chromium concentrations, although mean hair-chromium concentration in the total group of newborns did not differ significantly from that of the mothers. This group of 12 newborns probably received greater quantities of chromium from their mothers, thereby depleting the chromium pool of their mothers to a greater extent than did the newborns with low hair-chromium concentrations. Indeed, there was a significant difference in hair-chromium concentrations between the two groups of mothers. Whereas such parameters as mother's age and parity were not related to hair-chromium concentrations in this group of newborns, hair-chromium concentrations showed a significant correlation with birth weight. The group of newborns with high hair-chromium concentrations had significantly higher birth weights. A group of premature children whose gestational ages were 30 to 36 weeks, and a small group of infants with intrauterine growth retardation, were found to have significantly low hair-chromium concentrations compared to full-term babies (Hambidge and Baum, 1972). A similar gestational age-related change in the chromium content of tissues was shown by analysis of fetal ashed bones (Pribluda, 1963).

Hair-chromium concentrations were determined in a healthy population of children and adults. The infant and preschool age group consisted of cases that were followed in a private clinic and of infants and children in an orphanage. The school-age group was selected from a hospital outpatient population. Subjects with anthropometric measurements above 80% of accepted weight and height standards were considered to be in a normal nutritional state. Another group consisted of subjects between the ages of 20 and 35 with no known nutritional or metabolic disease; no women in this group had delivered children. Results on this group, including mean values and number of cases with hair-chromium concentrations below 200 ng/g, are indicated in Table V. Values below 200 ng/g were not encountered in any of the six private clinic infants. With the exception of two siblings, hair-chromium concentrations in all private-clinic preschool-age subjects were higher than 200 ng/g. The elder of these siblings had 233 ng/g of hair-chromium concentration when tested at 13 months; at 61 months he had an infant sister who had been subjected to a special long-term diet for her severe eczema. The hair-chromium values of the siblings assessed at that time were 151 ng/g and 134 ng/g, respectively.

The study on the orphanage children demonstrated a hair-chromium concentration lower than 200 ng/g in one infant and in 8 of the 12 preschool-age subjects. Three of the 11 cases of school-age children and 7 of the 17 young adults had values below 200 ng/g. These results would appear to agree with Hambidge, who suggested a value of 200 ng/g as the lower limit of normal for

Table V. Hair-Chromium Concentrations in Normal Subjects[a]

Group	Age	Mean ±SD	Hair-Cr concn. (<200 ng/g)
Normal			
Private clinic	Infants	408 ± 88	—
	(3–11 months)	(6)	
	Preschool	395 ± 142	134,161
	(1⅓–6½ years)	(10)	
Orphanage	Infants	377 ± 99	199
	(9–11 months)	(5)	
	Preschool children	190 ± 108	96,109,117,128,141
	(1⅓–6½ years)	(12)	144, 183, 187
Outpatient	Schoolchildren	322 ± 160	187, 188, 197
	(8–12 years)	(11)	
Staff	Adults	240 ± 123	79,126,129,135,135
	(23–35 years)	(17)	183,185

[a] Figures in parentheses are numbers of cases.

the U.S. population. These results lead one to speculate that marginal chromium deficiency may exist in children whose anthropometric measurements lie within normal limits, as well as in healthy-looking adults, and that hair is a suitable biological sample for the estimation of body chromium nutrition.

3.4. Chromium Metabolism in Protein-Energy Malnutrition (PEM)

Aberrations in carbohydrate metabolism in PEM of the marasmic or kwashiorkor type are well known (Gürson, 1972). Hypoglycemia in these patients has been an issue of conflict, reported as a frequent finding by a number of authors, whereas the experience of others has been different. A decrease in glucose removal rate following an i.v. GTT constitutes another important abnormality of carbohydrate metabolism in malnutrition. Majaj and Hopkins (1966) determined fasting blood glucose levels and removal rate following i.v. GTT in two groups of PEM subjects from two different regions in Jordan. The first group consisted of four patients from the mountainous area around Jerusalem who had been on a low-chromium diet. The five cases in the second group were from Jericho, which has a hot, dry climate and which leads to the ingestion of large amounts of water, hence to a relatively large intake of chromium. Hypoglycemia and low glucose removal rate following intravenous glucose load were established in the first group, whereas these aberrations of carbohydrate metabolism were nonexistent in the second group. When the first group of cases received a 250-μg dose of chromium chloride at 5 P.M. their fasting glycemia and removal rates converted to normal, as shown by tests performed the following day. These same workers, adding more cases to their study on PEM, established an increase in glucose removal rate from an initial mean value of 0.6–2.9 %-min following the administration of chromium. In a study from Nigeria, the glucose removal rate was reported to increase from 1.2–2.9 %-min after chromium, whereas no changes were observed in four control cases maintained on a hospital diet (Hopkins *et al.*, 1968). In contrast to these findings, Carter and associates (1968) established abnormal glucose tolerance tests in 33 out of 34 cases of kwashiorkor in Egypt, but noted no chromium effect on the diabetic curves. A high-protein, high-calorie diet caused the diabetic curves to revert to normal within a period of 1 to 2 weeks. These authors concluded that "Chromium deficiency was not responsible for the abnormal glucose utilization encountered in kwashiorkor cases in Egypt."

The effect of chromium on fasting glycemia and glucose removal rate was investigated in cases of marasmic PEM in this region (Gürson and Saner, 1971). Fourteen cases of marasmus with a mean age of 15 months were included in the study. An i.v. GTT was performed on all subjects within 24 hr of admission and following a fasting period of 10 to 12 hr. On the afternoon of the same day, at 5 P.M. a 250-μg dose of $CrCl_3 6H_2O$ dissolved in 2 ml water, was orally administered to each subject. GTT was repeated 15 hr later. Mean values for fasting blood glucose and increments in blood glucose, following GTT, both before and after chromium administration, are given in Table VI. Figure 3 shows the results on all 14 subjects. Significant increase in glucose removal rate occurred in nine of the cases; no change was observed in the remaining five cases. Fasting glycemia or glucose removal rates did not differ in GTT performed on two consecutive days under identical hospital conditions in a control group of PEM cases. These findings led to the conclusion that chromium was the factor responsible for the correction of glucose removal rate abnormality in the nine cases of PEM.

The effect of chromium on growth has been investigated in rats and mice. Retardation in growth was found to accompany chromium deficiency in these animals (Mertz and Roginski, 1969). Experimental data indicated that whereas a moderate degree of chromium deficiency led only to biological alterations, growth retardation also occurred when a number of stresses were superimposed on the deficiency state.

At present, information on the effect of chromium on growth in humans is lacking. We have established a significantly higher weight gain in PEM cases that responded to chromium by correction of impaired glucose tolerance as compared to controls and nonresponders, in subjects followed for a period of 30 days (Gürson and Saner, 1973). This observation suggests that in addition to protein and calories, chromium may be a necessary element for weight gain (Fig. 4).

It has been already mentioned that urinary chromium excretion or chromium concentration in hair can be accepted as criteria for evaluation of chromium nutrition. The PEM cases we have studied were classified according to their

Table VI. Mean Values for Fasting Blood Sugar, Increments in Blood Glucose after GTT, and Removal Rates before and after Chromium[a]

	Blood glucose (mg/100 ml)											
	Before Cr						After Cr					
	Fasting	(min postinjection)				Removal rate (%-min)	Fasting	(min postinjection)				Removal rate (%-min)
		15	30	45	60			15	30	45	60	
Admission (14)	58.8 ±13.7	242 ±44.6	210.7 ±47.7	169.9 ±32.7	144.9 ±23.6	1.71 ±0.72	61.4 ±11.9	209.9 ±43.8	177.0 ±39.3	124.2 ±32.4	97.9 ±33.1 (13)	3.91 ±2.56
After treatment (10)	67.9 ±9.1	246.8 ±52.3	188.4 ±50.0	116.4 ±53.3	86.3 ±53.2	5.51 ±2.12	76.8 ±16.7 (6)	234.8 ±32.7 (6)	164.3 ±43.4 (6)	99.3 ±35.1 (6)	81.1 ±28.4 (6)	7.02 ±3.7 (6)

[a]Removal Rate: At admission: Before chromium vs. after chromium, $p < 0.01$; admission vs. after treatment before chromium, $p < 0.001$; before chromium vs. after chromium (after treatment) 0.30 $p < 0.40$.

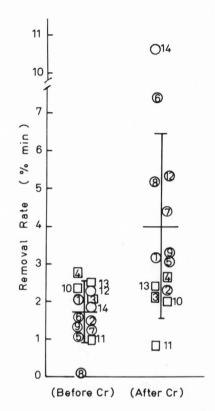

Fig. 3. Glucose removal rate in the individual cases. (O) Responders; (□) nonresponders (reproduced by permission of *American Journal of Clinical Nutrition,* from Gürson and Saner, 1971).

height and weight losses as either underweight or severe cases. A weight/height measurement 60–80% of the normal was considered underweight for any given age. Results of urinary chromium excretion in underweight infants and children are summarized in Table VII. The underweight group did not differ significantly from healthy controls in urinary chromium concentration values or in chromium excreted in the urine.

A significant negative correlation was found between glucose removal rate and urinary chromium concentrations and excretions both in underweight children and in normal children ($r = -0.7156, p < 0.01; r = -0.6373, p < 0.05; r = -0.8425, p < 0.001; r = -0.5494, p < 0.05$, respectively).

We have recently studied the urinary chromium excretion in 12 PEM cases. In these cases, the chromium excretion rate was significantly higher than in normal controls of the same age group (Table VIII). The relationship between urinary chromium excretion and glucose removal rate was investigated in nine cases. Urine was collected for 2–4-hr periods before and after i.v. GTT in these subjects. A significant negative correlation was found between the glucose removal rate and urinary chromium concentration ($r = -0.7602, p < 0.02$). When

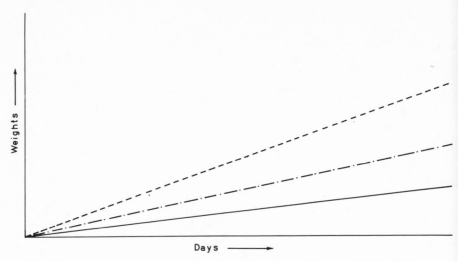

Fig. 4. Weight gain in the responder and nonresponder groups and in control infants. In the equation, y = body weight in grams and x = the number of days the patients were followed. (---) Responders: $y = 5262.83 + 34.88\,X$; $(-\cdot-)$ nonresponders: $y = 5737.56 + 20.60\,X$; (———controls: $y = 4743.06 + 11.50\,X$. Responders vs. nonresponders, $p < 0.02$; responders vs. controls, $p < 0.001$; nonresponders vs. controls, $0.1 < p < 0.2$ (reproduced by permission of *American Journal of Clinical Nutrition,* from Gürson and Saner, 1973).

the cases were classified as responders and nonresponders to chromium, according to the results of GTT, it was established that urinary chromium excretion was higher in those cases in which chromium failed to correct the GTT abnormality; however, the difference between the two groups was not statistically significant (Table IX). Similar findings were reported in insulin-dependent diabetic subjects (Doisy *et al.,* 1971; Hambidge, 1971). When values for urinary chromium excreted per minute before and after GTT were compared in PEM patients, no differences were noted in the responders whereas a small increase in urinary

Table VII. Urinary Chromium Excretion in Underweight Children[a]

Group	Urine vol. (ml/day)	Urinary Cr concn. (ng/ml)	Cr excretion (ng/day)
Infants (<2 years)	340 ± 149 (5)	4.24 ± 2.35 (5)	1382 ± 724 (5)
Preschool children (2–7 years)	333 ± 123 (12)	4.53 ± 1.44 (12)	1462 ± 696 (12)
Total group	335 ± 126 (17)	4.44 ± 1.68 (17)	1423 ± 684 (17)

[a]Results are expressed as mean ±SD. Figures in parentheses are numbers of cases.

Table VIII. Urinary Chromium Excretion in Normal
Controls and PEM[a]

Normal controls		PEM cases	
No.	ng/min	No.	ng/min
1	0.67	1	0.67
2	0.82	2	4.00
3	0.88	3	2.46
4	0.47	4	0.93
5	0.50	5	0.88
6	0.12	6	0.88
7	0.18	7	0.43
		8	0.49
		9	1.34
		10	1.63
		11	3.44
		12	0.87
Mean	0.52	Mean	1.50
±SD	0.29	±SD	1.18

[a] Normal controls vs. PEM, $p < 0.05$.

chromium occurred in the nonresponders. Further documentation and studies on a larger scale are needed to bring clarity to these inconclusive results.

Table X summarizes results on malnourished adults. Low values for urinary chromium concentration and excretion were obtained on this group, which consisted of subjects with sequelae of chronic disease and subjects who were drug addicts.

Results of hair-chromium concentrations in infants and children with malnutrition of varying degrees of severity and in malnourished adults (long-term hospitalized orthopedic cases and patients under treatment for drug addiction) are shown in Table XI.

Perhaps as a result of decreased rate of hair growth with deposition of available chromium in relatively greater amounts, no difference from normal

Table IX. Urinary Chromium Excretion in PEM Cases[a]

Group	Before oral GTT	After oral GTT
Responder (4)	1.81 ± 1.77	1.71 ± 1.16
Nonresponder (5)	0.65 ± 0.35	0.87 ± 0.73

[a] Results are expressed as mean ± SD. Figures in parentheses are numbers of cases.

Table X. Urinary Chromium Excretion in Different
Nutritional Backgrounds[a]

Subject group[b]	Urine vol. (ml/day)	Urinary Cr concn. (ng/ml)	Cr excretion (μg/day)
B1	593 ± 190 (19)	5.04 ± 2.27 (19)	2.77 ± 1.11 (19)
B2	440 ± 263 (10)	3.35 ± 1.76 (10)	1.40 ± 0.92 (10)
B3	950 ± 112 (5)	0.93 ± 0.30 (5)	0.91 ± 0.38 (5)

[a] Results are expressed as mean ± SD. Figures in parentheses are numbers of cases.
[b] B1 From the institution for the aged (45–85 years).
B2. Chronically disabled (30–70 years).
B3. Drug-addicted (45–70 years).

values could be established in hair chromium in the malnourished pediatric age group. It is known that in PEM hair growth may be as slow as one-eighth the normal rate (Sims, 1968). Another possibility is that the hair samples used in the analyses could be a reflection of the nutritional state prior to the development of malnutrition. In the group of malnourished adults, hair-chromium concentration was below 200 ng/g in 16 of the 22 cases. This group of subjects also showed lower values for urinary chromium concentration and excretion as compared to normal controls. These findings point to a relationship between chromium content of hair and that in the urine and indicate that hair chromium concentration may be accepted as a rather reliable criterion for chromium nutrition, with the exception of infants and children with protein deficiency.

3.5.   Chromium Metabolism in Diabetes Mellitus

Glinsmann and Mertz (1966) reported that oral GTT reverted to normal when chromium chloride was administered to diabetics in doses of 180–1000 μg

Table XI. Hair-Chromium Concentrations in Different Nutritional Backgrounds[a]

Group	Age	Hair-Cr concn. (ng/g)	Hair-Cr concn. (< 200 ng/g)
Underweight children	(1¼–6½ years)	335 ± 126 (17)	(98, 118, 146, 180)
PEM cases	(4–18 months)	278 ± 83	(176, 178, 180)
Chronically disabled patients	(30–70 years)	178 ± 58 (4) men 132 ± 58 (8) women	(102, 112, 193) (66, 82, 104, 109, 114, 150, 191)
Drug-addicted	(45–70 years)	228 ± 102 (10)	(102, 142, 172, 177, 189, 196)

[a] Results are expressed as mean ± SD. Figures in parentheses are numbers of cases.

under controlled experimental conditions. Subsequent observations led these authors to conclude that chromium chloride was not effective when used as a therapeutic agent in diabetics. Doisy *et al.* (1971) observed that in insulin-dependent diabetics there was a difference in intestinal absorption of chromium or its urinary excretion between normal controls and adult-type diabetic subjects. The administration of oral ^{51}Cr to these patients caused a greater elevation in plasma chromium as well as a higher urinary excretion than in normal individuals. Similar results were obtained with i.v. [51]Cr. Normal individuals excreted 25% of i.v. [51]Cr, whereas urinary excretion rose to 41% in insulin-dependent diabetics. As pointed out by Doisy and associates (1971), it is impossible to determine from these findings whether increased excretion of chromium in the urine is a function of enhanced intestinal absorption or a result of underutilization.

Additional evidence pointing to chromium loss is found in juvenile diabetics. Hair-chromium concentration in diabetic children was reported to be significantly low (Hambidge *et al.*, 1968). Doisy *et al.* (1971) found that urinary chromium excretion following i.v. [51]Cr was essentially normal in maturity-onset diabetics. However, in a study from Bangkok, it was reported that hair-chromium concentrations in patients with more than 9 years of history was significantly lower than in normal controls or diabetics in whom the onset was more recent (Benjanuvatra and Bennion, 1975). Signs of chromium deficiency in diabetes may be related more to duration of the disease than to its type. All these findings suggest that chromium deficiency is secondary to the diabetic state.

Determination of urinary chromium before and after oral GTT was carried out in eight diabetic subjects. The data in Table XII show that there is no

Table XII. Urinary Chromium Excretion in Diabetic Subjects before and after Oral GTT [a]

		Urinary Cr excretion			
		Before oral GTT		After oral GTT	
No.	Age (years)	ng/min	Cr(ng)/creatinine (mg)	ng/min	Cr(ng)/creatinine (mg)
1	11	8.68	15.95	9.31	16.57
2	13	6.56	8.04	4.00	7.81
3	11	5.86	17.28	3.05	9.52
4	5	5.04	20.14	4.51	11.84
5	14	6.95	14.30	5.71	12.82
6	20	4.52	10.51	13.35	23.50
7	25	5.08	9.80	9.83	13.11
8	22	23.69	13.53	10.00	9.25
Mean		8.30	13.69	7.47	13.05
±SD		6.36	4.09	3.65	5.04

[a]There is no significant difference between these values.

difference in the rate of Cr excretion per minute or the ratio before and after oral GTT. No difference was found in chromium excretion (ng/min) between normal and diabetic subjects, although a significant difference was noted in the chromium/creatinine ratio ($p < 0.05$). In these two groups, the urinary creatinine excretion was comparable. Therefore, the difference in chromium/creatinine does not depend on urinary creatinine excretion; the higher chromium excretion in diabetics was comparable to changes in PEM.

Doisy and associates (1976) in five insulin-dependent diabetics requiring insulin in doses of 60–130 U reported reduction in insulin requirements to as low as 20–45 U/day following the addition of brewer's yeast to the diet. The possible significance of these findings in keeping carbohydrate metabolism under control in chemical diabetes, which may occur in the offspring or siblings of diabetic patients, by a so-called "GTF prophylaxis," needs particular emphasis. The possibility of preventing diabetes by GTF requires further intensive research.

3.6. Chromium Metabolism in Pregnancy

A reduction in body chromium pool and a placental transfer of chromium to the fetus is thought to occur in pregnancy. Body chromium content before pregnancy, chromium intake during pregnancy, and gestational age at delivery are considered factors that regulate maternal–fetal chromium homeostasis. Hair-chromium concentrations in the newborn infant may be used as an indicator of fetal body-chromium content.

Transfer of chromium to the fetus during pregnancy leads to alterations in hair-chromium concentrations in the mother, a decrease in hair chromium being most marked in multiparae. In this respect, a distinct difference was observed when hair-chromium concentrations in women with a history of multiple births were compared with those in women with no children (Wolf et al., 1971, unpublished data). Changes in hair-chromium concentrations pertaining to different stages of pregnancy can be estimated by measuring chromium concentrations in hair samples taken from varying distances from the scalp. Results obtained on these samples indicate that a fall in hair chromium exists in the early stages of pregnancy but that it becomes most accentuated during the last months before delivery, which shows a relationship to the increasing requirements of the fetus (Saner, 1975b).

Davidson and Burt (1973) studied the effect of GTT on plasma-chromium levels in nonpregnant women. Significantly lower fasting plasma chromium levels were observed in the pregnant group (2.97 ± 0.11 ng/ml) than in the nonpregnant group (4.70 ± 0.15 ng/ml). Whereas GRR following i.v. GTT was the same in both groups, a sudden and persisting fall was noted in plasma-chromium levels in the nonpregnant subjects. This fall did not occur in subjects in their 37th to 40th week of pregnancy. A similar fall was observed following an oral glucose tolerance load in the nonpregnant group. The authors interpreted the drop in plasma level in nonpregnant subjects as indicative of a rapid transfer of chromium to the tissues. Whereas immunoreactive insulin (IRI) levels were essentially comparable in the fasting state in pregnant and nonpregnant subjects, IRI response to GTT was found to be more exaggerated in the pregnant group.

Sutherland and Stowers (1975) calculated the increment index by performing repeat GTT on women in their first (11–13 weeks), second (25–27 weeks), and third (37–39 weeks) trimesters of pregnancy. Consecutive mean increment indexes or GRR for the different stages of pregnancy were 6.29 ± 1.26, 5.0 ± 1.01, and 3.64 ± 0.42 %-min. Significant differences were found in increment indexes (GRR) between the first and second as well as between the second and third trimesters ($p < 0.05$ and $p < 0.001$, respectively).

In a recent study, GRR and urinary chromium were determined in seven women in their third trimester of pregnancy (Saner and Gürson, 1976b, unpublished data). The results are indicated in Fig. 5. A significant negative correlation between GRR and urinary chromium concentrations in these women was established ($r = -0.7768, p < 0.05$). In chromium deficiency a low GRR may be accepted as a constant finding. On the other hand, an increased urinary chromium excretion in chromium deficiency, such as diabetes and PEM, has been found. Thus, the significant negative correlation between GRR and urinary chromium in these cases may be accepted as evidence of chromium deficiency. This deficiency became more evident in women who had successive pregnancies. Sutherland and Stowers (1975) drew attention to the impairment of GRR and summarized their results as follows: "Many questions remain unanswered and one of the most challenging is why many women with an abnormal IV GTT in a preceding

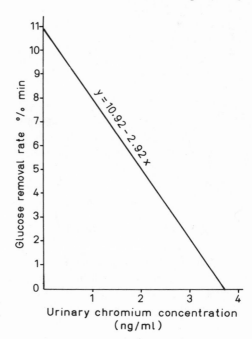

Fig. 5. The correlation between glucose removal rate and urinary chromium concentration.

pregnancy may have normal IV GTT results in successive pregnancies in spite of the increased age and parity.'' These observations suggest that chromium deficiency may occur among the multiparae and that marginal chromium deficiency is a widespread problem in some parts of the world.

3.7. Chromium in the Aged

The observations of Schroeder and associates (1962) in the United States indicate a decrease in tissue-chromium content in aged individuals. Streeten *et al.* (1965) reported significant impairment in GTT in aged residents of an institution. Of the subjects over 70 years of age, 86% showed abnormal GTT. These findings led to investigations aimed at establishing a relationship between chromium deficiency and impaired GTT. It was observed that chromium supplementation, in daily doses of 150 μg $CrCl_3 \cdot 6H_2O$ caused a restoration of GTT to normal in 4 out of 10 subjects, but that it had no effect when the GTT abnormality was severe. Canfield and Doisy (1976) have demonstrated a correlation between decrease in urinary chromium concentrations and increasing age. These findings are in agreement with our experience in older subjects (aged 45–85) residing in an old-age home (Gürson *et al.*, 1975). Canfield and Doisy (1976) relate the lower urinary chromium concentrations to a decrease in volatile and nonvolatile chromium in the urine.

Hair-chromium concentrations determined on 19 subjects in a home for the aged gave a mean value of 296 ± 242 ng/g (Gürson *et al.*, 1975). In eight of these cases, hair-chromium concentrations were lower than 200 ng/g. These relatively low values may be considered an indication of a marginal deficiency in chromium in old-age subjects.

Recently, Doisy and associates (1976) studied the effect of brewer's yeast extracts on impaired GTT in aged subjects. This work was done in an institution for the aged where the individuals had a choice for selecting the items in their diet, rather than being offered a fixed menu. GTTs were found to be abnormal in 45% of the subjects, which indicated that even under favorable circumstances that provide for a selection of diet, chromium deficiency may occur in old age. Supplementation of the diet with chromium resulted in correction of GTT in 50% of these subjects. These findings may be interpreted to be an indication of an incapacity to convert inorganic chromium to GTF in the aged.

4. Chromium Requirements

Because of differences in the rate of absorption of different chromium salts, it has not yet been possible to establish the daily requirements for chromium. Although Mertz (1969) advocated that the daily requirement should be in the range of 10–30 μg, provided that the ingested chromium is converted into biologically active GTF during the process of digestion, in a recent publication the same author states that "although chromium is an essential element in human nutrition, the daily requirement for chromium is not known" (Mertz, 1974b). Indeed, neither of the two different committees of experts was able to suggest figures for daily chromium requirements (WHO, 1973; National Academy of Science, 1974).

5. Conclusion

Present evidence indicates that marginal chromium deficiency exists as a nutritional problem in a number of regions throughout the world. Investigations that demonstrate the relationships between chromium and carbohydrate and lipid metabolism suggest that chromium may also have an important role in the pathogenesis of abnormalities of carbohydrate and lipid metabolism as well as evaluating some disease states that constitute a health hazard in all parts of the world. Existing methods for the determination of the chromium nutritional state are not entirely satisfactory, and at the same time the necessity for laboratory facilities for methods of determinating chromium in organic substances poses altogether different problems. In an attempt to evaluate the different methods used for the assessment of chromium nutrition, it may be said that determinations of urinary chromium, particularly urinary chromium following oral GTT, appears to be of great value in certain age groups and in a number of pathological states to the extent that it may replace the use of i.v. GTT. Despite a number of paradoxical results encountered in some publications, increase in urinary chromium excretion following an oral gluose load can be taken as an index for the assessment of the body chromium pool. This issue is of great significance for future field studies.

The present author is inclined to believe that hair-chromium concentrations are of limited value in the assessment of body-chromium content, particularly in individuals who have other deficiency states, such as protein and energy malnutrition. Hair-chromium concentrations may have a place in the assessment of the overall state of chromium nutrition in a population or in the estimation of chromium losses during pregnancy.

It is difficult to offer an explanation for the metabolic aberrations that lead to urinary chromium excretion observed in states of chromium deficiency, such as PEM and diabetes. It may be speculated that in the early phase of chromium deficiency in these conditions, the peripheral tissue response to insulin decreases and plasma insulin level is slightly increased. Elevated plasma insulin levels always lead to an increase in chromium excretion. With continuing chromium loss, body stores are gradually depleted, and finally a state of unresponsiveness to GTT is reached. Increased urinary excretion of chromium, despite the existence of chromium deficiency, as observed in conditions such as PEM and diabetes, may seem paradoxical. It is highly probable that the abnormal urinary chromium loss in these diseases is a result of defective tubular reabsorption of this element. Needless to say, this hypothesis requires documentation.

A significant negative correlation between the efficiency of glucose utilization and urinary chromium in PEM and pregnancy suggests that chromium deficiency could occur as an isolated deficiency. Speculations on the value of organic chromium salts of GTF character in "glucose tolerance factor prophylaxis" as a possible preventive measure in diabetes and diseases states related to hyperlipidemia, need particular emphasis and further research.

References

Baetjer, A. M., 1950, Pulmonary carcinoma in chromate workers, I. A review of literature and report of cases, *Arch. Indust. Hyg.* **2**:487.

Behne, D., Braetter, P., Gessner, H., Mertz, W., and Roesick, U., 1976, Problems in the determination of chromium in biological materials. Comparison of flameless atomic absorption spectrometry and neutron activation analysis, *Z. Anal. Chem.* **278**:269.

Behne, D., and Diehl, F., 1972, Relations between carbohydrate and trace element metabolism investigated by neutron activation analysis, in *Nuclear Activation Techniques in the Life Sciences*, IAEA LAEA-SM-157/11 Vienna.

Benjanuvatra, N. K., and Bennion, M., 1975, Hair chromium concentration of subjects with and without diabetes mellitus, *Nutr. Rep. Int.* **12**:325.

Campbell, W. J., and Mertz, W., 1963, The interaction of insulin and chromium (III) on mitochondrial swelling, *Am. J. Physiol.* **204**:1028.

Canfield, W. K., and Doisy, R. J., 1976, Chromium and diabetes in the aged, in *Biomedical Role of Trace Elements in Aging* (J. M. Hsu, R. L. Davis, and R. W. Nerthamer, eds.) pp. 117–126. Eckerd College Gerontology Center, St. Petersburg, Fla.

Carter, J. P., Kattab, A., Abd-El-Hadi, K., Davis, J. T., El Gholmy, A., and Patwardhan, V. N., 1968, Chromium III in hypoglycemia and in impaired glucose utilization in kwashiorkor, *Am. J. Clin. Nutr.* **21**:195.

Collins, R. J., Fromm, P. O., and Collings, W. D., 1961, Chromium excretion in the dog, *Am. J. Physiol.* **201**:795.

Curran, G. L., 1954, Effect of certain transition group elements on hepatic synthesis of cholesterol in the rat, *J. Biol. Chem.* **210**:765.

Davidson, I. W. F., and Blackwell, W. L. 1968, Changes in carbohydrate metabolism of squirrel monkeys with chromium dietary supplementation, *Proc. Soc. Exp. Biol. Med.* **127**:66.

Davidson, I. W. F., and Burt, R. L., 1973, Physiologic changes in plasma chromium of normal and pregnant women: Effect of a glucose load, *Am. J. Obstet. Gynecol.* **116**:601.

Davidson, I. W. F., and Parker, J. C., 1974, Renal excretion of chromium and copper, *Proc. Soc. Exp. Biol. Med.* **147**:720.

Doisy, R. J., Streeten, D. H. P., Freiberg, J. M., and Schneider, A. J., 1976, Chromium metabolism in man and biochemical effects, in *Trace Elements in Human Health and Disease* (A. S. Prasad, ed.), Nutrition Foundation Monograph, Academic Press, New York.

Doisy, R. J., Streeten, D. H. P., Souma, M. L., Kalafer, M. E., Rekant, S. I., and Dalakos, T. G., 1971, Metabolism of 51 chromium in human subjects, in *Newer Trace Elements in Nutrition* (W. Mertz and W. Cornatzer, eds.), pp. 155–68, Marcel Dekker, New York.

Donaldson, R. M., and Barreras, R. F., 1966, Intestinal absorption of trace quantities of chromium, *J. Lab. Clin. Med.* **68**:484.

Durfor, C. N., and Becker, E., 1962, Public Water Supplies of the 100 Largest Cities in the United States, Geological Survey, Water-Supply Paper 1812.

Franzen, E., Pohle, R., and Knoblich, K., 1970, Arbeitshygienische Untersuchungen in Betrieben der Galvatechnik. III. Chrom in Urin, *Z. Ges. Hyg.* **16**:657.

Glinsmann, W. H., and Mertz, W., 1966, Effect of trivalent chromium on glucose tolerance, *Metabolism* **15**:510.

Gürson, C. T., 1972, The biochemical aspects of protein-calorie malnutrition in *Newer Methods of Nutritional Biochemistry* (A. A. Albanese, ed.), pp. 66–124, Academic Press, New York.

Gürson, C. T., and Saner, G., 1971, Effect of chromium on glucose utilization in marasmic protein-calorie malnutrition, *Am. J. Clin. Nutr.* **24**:1313.

Gürson, C. T., and Saner, G., 1972, Unpublished data.

Gürson, C. T., and Saner, G., 1973, Effect of Chromium supplementation on growth in marasmic protein-calorie malnutrition, *Am. J. Clin. Nutr.* **26**:988.

Gürson, C. T., Saner, G., Mertz, W., Wolf, W. R., and Sökücü, S., 1975, Nutritional significance of chromium in different chronological age groups and in populations differing in nutritional backgrounds, *Nutr. Rep. Int.* **12**:9.

Hambidge, K. M., 1971, Chromium nutrition in the mother and the growing child, in *Newer Trace Elements in Nutrition* (W. Mertz and W. Cornatzer, eds.), pp. 169–94, Dekker, New York.

Hambidge, K. M., and Baum, J. D., 1972, Hair chromium concentrations of human newborn and changes during infancy, *Am. J. Clin. Nutr.* **25**:376.

Hambidge, K. M., Rodgerson, D. O., and O'Brien, D., 1968, The concentration of chromium in the hair of normal and children with juvenile diabetes mellitus, *Diabetes* **17**:517.

Hopkins, L. L., Jr., Ransome-Kuti, O., and Majaj, A. S., 1968, Improvement of impaired carbohydrate metabolism by chromium III in malnourished infants, *Am. J. Clin. Nutr.* **21**:203.

Jeejeebhoy, K. N., Shu, R., Marliss, E. B., Greenburg, G. R., and Robertson, A. B. 1975, Chromium deficiency, diabetes and neuropathy reversed by chromium infusion in a patient on total parenteral nutrition (TPN) for three and one half years, *Clin. Res.* **23**:636.

Jones, G. B., Buckley, R. A., and Chandler, C. S., 1975, The volatility of chromium from brewer's yeast during assay, *Anal. Chim. Acta* **80**:389.

Levine, R. A., Streeten, D. H. P., and Doisy, R. J., 1968, Effects of oral chromium supplementation on the glucose tolerance of elderly human subjects, *Metabolism* **17**:114.

Majaj, S. A., and Hopkins, Jr., L.L., 1966, Human chromium deficiency. The response of hypoglycemia and impaired glucose utilization to chromium (III) treatment in protein calorie malnourished infants, *Lebanase Med. J.* **19**:177.

Mathur, R. K., and Doisy, R. J., 1972, Effect of diabetes and diet on the distribution of tracer doses of chromium in rats, *Proc. Soc. Exp. Biol. Med.* **139**:836.

Maxia, V., Meloni, S., Rollier, M. A., Brandone, A., Patwardhan, V. N., Waslien, C. I., and El-Shami, S., 1972, Selenium and chromium assay in Egyptian foods and in blood of Egyptian children by activation analysis, in *Nuclear Activation Techniques in the Life Sciences*, IAEA, Vienna.

Mertz, W., 1969, Chromium occurrence and function in biological systems, *Physiol. Rev.* **49**:169.

Mertz, W., 1974a, Chromium as a dietary essential in *Trace Element Metabolism in Animals. 2* (W. G. Hoekstra, J. W. Suttie, H. E. Ganther, and W. Mertz, eds.), pp. 185–98, University Park Press, Baltimore.

Mertz, W., 1974b, Recommended dietary allowances up to date trace minerals, *J. Am. Diet. Assoc.* **64**:163.

Mertz, W., and Roginski, E. E., 1963, The effect of trivalent chromium on galactose entry in rat epididymal fat tissue, *J. Biol. Chem.* **238**:868.

Mertz, W., and Roginski, E. E., 1969, Effects of chromium (III) Supplementation on growth and survival under stress in rats fed low protein diets, *J. Nutr.* **97**:531.

Mertz, W., and Roginski, E. E., 1971, Chromium metabolism: The glucose tolerance factor, in *Newer Trace Elements in Nutrition* (W. Mertz and W. Cornatzer, eds.), pp. 123–53, Dekker, New York.

Mertz, W., and Schwarz, K., 1955, Impaired intravenous glucose tolerance as an early sign of dietary necrotic liver degeneration, *Arch. Biochem. Biophys.* **58**:504.

Mertz, W., and Schwarz, K., 1959, Relation of glucose tolerance factor to impaired glucose tolerance in rats on stock diets, *Am. J. Physiol.* **196**:614.

Mertz, W., Toepfer, E. W., Roginski, E. E., and Polansky, M. M., 1974, Present knowledge of the role of chromium, *Fed. Proc.* **33**:2275.

Mitman, F. W., Wolf, W. R., Kelsay, J. L., and Prather, E. S., 1974, Urinary chromium levels of nine young women eating freely chosen diets, *J. Nutr.* **105**:64.

National Academy of Sciences, 1974, *Recommended Dietary Allowances*, 8th ed., p. 101, Washington, D.C.

Pribluda, L. A., 1963, Chromium content of the long bones of rats at different stages of pregnancy, *Chem. Abstr.* **59**:4331.

Saner, G., 1975a, Urinary chromium excretion in the newborn and its relation to intravenous glucose loading, *Nutr. Rep. Int.* **11**:387.

Saner, G., 1975b, Chrom bei Neugeborenen, in *Spurelemente in der Entwicklung von Mensch und Tier* (K. Betke und F. Bidlingmaier, eds.), pp. 199–206, Urban und Schwarzenberg, Munich–Berlin–Vienna.

Saner, G., and Gürson, C. T., 1976a, Hair chromium concentration in newborns and their mothers, *Nutr. Rep. Int.* **14**:155.

Saner, G., and Gürson, C. T., 1976b, Unpublished data.

Saner, G., Wolf, W. R., and Gürson, C. T., 1974, The relationship of glucose removal rate and urinary chromium excretion in normal and underweight children from Turkey, *Fed. Proc.* **33**:660, Abst. 2541.

Schroeder, H. A., 1968, The role of chromium in mammalian nutrition, *Am. J. Clin. Nutr.* **21**:230.

Schroeder, H. A., and Balassa, J. J., 1965, Influence of chromium, cadmium, and lead on rat aortic lipids and circulating cholesterol, *Am. J. Physiol.* **209**:433.

Schroeder, H. A., Balassa, J. J., and Tipton, I. H., 1962, Abnormal trace metals in man: Chromium, *J. Chron. Dis.* **15**:941.

Schroeder, H. A., Nason, A. P., and Tipton, I. H., 1970, Chromium deficiency as a factor in atherosclerosis, *J. Chron. Dis.* **23**:123.

Schwarz, K., and Mertz, W., 1959, Chromium III and the glucose tolerance factor, *Arch. Biochem. Biophys.* **85**:292.

Sims, R., Quoted by Wayburne, S., 1968, Malnutrition in Johannesburg, in *Calorie Deficiencies and Protein Deficiencies* (R. A. McCance and E. M. Widdowson, eds.), pp. 7–21, Churchill, London.

Staub, H. W., Reussner, G., and Thiessen, R. T., Jr., 1969, Serum cholesterol reduction by chromium in hypercholesterolemic rats, *Science* **166**:746.

Streeten, D. H. P., Gerstein, M. M., Marmor, B. M., and Doisy, R. J., 1965, Reduced glucose tolerance in elderly human subjects, *Diabetes* **14**:579.

Sutherland, H. W., and Stowers, J. M., 1975, The detection of chemical diabetes during pregnancy using the intravenous glucose tolerance test, in *Carbohydrate Metabolism in Pregnancy and the Newborn* (H. W. Sutherland and J. M. Stowers, eds.), pp. 153–166, Churchill Livingstone, Edinburgh–London–New York.

Tipton, I. H., 1960, Distribution of trace metals in the human body, in *Metal Binding in Medicine* (M. J. Seven, ed.), pp. 27–42, Lippincott, Philadelphia.

Tipton, I. H., and Cook, M. J., 1963, Trace elements in human tissue. Part II. Adult subjects from the United States, *Health Phys.* **9**:103.

Tipton, I. H., Schroeder, H. A., Perry, H. M., Jr., and Cook, M. J., 1965, Trace elements in human tissue. III. Subjects from Africa, the Near and Far East and Europe, *Health Phys.* **11**:403.

Toepfer, E. W., Mertz, W., Roginski, E. E., and Polansky, M. M., 1973, Chromium in foods in relation to biological activity, *J. Agr. Food Chem.* **21**:69.

Wacker, W. E. C., and Vallee, B. L., 1959, Nucleic acids and metals. I. Chromium, manganese, nickel, iron, and other metals in ribonucleic acid from diverse biological sources, *J. Biol. Chem.* **234**:3257.

Wolf, W. R., 1975, Uncertainties in the analysis of chromium in biological materials, *Fed. Proc.* **34**:927.

Wolf, W. R., Mertz, W., Saner, G., and Gürson, C. T., 1971, Unpublished data.

Wool, I. G., Rampersad, O. R., and Moyer, A. N., 1966, Effect of insulin and diabetes on protein synthesis by ribosomes from heart muscle, *Am. J. Med.* **40**:716.

World Health Organization, 1973, Trace Elements in Human Nutrition, WHO Technical Report Series, No. 532, Geneva, 1–65.

Chapter 3

The Significance of Folate Binding Proteins in Folate Metabolism

Samuel Waxman, Carol Schreiber, and Mitchell Rubinoff

1. Folate Binding Proteins in Folate Metabolism

1.1. Introduction

Various aspects of folate metabolism in the microorganism and in the multicellular organism call for functional, specific folate binding proteins. These folate binding protein functions would include extracellular, intercellular, and intracellular folate transport, distribution, and storage.

Until recently it was generally thought that the folates were carried in the serum and other body fluids in either an unbound form or loosely associated with albumin. This was supported by the finding that serum folate was for the most part removed by dialysis (Hampers *et al.*, 1967) and was rapidly cleared *in vivo* from the serum following infusion (Chanarin *et al.*, 1958). Attempts to establish specific folate binding proteins either by fractionation of endogenous folate activity into free and bound (Johns *et al.*, 1961) or by addition of exogenous tritiated folic acid (^{3}H-PteGlu) (Metz *et al.*, 1968) were at first equivocal.

Samuel Waxman, Carol Schreiber, and Mitchell Rubinoff • Division of Medical Oncology, Cancer Chemotherapy Foundation Laboratory, Department of Medicine, The Mount Sinai School of Medicine of The City University of New York, New York, New York 10029. The research was supported by National Institutes of Health Grant AM 16690 and the Cancer Chemotherapy Foundation. Address all correspondence to Dr. Samuel Waxman.

During the past few years both direct and indirect evidence for the existence of folate binders has been reported. These folate binding proteins can be divided into high-affinity binding proteins and low-affinity, reversible binding proteins. The low-affinity binding proteins probably represent a folate–macromolecular complex which is nonspecific and is analagous to many drug–albumin interactions (Markkanen, 1968; Elsborg, 1972). This may result from the negative charge produced by the ionized carboxyl groups of folate at physiologic pH as well as the resulting tendency for this end of the molecule to form weak electrostatic attractions with positively charged moieties of macromolecules. This type of complex is characteristic of nonspecific, low-affinity binding and does not demonstrate saturation kinetics. This nonspecific binding may not be dissociated by short periods of dialysis or gel filtration chromatography, but the complex can be separated by adsorbing this weakly bound folate to a higher-affinity attracting surface, such as charcoal.

Low-affinity, nonspecific binding interaction with folate and serum proteins such as albumin, probably accounts for the large fraction of nondialyzable folate reported by Retief and Huskisson (1969) in serum, red blood cells, bile, and saliva, but not in urine (Retief and Huskisson, 1970). Similarly, the endogenous protein-bound folate activity patterns described by Markkanen *et al.* (1968, 1970–1974), for the most part represent low-affinity binding to albumin. They have reported that in human serum almost one-half of folate activity is bound to various proteins and the balance is free. These findings were confirmed by Elsborg (1972) who also demonstrated that folate binding was pH dependent with an optimum at 7.4. These investigators found that the endogenous protein-bound folate activity was mainly found in the albumin fraction using Sephadex G-150 column chromatography and microbiologic assay of each fraction. In addition, endogenous folate-bound protein activity was recovered in the transferrin, γ-globulin and α_2-macroglobulin regions to a lesser extent and that these zones changed in relationship to various clinical situations, such as pregnancy and folate deficiency. The totality of this work probably represents the combination of nonspecific low-affinity binding, such as in the albumin fraction, and in addition, a small amount of specific binding perhaps present in the other zones. Although most of the endogenous folate appears to be carried or associated with nonspecific binding to albumin it does not necessarily imply that this serves no physiologic function. It may be that this nonspecific, low-affinity binding could retard the ultrafiltration of folate into the glomerular fluid (Johns *et al.*, 1961) or facilitate the transport of folate from the intestinal epithelial cells into the blood and preserve it by increasing the unidirectional flux of absorbed folate into the portal system.

In contrast to the nonspecific binding, the specific, high-affinity binders of folate are coupled to the pteridine moiety of the folate molecule, having binding constants estimated to be in the range of 1×10^{12} liters/mol at pH 7.4 and demonstrate saturation kinetics. High-affinity folate binding can be demonstrated by reacting the proposed binder source with increasing concentrations of folate. When the saturating concentration is reached, the ratio of bound to free folate will decrease as the concentration of folate is further increased. The high-affinity binding cannot be dissociated by exposure to low concentrations of charcoal and is strong enough to prevent displacement of bound folate by great excesses of unbound folate. The data presented and discussed here are chiefly concerned with the high-affinity, specific types of folate binding proteins. The physico-chemical properties of these folate binding proteins from several sources are compared and discussed with relationship to the various roles they may play in normal and abnormal folate metabolism.

2. Extracellular Folate Binding Protein

2.1. Serum

There is now evidence for circulating high-affinity, specific folate binding proteins that are unsaturated or partially saturated with folate in normal serum (Waxman and Schreiber, 1973a; Zettner and Duly, 1974; Eichner *et al.*, 1975). These high-affinity folate binding proteins have been identified in the sera from patients with folate deficiency (Waxman and Schreiber, 1971, 1973a,b), liver disease (Colman and Herbert, 1974; Waxman, 1975), uremia (Hines *et al.*, 1973; Eichner *et al.*, 1975), leukemia (Rothenberg and daCosta, 1971), and from women who are pregnant or taking oral contraceptives (daCosta and Rothenberg, 1974a), patients with prostatic carcinoma taking estrogen (Waxman and Schreiber, 1976, unpublished observations), and in the sera derived from um-bilical cord blood (Kamen and Caston, 1975a).

In these studies, folate binding protein was measured by the addition of exogenous ^{3}H-PteGlu, which measures the unsaturated folate binding protein capacity. This unsaturated folate binding protein capacity is increased in folate deficiency, and in some of the above conditions ($>$ 100 pg ^{3}H-PteGlu bound/0.4 ml) as compared to most normal sera ($<$ 20 pg ^{3}H-PteGlu bound/0.4 ml) (Waxman and Schreiber, 1973a). Reproducible measurements of serum unsaturated folate binding protein by exogenous ^{3}H-PteGlu binding requires a fixed propor-tion of ^{3}H-PteGlu (500 pg) to serum (0.4 ml) and sufficiently protein-coated charcoal for adsorption (Waxman and Schreiber, 1973a; Leonard and Beckers, 1976). Excessive ^{3}H-PteGlu or insufficiently coated charcoal will give spurious folate binding protein levels and nonspecific reversible binding to low-affinity serum proteins, such as albumin. This may explain the higher levels of serum unsaturated folate binding protein reported by Zettner and Duly (1974) in whose studies 10 times the above ^{3}H-PteGlu/serum ratio was used. It is probable that some of the ^{3}H-PteGlu binding was nonspecific binding. Specific ^{3}H-PteGlu binding to folate binding protein must be noncoated charcoal adsorbable and not displaced by stable PteGlu.

The apparent discrepancy of finding an unsaturated folate binding capacity of such a small quantity of ^{3}H-PteGlu in relation to serum, which contains 100–500 times greater concentrations of folate in the form of 5-methyltetra-hydrofolic acid (5-CH$_3$H$_4$ PteGlu) (Herbert *et al.*, 1962), can be explained by the much lower affinity of this folate binding protein for reduced folate, as compared to oxidized folate (Rothenberg and daCosta, 1971; Waxman and Schreiber, 1973b). Thus, if the folate binding protein is not saturated by a form of oxidized folate it may be weakly associated with circulating 5-CH$_3$H$_4$ PteGlu, and this will be displaced by the exogenous addition of ^{3}H-PteGlu used in these *in vitro* assay systems. Therefore, a low unsaturated folate binding protein capacity could be attributable to two factors—either it is saturated with oxidized folate or the protein is actually present in lesser amounts.

Unsaturated folate binding protein has been characterized in various sera and appears to be of two types. The most usual folate binding protein (small folate binding protein) has a molecular weight of approximately 40,000 and is present in all sera with folate binding capacity (Waxman and Schreiber, 1973b). In addition, sera from some patients with alcoholism, liver disease, and various malignancies, but not complicated by uremia or nutritional folate deficiency have, in addition, a folate binding protein with a molecular weight of greater than 200,000 (large folate binding protein) (Fig. 1) (Waxman, 1975). Large folate binding protein is also found in human milk and placenta (Waxman and Schreiber, 1975a). Large folate binding protein is present in larger amounts in plasma than in serum and accounts for one-sixth the unsaturated folate binding capacity, the remainder being small folate binding protein. Large folate binding protein has been separated from small folate binding protein by affinity chromatography and has been partially purified by Sephadex G-200, DEAE cellulose, and concanavalin-A Sepharose chromatography (Waxman *et al.*, 1976). The large folate binding protein is eluted in one peak from DEAE cellulose with 0.15 M phosphate buffer pH 7.0 and can be partially separated from the small folate binding protein that elutes from DEAE cellulose with 0.05 *M* phosphate buffer pH 7.0 (Sharon *et al.*, 1976). The large folate binding protein does not adhere to concanavalin-A Sepharose, and folate binding activity is lost by treatment with 0.1% Triton X-100, but not by neuraminidase. Large folate binding protein purified from serum is recovered with the α_2-macroglobulin fraction. These characteristics appear to differentiate large folate binding protein from the small folate binding protein, and because it does occur in some sera derived from patients with liver disease, suggest that it may be partly liver derived. The large folate binding protein in hog plasma, in contrast to that in human plasma, gives rise to the 40,000 mol. wt. folate binding protein when isolated and subjected to fractionation by Sephadex gel filtration (Kamen, 1976).

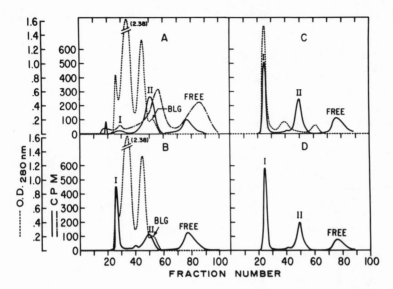

Fig. 1. Determination of the apparent molecular weight of various folate binding proteins labeled with ³H-PteGlu by gel filtration on a column (2.2 × 85 cm) of Sephadex G-200 equilibrated with 0.1 M phosphate buffer pH7.4 containing 0.15 M NaCl and 0.02% sodium azide. Protein markers are as previously described (Waxman and Schreiber, 1975a). Peak I: large folate binding protein (> 200,000); peak II: small folate binding protein (40,000); free: unbound ³H-PteGlu. (A) (———) Folate-deficient serum; uremic serum (· — ·); BLG: marker for small folate binding protein in cow's milk; (B) obstructive jaundice serum; (C) human milk; (D) human placental extract.

However, concentrated 40,000 mol. wt. folate binding protein subjected to Sephadex gel filtration does not result in the appearance of large folate binding protein (Kamen, 1976; Waxman et al., 1976). We have found that the large folate binding protein in goat serum is immunologically related to the small folate binding protein isolated from goat milk using rabbit anti-goat small folate binding protein (Waxman et al., 1975b). It may be that the large folate binding protein represents the small folate binding protein bound to a carrier or membrane protein. This is suggested by the finding that large folate binding protein is present in large amounts in human milk.

The small folate binding protein was first measured and characterized in some patients with chronic granulocytic leukemia (Rothenberg, 1970) and thereafter in the serum of patients with folate deficiency. The characteristics of small folate binding protein in human milk are outlined in Table I. With minor exceptions, these characteristics are remarkably consistent in studies reported for human, umbilical cord, and hog serum (Rothenberg and daCosta, 1971; Waxman and Schreiber, 1973b; Kamen and Caston, 1975a; Kamen, 1976). Small folate binding protein has a molecular weight of 35,000 to 40,000 and is clearly separable from dihydrofolate reductase activity. The binding constant for folic

Table I. Characteristics of ^{3}H-PteGlu Binding to Human Small Folate Binding Protein

pH range[a]	5–10
pH maximum[a]	7.6–10
pH for dissociation[a]	< 3
Temperature[a]	0–56°C
Boiling[a]	Destroys
8 M Urea[a]	Inhibits binding
SDS (sodium dodecylsulfate)	Inhibits binding
Addition of PteGlu	Protects
Effect of folate analogues[a]	PteGlu = 10 formyl PteGlu = PteGlu$_{3-7}$ >DHFA >Pteroic acid > Methyl H$_4$-PteGlu = methyl H$_4$PteGlu$_{3-7}$ > MTX > 5 formyl H$_4$-PteGlu
DEAE-cellulose	Elutes in 0.05 M phosphate buffer pH 7.0
Buffer (at pH 4.8)	0.1 M monobasic potassium phosphate > Na phosphate
Sephadex G-200[a]	35–40,000 MW
SDS gel electrophoresis	87,000; 35,000; 19,000
Isoelectric points	6.8, 7.5, 8.2
Adheres to Con-A Sepharose	(87%)
PAS stain	+

Methyl H$_4$-PteGlu dissociated by PteGlu[a] and PteGlu$_{3-7}$
Folate binding protein-bound PteGlu not a substrate for dihydrofolate reductase[a]
Folate binding protein-bound PteGlu$_{3-7}$ not a substrate for conjugase
Folate binding protein-bound PteGlu or methyl H$_4$-PteGlu not taken up by cells

[a] Similar in serum, milk, and leukemic lysates.

acid at pH 7.6 is approximately 5×10^{12} liters/mol, which is reduced at pH 9.6 to 2×10^9 liters/mol (Kamen, 1976). This binder has a greater preference for PteGlu than for 5-CH$_3$H$_4$ PteGlu at pH 7.6, but this preference disappears as the pH increases to 9.4. At pH 3–3.5 the folic acid–folate binding protein complex dissociates; it reverses when the pH is returned to 7.6.

At the same pH, human small folate binding protein binding of ^{3}H-PteGlu is inhibited more by citrate and acetate than with phosphate buffer (Table I). The binding at low pH is greater in monobasic potassium buffer than in monobasic sodium buffer and intermediate in a mixed Na–K phosphate buffer. This may relate to the observation that extracellular fluid high in sodium, such as serum, has predominantly unbound folate, whereas folate is mainly protein bound in the intracellular fluid which is high in potassium. In addition, this lesser effect of sodium and greater effect of potassium on the binding of PteGlu to folate binding protein suggests that the ATP-dependent Na–K pump mechanism may be involved in the membrane transfer of folates.

Folate binding protein binding of PteGlu does not differ significantly from 0°–56°C, whereas boiling irreversibly destroys binding. Pure folate binding protein is stable at -10°C but loses binding activity at -70°C and at room temperature. It is more stable when saturated with PteGlu. The addition of 8 M urea, guanidine, and sodium dodecylsulfate reversibly inhibits the binding of PteGlu to folate binding protein. Treatment with trypsin, but not ribonuclease or deoxyribo-nuclease, completely destroys hog folate binding protein (Kamen and Caston, 1975b). These data suggest that folate binding protein represents one or more proteins that bind folates by noncovalent bonding.

There is a similar pattern of competition by folate analogues for the binding of ^{3}H-PteGlu to various folate binding proteins in human and other animal tissues. The preference is for oxidized folyl mono (including formyl PteGlu) and poly-glutamates (glutamate chain length 3 to 7) over reduced folyl mono- and poly-glutamates. Pteroic acid, but not pteridine-6-carboxylic acid or glutamic acid inhibits ^{3}H-PteGlu binding to folate binding protein. The presence of a methyl group in the N-5 position diminishes competition. The N-5 formyl group totally inhibits competition for binding to folate binding protein, whereas N-10 formyl H_4-PteGlu exhibits some competition (Waxman and Schreiber, 1973b; Kamen and Caston, 1975a,b). Methyl-H_4-PteGlu is dissociated from folate binding protein by PteGlu or PteGlu (3–7) equally at 23° or 37°C (Waxman and Schreiber, 1975c). Thus, the oxidative state of the pteridine portion and the linkage to the p-amino benzoic acid moiety, but not the glutamyl portion, are the determinants for the binding of the folate to folate binding protein.

Evidence is developing for the existence of saturated folate binding protein in serum (Mantzos et al., 1974; Kamen and Caston, 1975a). These workers noted that serum folate concentration measured by radioligand binding assay was higher when serum was extracted with boiling or acid rather than when unex-tracted serum was assayed. This suggests that some of the serum folate was complexed to a high-affinity binder and was therefore not measurable in a com-petitive binding radioassay until the binding factor was destroyed. This observa-tion led to the finding and subsequent purification of saturated folate binder in umbilical cord serum (Kamen and Caston, 1975a). The effect of the acid treat-ment for dissociation of the folate–folate binding protein complex on the prop-erties of the cord serum factor has not yet been determined. The folate binding protein uncovered in cord serum has binding determinants similar to that found in the unsaturated folate binding protein of the 40,000-mol. wt. size. Cord serum contains only saturated folate binding protein, as opposed to only unsaturated folate binding protein in maternal serum. The cord folate binding protein can bind 5-CH_3H_4 PteGlu, and this may be a mechanism whereby the fetus can maintain a 5 : 1 serum folate advantage and sequester folate for its own use at maternal expense.

Hog plasma contains both saturated and unsaturated folate binding proteins (Mantzos et al., 1974). These folate binding protein fractions were separated by adsorption of the unsaturated binder onto Methotrexate–Sepharose affinity col-umns (Kamen et al., 1976a). Thereafter, the saturated binder was purified from the plasma by acid charcoal treatment of the fluid to remove endogenous folate, neutralize the clarified preparation, and finally adsorb the now unsaturated binding protein onto Methotrexate–Sepharose. The properties of the two fractions (saturated and unsaturated folate binding protein) purified 1000-fold were found to be essentially identical to each other.

A radioimmunoassay for the measurement of folate binding protein in fluids and tissues has been developed by Rothenberg and daCosta (1976). These investigators used an antiserum to crude folate binding protein derived from chronic granulocytic leukemia cell lysates. Using this methodology, these investigators state that samples of human serum, urine, milk, extracts of human tissues, as well as extracts of L1210 leukemic cells contain a protein immunologically similar to the folate binding protein of the chronic granulocytic leukemia cell lysate. However, it has not been possible to do a quantitative evaluation of these materials with their present assay, and a more sensitive radioassay using folate binding protein labeled with ^{125}I is now being developed.

We have attempted to measure saturated folate binding proteins in various sera by dissociating the folic acid–folate binder complex with citric acid, guanidine, and urea, and by boiling to release folate from presumed binders followed by radioassay. For the most part, we have not been able to uncover more than 0.5–0.15 ng/ml additional folate binding capacity to represent a saturated pool. In some human cord serum we were able to uncover 0.2–0.5 ng/ml additional folate binding capacity following some of these dissociation procedures. This is similar to the level of serum-saturated folate binding protein reported by Colman and Herbert (1976). Kamen and Caston (1975a) and Hines *et al.* (1973) have reported that they can find as much as 5–10 ng/ml more radioassayable folate following destruction of folate binding protein or dissociation of the folate from the folate binding protein. These discrepancies may be explained by the observation that circulating folate binding protein in native serum appears to bind 0.1–0.5 ng oxidized folate as determined by radioassay (Rothenberg and da Costa, 1971; Waxman and Schreiber, 1972). The release of this oxidized folate will give a falsely elevated radioassayable folate when the standard curve is prepared at pH 7.4 with 5-CH_3H_4 PteGlu, since oxidized folate competes five to eight times greater for the folate binding protein in the assay system. If the standard curve is prepared with PteGlu, the net increase in radioassayable folate following dissociation is in the 0.1–0.5-ng level, rather than in the 5–10-ng range. Moreover, serum with purported saturated folate binding protein of 5 ng folate as measured by dissociation and radioassay actually contains < 1 ng folate following dialysis when measured by *L. casei* microbiologic assay.

2.2. Milk

Folate binding protein was first observed in bovine milk by Ghitis (1967), partially studied by Metz *et al.* (1968) and Ford *et al.* (1969), and then purified and characterized by Salter *et al.* (1972). A source of bovine milk folate binding protein was isolated by Waxman and Schreiber (1974a) from the commercially available β-lactoglobulin preparations used as the folate binder in many radioassay methods. Human milk, an apocrine secretion, is much higher in folate binding protein content than is serum (Waxman, 1975).

Human milk, like some sera, contains two folate binding proteins (a large- and small-molecular-weight binding protein) (Fig. 1). Small folate binding protein has been purified 10,000-fold from human milk using affinity chromatography as a major purification technique (Waxman and Schreiber, 1975a). An initial preparation using acid–charcoal improves specific activity by removing endogenous bound folate as well as denaturing and precipitating some of the non-folate binding milk protein. The large-molecular-weight folate binding protein (> 200,000) was separated from the smaller-molecular-weight folate binding protein because it was excluded by the PteGlu–Sepharose affinity column and recovered in the milk eluant.

The remainder of the milk folate binding protein representing small folate binding protein binds tightly to the PteGlu–Sepharose and remains bound while the PteGlu–Sepharose is washed with large volumes of a number of solutions of varying salt concentrations. The folate binding protein is rapidly removed by the addition of 0.2 M acetic acid without significant loss of folate binding activity. The total recovery in the purification procedure is 32%. It is possible to use Methotrexate–Sepharose affinity chromatography for the same purpose (Selhub and Grossowicz, 1973), and this has been done successfully in the purification and isolation of hog plasma folate binding protein (Kamen et al., 1976a). These authors claim that this methodology is more effective because it allows for easier elution of the bound folate from the methotrexate, as compared to the PteGlu–Sepharose affinity column.

The small folic acid binding protein from human milk was further purified by DEAE cellulose chromatography where it elutes in 0.05 M phosphate buffer. This folate binding protein (7.2 μg PteGlu-bound/mg protein) was characterized and found to show two bands of binding activity by polyacrylamide disk gel electrophoresis, which stained positively for both protein and sugar. Most of the folate binding protein activity adsorbs to concanavalin-A-Sepharose where it can be eluted with α-methyl mannoside. Sodium dodecylsulfate polyacrylamide disk gel electrophoresis of the purified binding protein indicated three bands of protein, which suggests that folate binding protein may contain two subunits or monomers held together by hydrogen bonds. Larger species were found and thought to be caused by reversible conformational changes such as reported by Salter et al. (1972).

Isoelectric focusing of the pure folate binding protein indicated three major peaks and three minor peaks of folate binding protein activity at near-neutral or alkaline pH (pI6.8, 7.5, and 8.2). This is similar to the pattern found for goat milk (Rubinoff et al., 1975) (Fig. 2) and the hog-kidney folate binding protein (Kamen and Caston, 1975b). Three peaks of folate binding protein with similar isoelectric points have been found in cow's milk (Waxman and Schreiber, 1975a) and goat's milk, even when the latter was immunologically homogeneous and

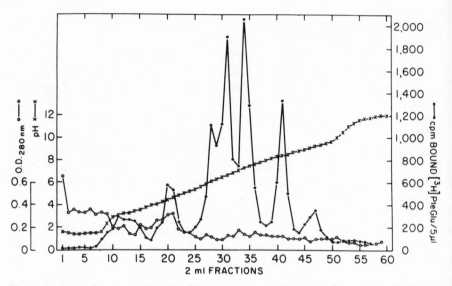

Fig. 2. Isoelectric focusing (pH 3–10) of folate binding protein isolated from goat's milk labeled with ^{3}H-PteGlu. The radioactive peak at pH 4.7 is charcoal-adsorbable and represents free ^{3}H-PteGlu.

demonstrated a single terminal amino acid and a single amino acid in the third position (Rubinoff *et al.*, 1975). The multiple peaks of folate binding protein on isoelectric focusing may represent isoproteins and may be related to a varying amount of sugar moieties in the glycoprotein similar to that described for the R-type vitamin B_{12} binding proteins (Stenman *et al.*, 1968). A majority of the contaminating protein recovered from the PteGlu–Sepharose affinity column with the folate binding protein separated at a lower pH, contained iron-binding capacity, and was, presumably, lactoferrin. No iron-binding activity was recovered in the three peaks of folate binding protein.

There does not appear to be a functional difference between the various pure folate binding proteins separated by isoelectric focusing, as all three peaks retard the uptake of ^{3}H-PteGlu into HeLa cells (Waxman, 1976). This inhibition of uptake was previously reported for crude human milk and serum folate binding protein and indicates that the final preparation of pure folate binding protein retains its native characteristics (Waxman and Schreiber, 1974a). There is preliminary evidence that the folate binding protein preincubated with the cell may enhance attachment of folic acid to the membrane of the HeLa cell within the first 5 min of exposure (Waxman and Schreiber, 1974a).

Folic acid can be separated from the purified folate binding protein by urea, guanidine, low pH, and negative ions. This may result from disruption of the hydrogen bonds, alteration of protein conformation, or indirect alteration of hydrophobic bonding. PteGlu has a greater affinity for pure folate binding protein and can dissociate or exchange with 5-CH₃H₄ PteGlu previously bound to folate binding protein (Waxman and Schreiber, 1975c).

Goat's milk is high in unsaturated small folate binding protein (7.8 ng bound/mg protein). It does not contain large folate binding protein. The small folate binding protein from goat's milk was isolated by the techniques outlined earlier (Rubinoff et al., 1975). Goat's milk has been shown to be homogeneous by means of various techniques including SDS polyacrylamide disk gel electrophoresis (Fig. 3), amino acid sequencing, and the production of a monospecific antibody. The characteristics are similar to those of the human small folate binding protein found in milk and in serum. Pure small folate binding protein from goat's milk has been analyzed as to amino acid composition and has been shown to contain 22% carbohydrate (Table II). The small folate binding protein isolated from goat's milk represents the first example of totally characterized folate binding protein in a homogeneous form available in large amounts. This source of folate binding protein has been used to do folate binding

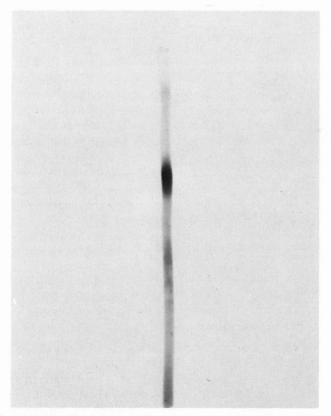

Fig. 3. Sodium dodecyl sulfate polyacrylamide gel electrophoresis of 30μg folate binding protein isolated from goat's milk.

Table II. Amino Acid and Carbohydrate Composition
of Goat Milk Folate Binding Protein

Substance	Residues/mol bound PteGlu
Amino acid[a]	
Lysine	17
Histidine	9
Arginine	12
Aspartic acid	26
Threonine	12
Serine	21
Glutamic acid	24
Proline	22
Glycine	29
Alanine	22
Half-cystine	4
Valine	16
Methionine	3
Isoleucine	10
Leucine	15
Tyrosine	7
Phenylalanine	7
Total	256
Carbohydrate	
Fucose	4
Galactose	3
Mannose	11
Galactosamine	6
Glucosamine	9
Sialic acid	3
Total	36

[a] Tryptophan not determined.

protein clearance studies as well as to develop immunoassays of folate binding protein.

Immunologic studies on the various folate binding proteins were done with a monospecific antibody produced in rabbits against the small folate binding protein from goat's milk (Waxman et al., 1975b). This antiserum forms a single precipitin line on Ouchterlony plates in the presence of the isolated folate binding protein (Fig. 4). PteGlu, PteGlu$_3$, or 5-CH$_3$H$_4$PteGlu preincubated with folate binding protein does not inhibit the subsequent precipitin line formed with anti-folate binding protein serum. Anti-goat folate binding protein diluted 1 : 1280 will precipitate (37°C for 2 hr followed by 24 hr at 4°C) the small folate binding protein and large folate binding protein in goat's milk and serum or bovine milk, but not the folate binding protein from human milk. Anti-goat folate binding protein blocks the binding of ^{3}H-PteGlu to purified goat, bovine, and human

Fig. 4. Precipitin reaction of rabbit anti-goat milk folate binding protein serum (center well) to the folate binding protein purified from (1) human milk small folate binding protein; (2) human serum folate binding protein; (3) human placental folate binding protein; (4) goat milk foalte binding protein; (5) goat milk folate binding protein saturated with PteGlu; and (6) bovine milk folate binding protein.

folate binding protein. Thus, the folate binding protein structure appears to be more species-specific, whereas the folate binding protein binding site for folic acid is immunologically similar in several species, including humans. Anti-goat folate binding protein-blocking and -precipitating antibodies are now being used in a radioimmunoassay for the measurement of total folate binding protein (saturated and unsaturated with folic acid) in various tissues and cell fractions.

3. Tissue Folate Binding Proteins

3.1. Intact Tissue

Low-affinity folate binding proteins have been found in the rat small intestinal brush border (Leslie and Rowe, 1972) and in some bacterial transport systems (Cooper, 1970; Heunnekens and Henderson, 1975; Shane and Stokstad, 1975). These binders demonstrate a rapid, saturable capacity for folic acid with a broad pH optimum; however, the affinity constant is 10^4 or 10^5 less than the high-affinity folate binders present in serum and milk.

Evidence for the presence of folate binding protein in the liver has been reported (Corrocher et al., 1974) but not well characterized. Zamierowski and

Wagner (1974) found intravenously injected folate activity associated with high-molecular-weight fractions in the liver, kidney, and intestine. The bound folate was predominantly in the folyl polyglutamate form in complexes of approximately 350,000, 150,000, and 25,000 daltons in the liver cell supernatant and 90,000 daltons in the liver nuclear fraction. The affinity of these binding proteins for folic acid has not been determined.

The first direct evidence of an unsaturated intracellular folate binding protein was identified in chronic granulocytic leukemia cells (Rothenberg, 1970). The crude and slightly purified folate binding proteins from these cells have properties similar to the serum folate binding proteins. The folate binding activity can be separated by DEAE cellulose chromatography into two fractions of different molecular weights—one eluting from the column with 0.005 M phosphate buffer pH 6.0 and a second more acidic binder eluting with 0.1 M phosphate buffer pH 7.0 (Fischer *et al.*, 1975).

The folate binding protein extracted from the hog kidney represents a binder that was initially saturated endogenously with folate (Kamen and Caston, 1975b). The nature of the form of folate in its naturally bound state was not determined. This folate binding protein was acid–charcoal dissociated and purified by Methotrexate–Sepharose affinity chromatography. Analysis of binding on polyacrylamide gel showed one major protein band, whereas isoelectric focusing indicated three major bands of PteGlu binding at pI 5.7, 6.6, and 6.8. The results of this experiment correlate well with human milk folate binding protein (Waxman and Schreiber, 1975a). The folate binding protein in hog kidney was found to have a molecular weight of 40,000 and a binding constant for folic acid of 5×10^{12} liters/mol at pH 7.6. At pH 9.6, a lower value of 1.4×10^{10} liters/mol was determined. The binding activity was destroyed by incubating the binder with proteolytic enzymes, but not with DNase, RNase, neuraminidase, phospholipase C, or β-galactosidase.

Folate binding protein has been purified from human placenta using either acetone extract, macro-Sephadex G-200 column chromatography, or PteGlu-Sepharose affinity chromatography (Liu *et al.*, 1976). Large and small folate-binding proteins were found with DEAE cellulose elution profiles and affinities for folic acid and folate analogues similar to those in serum and milk.

3.2. Subcellular Fractions

3.2.1. Membrane Fractions

Folate binding protein with a molecular weight of greater than 100,000 has been extracted from the brush-border membrane of the rat intestinal epithelial cell and found to have a binding constant (K_b) of 3.98×10^{-5} with broad specificity of the pteroic acid moiety (Leslie and Rowe, 1972). Similarly, a folate binding protein has been solubilized in high yield by sonic disruption of lysozyme-treated *L. casei* in the presence of ^{3}H-PteGlu and Triton X-100 (Huennekens and Henderson, 1975). The protein Triton complex (230,000 MW as estimated by Sephadex G-150 filtration) is stable to dialysis but dissociates on heating. The properties of this binder suggest that it may function as the carrier of folate during its transport into the cells.

Membrane fractions have been prepared from human lymphocytes previously pulsed with ^{3}H-PteGlu and extracted with 0.1% SDS (Fig. 5). Passage of this SDS membrane extract through a Sephadex G-200 column revealed a large and small folate binding protein similar to that found in milk and some sera. Attempts to find similar folate binding activity in red blood cell ghosts were unsuccessful, although the red blood cells were not pulsed with ^{3}H-PteGlu prior to preparation of the membrane fraction.

3.2.2. Nonmembrane Fractions

Folate binding protein has been studied in the subcellular fractions of chronic granulocytic cells (daCosta and Rothenberg, 1974b). The highest binding capacity was found in the nuclear and mitochondrial fractions. It was found that those cells with high intracellular concentration of unsaturated folate binding protein released this protein into the culture medium during the course of incubation.

4. Proposed Roles of Folate Binding Proteins

The proposed roles for folate binding proteins are outlined in Table III. Present knowledge of the folate binding proteins permit only inferences as to their biologic significance. There are several classes of folate binders, which differ in their molecular weight and which have some differences in binding specificity and affinities for the natural reduced folates. Folate binding proteins

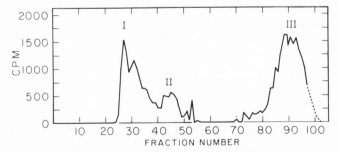

Fig. 5. Sephadex G-200 gel filtration of SDS extracted folate binding protein from the membrane of human lymphocytes pulsed with ^{3}H-PteGlu. (Peak I) Large folate binding protein ($>$ 200,000); (Peak II) small folate binding protein ($\approx$ 40,000); and (Peak III) free ^{3}H-PteGlu.

Table III. Proposed Roles for Folate Binding Protein in Folate Metabolism

---

1. Extracellular binding function
   Binds small amount of *oxidized folate* in serum
   Trapping of folate such as in cord serum for the fetus, milk for neonates, and in the choroid plexus for the brain
   Bacteriostatic property, such as in gut.
   Enterohepatic recirculation of folate
2. Intracellular transport
   Specific carrier protein for folate transport into hematopoietic cells, intestine, liver, and kidney.
3. Intracellular function
   Stores *oxidized folyl mono-* and *polyglutamates* but not reduced forms. Bound folates protected from action of *dihydrofolate reductase and conjugase.*

---

have been found in extracellular and intracellular compartments and in secretions. However, despite some minor differences, these binding proteins exhibit distinct similarities such as trans-species immunologically similar binding sites for folate and a preference for oxidized folyl mono- or polyglutamates.

4.1. Extracellular Binding Function

Homogeneous folate binding protein isolated from goat's milk has been used to determine the effect of extracellular folate binding protein on folate metabolism (Rubinoff *et al.*, 1976). Studies using ^{75}Se-selenofolate (30 ng) free or bound to goat folate binding protein injected intravenously into rats demonstrated that folate binding protein prolonged the plasma survival of ^{75}Se-selenofolate. ^{125}I-albumin was simultaneously injected to monitor plasma volume. Total urinary excretion of the free ^{75}Se-selenofolate was greater than that of the bound ^{75}Se-selenofolate. The bound ^{75}Se-selenofolate rapidly goes to the liver within the first 5 min in greater amounts than the free ^{75}Se-selenofolate. After 30 min, 40% of the ^{75}Se-selenofolate radioactivity was recovered in the bile from the bound folate as compared to 20% when the folate was injected in its free form. The ^{75}Se-selenofolate radioactivity was greatest in the liver, less in the kidney and bone marrow, and less than 1% in heart, muscle, spleen, and stomach at 2 hr. Recovery was greater when the ^{75}Se-selenofolate was injected bound, as compared to free, and probably represents a more rapid cellular uptake of the free folate as compared to the bound folate. ^{75}Se-radioactivity was concentrated in the middle third of the small intestine, increased with time, and was threefold more following injection of bound ^{75}Se-selenofolate, as compared to free ^{75}Se-selenofolate. ^{75}Se-radioactivity was no longer recovered in the small intestine when the bile duct was cannulated. These studies suggest that circulating folate binding protein prolongs the plasma survival of folate, inhibits its urinary excretion, and directs folate to the liver. These data suggest that the folate binding protein maintains small amounts of circulating folate, which may explain the finding of an elevated folate binding protein in folate deficiency.

Extracellular folate binding protein may be important in the entero-hepatic circulation and in the intestinal reabsorption of various folates.

[3]H-PteGlu bound to goat or cow milk folate binding protein studied over 24 hr was associated with a substantial reduction in its absorption from rate jejunum, whereas the ileal absorption showed a marked increase (Izak et al., 1972). Folate depletion or overload had no effect on the amount of milk-bound folate absorbed, whereas such treatments decreased the absorption of free folate. Thus, these data support the concept that folate bound to binder is absorbed differently from free folate. Further differences might have been found if the absorption comparison had been made earlier, as it appears that there is a significant and early enterohepatic recirculation for folate that differs depending on whether the folate is free or bound (Rubinoff et al., 1976).

Clearly, the binding of oxidized folate to folate binding protein at 10^{-9} M is essentially irreversible. It may be that this avid affinity is used to remove an unwanted form of oxidized folate where it cannot be metabolized, such as in the brain, where there is no dihydrofolate reductase (Makuld et al., 1973). Folate binding protein may control the concentration of a physiologic form of oxidized folate that is present in extracellular fluid, such as formyl folic acid, which it avidly binds (Table I).

Functionally, secretory folate binding protein can sequester folates for the fetus (cord serum), neonates (milk), as well as to organs, such as the brain (choroid plexus), which have a greater demand for folates (Spector and Lorenzo, 1975). Similarly, extracellular, secretory folate binding proteins may prevent bacterial utilization of bound folates and consequently serve as a bacteriostatic property in the intestinal lumen (Ford, 1974).

4.2. Intracellular Transport Function

A review of the literature concerned with folate transport in mammalian cell lines yields data that demonstrate the operation of probably two carrier-mediated saturable transport systems for the folates (one for oxidized and another for reduced forms of folates) (Huennekens et al., 1974). Studies in bacteria, intestinal brush border membrane, and lymphocyte membranes support that a folate binding protein similar to the extracellular folate binding protein may be part of this carrier-transport system. It has been proposed that the large folate binding protein may represent the small folate binding protein attached to a carrier. The cellular uptake of PteGlu in the L1210 leukemia system is biphasic with an immediate attachment of folate followed by a slower linear uptake (Lichtenstein et al., 1969; Bruckner et al., 1975). The immediate attachment phase is observed with ^{3}H-PteGlu, but not with ^{14}C-5-CH$_3$H$_4$ PteGlu or ^{3}H-Methotrexate, which suggests that there are differences in the dynamics of transport (Bruckner et al., 1975).

The possibility that there is a receptor or carrier for the folate–binder complex becomes very likely when one examines related literature. For example, the role of the retinol-binding protein and a specific cell receptor for the retinol–retinol-binding protein complex was recently described (Heller, 1975). Significantly, the affinity of the receptor for the apobinder was lower so that the retinol–binder complex would displace it from the cell receptor. This may relate to a possible mechanism of release of folate bound to folate binding protein whereby the unbound folate binding protein attached to a membrane receptor is displaced by a saturated folate binding protein that is capable of delivering to the cell. Clearly, much additional work is needed to establish this model, such as the mechanism of folate release from the binder. Thus far, dissociation has been accomplished by low pH and high ionic concentration. It is possible that the intramembranous conditions meet these requirements (Blair and Matty, 1975).

4.3. Intracellular Metabolic Function

The presence of folate binding proteins in the cell (cytoplasmic or nuclear) can significantly influence folate storage, interconversions, and utilization. Folates in the cell are predominantly in the form of $5\text{-}CH_3H_4$ PteGlu polyglutamates and can carry out metabolic functions either in the mono- or polyglutamate form. Folate binding protein can bind oxidized folyl mono- or polyglutamates with equal affinity. Consequently, when bound to folate binding protein, dihydrofolate may not be available as a substrate for dihydrofolate reductase and may be important in modulating *de novo* thymidylate synthesis. Oxidized folyl polyglutamate, when bound to folate binding protein, is not accessible to the action of conjugase (Waxman and Schreiber, 1975c); this may be important in the intracellular storage of folates. In contrast, $5\text{-}CH_3H_4PteGlu_3$, $-PteGlu_5$, or $-PteGlu_7$ are poorly bound by folate binding proteins, which may make these forms of folate available for coenzymatic functions (Waxman and Schreiber, 1975, unpublished observations).

The formyl derivates of reduced folate are the natural folate analogues with the lowest affinity for the various folate binding proteins studied. It may be that the formyl derivatives of folate are the intracellular forms of folate, which are free for metabolic interconversions, and when converted to other folyl mono- or polyglutamate derivatives become available for storage by becoming bound to folate binding proteins.

5. Clinical Significance of Folate Binding Proteins

5.1. Clinical

The finding of folate binding protein in serum, cells, and secretions suggests new parameters for studying serum folate in the normal and various clinical situations. A multifaceted radioassay can be used to measure total folate, bound folate, unbound folate, total folate binding protein, saturated and unsaturated folate binding protein, and oxidized and reduced folates in serum. These parameters can be studied using radiofolate, folate binding protein, antibody to folate binding protein, and radioiodinated folate binding protein. The use of such multifaceted radioassays may provide new information concerning folate metabolism in the serum, milk, cerebrospinal fluid, intestinal secretions, and various tissues in patients with cirrhosis, sprue, folate deficiency, uremia, and neoplastic diseases, as well as those who are pregnant, taking oral contraceptives, anticonvulsants, and antifols.

Recent studies show that uremic patients have levels of unsaturated folate binding protein that increase upon dialysis, as well as having an apparent pool of bound folate in the serum. An unusual folate binding protein present in serum has been found in patients with certain types of carcinoma (Kamen *et al.*, 1976b). Cerebrospinal fluid appears to contain a folate binding factor that facilitates the uptake of folate into lymphocytes (Kamen *et al.*, 1976c). Malignant ascites obtained from a patient with advanced adenocarcinoma contain folate binding protein that can bind significant amounts of folate (Waxman and Schreiber, 1976). Thus far, anticonvulsants, such as diphenylhydantoin (Dilantin) do not appear to interfere with the binding of folates to folate binding proteins. Thus, it may be anticipated that new disorders of folate metabolism will be discovered in measuring these new parameters of folate and folate binding protein in the serum.

Folate binding protein dysfunction could produce clinical disorders of folate metabolism. Increased folate binding protein in the serum could produce a state of intracellular folate deficiency despite adequate serum folate levels. Such a condition has been suggested by Hines *et al.* (1973) as a cause of the megaloblastic marrow maturation in some uremic patients whose serum contains folate binding protein. Absence (congenital or acquired) of cellular (membrane) folate binding protein could result in folate malabsorption and may explain the isolated folate intestinal malabsorption and transport defect described by Lanzkowsky *et al.* (1965). This may also contribute to deranged enterohepatic folate reutilization and to the folate malabsorption in the sprue syndromes, in which there is usually membrane injury. Lack of folate binding proteins in the choroid plexus may result in defective brain development and function as a result of central nervous system folate deficiency in the presence of a normal serum folate level.

The failure to retain intracellular folates in a patient with acquired aplastic anemia who has a normal serum folate and low red blood cell folate level that responded to folate administration may be attributable to lack of intracellular folate binding protein (Branda *et al.*, 1975). Another possibility for deranged folate metabolism is defective intracellular release of folates bound to folate binding protein, which could manifest as a state of intracellular metabolic folate deficiency despite normal serum and cellular folate levels. Whether agents (diphenylhydantoin, ethanol, oral contraceptives) that provoke folate malabsorption influence the binding of folate to folate binding protein remains to be established.

References

Blair, J. A., and Matty, A. J., 1975, Acid microclimate in intestinal absorption, *Clin. Gastroenterol.* **3**:183.

Branda, R. F., Moldow, C. F., and Jacob, H. S., 1975, Folate induced remission in aplastic anemia with impaired cellular folate uptake, *Clin. Res.* **23**:270a.

Bruckner, H. W., Schreiber, C., and Waxman, S., 1975, The interaction of chemotherapeutic agents with methotrexate and 5-fluorouracil and its effect on de novo DNA synthesis, *Cancer Res.* **35**:801.

Chanarin, I., Mollin, D., and Anderson, B. B., 1958, The clearance from plasma of folic acid injected intravenously in normal subjects and patients with megaloblastic anemia, *Br. J. Haematol.* **4**:435.

Colman, N., and Herbert, V., 1974, Release of folate binding protein FBP from granulocytes. Enhancement by lithium and elimination by fluoride. Studies with normal pregnant, cirrhotic and uremic persons, *Program, 17th Annual Meeting of the American Society Hematology,* p. 155.

Colman, N., and Herbert, V., 1976, Total folate binding capacity of normal human plasma, and variations in uremia, cirrhosis and pregnancy, *Blood* **48**:911.

Cooper, B. A., 1970, Studies of ^{3}H-folic acid uptake by Lactobacillus casei, *Biochim Biophys. Acta* **208**:99.

Corrocher, R., DeSandre, G., Pacor, M. L., and Hoffbrand, V., 1974, Hepatic protein binding of folate, *Clin. Sci. Mol. Med.* **46**:551.

daCosta, M., and Rothenberg, S. P., 1974a, Appearance of a folate binder in leukocytes and serum of women who are pregnant or taking oral contraceptives, *J. Lab. Clin. Med.* **83**:207.

da Costa, M., and Rothenberg, S. P., 1974b, Studies of the folate binding factor in cultures and subcellular fractions of chronic myelogenous leukemia cells, *Clin. Res.* **22**:486.

Eichner, E. R., Paine, C. J., Dickson, V. L., and Hargrove, M. D., 1975, Clinical and laboratory observations on serum folate-binding protein, *Blood* **46**:599.

Elsborg, L., 1972, Binding of folic acid to human plasma proteins, *Acta Haematol.* **48**:207.

Fischer, C., daCosta, M., and Rothenberg, S. P., 1975, Heterogeneity and properties of folate binding protein from chronic myelogenous leukemia cells, *Blood* **46**:855.

Ford, J. E., 1974, Some observations on the possible nutritional significance of vitamin B_{12} and folate binding proteins in milk, *Br. J. Nutr.* **31**:243.

Ford, J. E., Salter, D. N., and Scott, K. J., 1969, The folate binding protein in milk, *J. Dairy Res.* **36**:435.

Ghitis, J., 1967, The folate binding in milk, *Am. J. Clin. Nutr.* **20**:1.

Hampers, C. L., Streiff, R., Nathan, D. C., Snyder, D., and Merrill, J. P., 1967, Megaloblastic hematopoesis in uremia and in patients on long term hemodialysis, *New Engl. J. Med.* **276**:551.

Heller, J., 1975, Interactions of plasma retinol-binding protein with its receptor. Specific binding of bovine and human retinol-binding protein to pigment epithelium cells from bovine eyes, *J. Biol. Chem.* **250**:3613.

Herbert, V., Larrabee, A. R., and Buchanan, J. N., 1962, Studies on the identification of a folate compound of human serum, *J. Clin. Invest.* **41**:1134.

Hines, J. D., Kamen, B. A., and Caston, J. D., 1973, Abnormal folate binding proteins in azotemic patients, *Blood* **42**:997.

Huennekens, F. M., and Henderson, G. B., 1975, Transport of folate compounds into mammalian and bacterial cells, in *Chemistry and Biology of Pteridines* (W. Pfleiderer, ed.) pp. 179–196, Walter de Gruyter, Berlin.

Huennekens, F. M., DiGirolamo, P. M., Fujii, K., Henderson, G. B., Jacobsen, D. W., Neef, V. G., and Rader, J. I., 1974, Folic acid and vitamin B_{12}: transport and conversion to coenzyme forms, *Advan. Enzyme Reg.* **12**:131.

Izak, G., Galewski, K., Rachmilewitz, M., and Grossowicz, N., 1972, The absorption of milk-bound pteroylglutamic acid from small intestine segments, *Proc. Soc. Exp. Biol. Med.* **140**:248.

Johns, D. J., Sperti, S., and Gurgen, A. S. V., 1961, The metabolism of tritiated folic acid in man, *J. Clin. Invest.* **40**:1684.

Kamen, B. A., 1976, Characterization of Folate Binding Proteins, *Doctoral dissertation*, Case Western Reserve Univ., Department of Anatomy.

Kamen, B. A., and Caston, J. D., 1975a, Purification of folate binding factor in normal umbilical cord serum, *Proc. Natl. Acad. Sci. USA* **72**:4261.

Kamen, B. A., and Caston, J. D., 1975b, Identification of a folate binder in hog kidney, *J. Biol. Chem.* **250**:2203.

Kamen, B. A., Tackach, P. T., Vatev, R., and Caston, J. D., 1976a, A rapid radio-chemical ligand binding assay for methotrexate, *Anal. Biochem.* **70**:54.

Kamen, B. A., Anthony, D. A., and Caston, J. D., 1976b, Properties of a folate binding protein in the serum from a patient with an adenocarcinoma, in preparation.

Kamen, B. A., Gross, S., and Caston, J. D., 1976c, Role of cerebrospinal fluid folate binder in the uptake of folate by lymphoblasts.

Lanzkowsky, P., Erlandson, M. E., and Bezan, A. I., 1965, Isolated defect of folic acid absorption associated with mental retardation and cerebral calcification, *Blood* **34**:452.

Leonard, J. P., and Beckers, C., 1976, Limits of the estimate of serum folate binding protein by radioassay, *Clin. Chim. Acta* **70**:119.

Leslie, G. I., and Rowe, P. B., 1972, Folate binding by the brush border membrane proteins of small intestinal epithelial cells, *Biochemistry* **11**:1696.

Lichtenstein, N. S., Oliverio, V. T., and Goldman, I. D. 1969, Characteristics of folate transport in the L1210 leukemia cell, *Biochim. Biophys. Acta* **193**:456.

Liu, C. K., Rubinoff, M., Schreiber, C., and Waxman, S. 1976, unpublished observations.

Makuld, D. R., Smith, E. F., and Bertino, J. R., 1973, Lack of dihydrofolate reductase activity in brain tissue of mammalian species: Possible implications, *J. Neurochem.* **21**:241.

Mantzos, J. D., Alevizou-Terzaki, V., and Gyftaki, E., 1974, Folate binding in animal plasma, *Acta Haematol.* **51**:204.

Markkanen, T., 1968, Pteroylglutamic acid (PGA) activity of serum in gel filtration, *Life Sci.* **7**:887.

Markkanen, T., and Peltola, O., 1970, Binding of folic acid activity by body fluids, *Acta Haematol.* **43**:272.

Markkanen, T., and Peltola, O., 1971, Carrier proteins of folic acid activity in human serum, *Acta Haematol.* **45**:176.

Markkanen, T., Pajula, R. L., Virtanen, S., and Himanen, P., 1972, New carrier protein(s) of folic acid in human serum, *Acta Haematol.* **48**:145.

Markkanen, T., Pajula, R. L., Himanen, P., and Virtanen, S., 1973, Serum folic acid activity (L. casei) in Sephadex gel chromatography, *J. Clin. Pathol.* **26**:486.

Markkanen, T., Pajula, R. L., Himanen, P., and Virtanen, S., 1974, Binding of folic acid to serum proteins IV. In some animal species, *Int. J. Vit. Nutr. Res.* **44**:347.

Metz, J., Zalusky, R., and Herbert, V., 1968, Folic acid binding by serum and milk, *Am. J. Clin. Nutr.* **21**:259.

Retief, F. P., and Huskisson, Y. J., 1969, Serum and urinary folate in liver disease, *Br. Med. J.* **2**:150.

Retief, F. P., and Huskisson, Y. J. 1970, Folate binders in body fluids, *J. Clin. Pathol.* **23**:703.

Rothenberg, S. P., 1970, A macromolecular factor in some leukemic cells which binds folic acid, *Proc. Soc. Exp. Biol. Med.* **133**:428.

Rothenberg, S. P., and daCosta, M. 1971, Further observations on the folate binding factor in some leukemic cells, *J. Clin. Invest.* **50**:719.

Rothenberg, S. P., and daCosta, M., 1976, Folate binding proteins and radioassay for folate, *Clinics Hematol.* **5**(3):569–587.

Rubinoff, M., Schreiber, C., and Waxman, S., 1975, Isolation and characterization of folate binding protein (FABP) from goat milk by affinity chromatography, FEBS Lett. **75**:244.

Rubinoff, M., Abramson, R., Schreiber, C., and Waxman, S., 1976, The effect of FABP on plasma transport of folates, *Prog. Am. Soc. Hematol.* Abst. 506.

Salter, D. N., Ford, J. E., Scott, K. J., and Andrews, P., 1972, Isolation of the folate binding protein from cow's milk by the use of affinity chromatography, *FEBS Lett.* **20**:302.

Selhub, J., and Grossowicz, N., 1973, Chemical fixation of folate binding protein to activated sepharose, *FEBS Lett.* **35**:76.

Shane, B., and Stokstad, E. R. L., 1975, Transport of folates by bacteria, *J. Biol. Chem.* **250**:2243.

Sharon, B., Liu, C. K., Schreiber, C., and Waxman, S., 1976, Unpublished observations.

Spector, R., and Lorenzo, A. V., 1975, Folate transport by the choroid plexus in vitro, *Science* **187**:540.

Stenman, U. H., Simons, K., and Grasbeck, R., 1968, Vitamin B_{12} binding protein in normal and leukemic human leukocytes and sera, *J. Lab. Clin. Invest.* **21**:202.

Waxman, S., 1975, Annotation, folate binding proteins, *Br. J. Haematol.* **29**:23.

Waxman, S., 1976, The role of folate binding proteins in folate metabolism in *Chemistry and Biology of Pteridines* (W. Pfleiderer, ed.), pp. 165–178, Walter de Gruyter, Berlin.

Waxman, S., and Schreiber, C., 1971, Characteristics of folic acid binding macromolecular factor (FABF) in folic acid (FA) deficient serum, *Prog. Am. Soc. Hematol.*

Waxman, S., and Schreiber, C. 1972, Measurement of serum folate and folic acid binding protein by [3]HPGA radioassay, *Clin. Res.* **20**:572.

Waxman, S., and Schreiber, C., 1973a, Measurement of serum folate levels and serum folic acid binding protein by [3]HPGA radioassay, *Blood* **42**:281.

Waxman, S., and Schreiber, C., 1973b, Characteristics of folic acid binding protein in folate deficient serum, *Blood* **42**:291.

Waxman, S., and Schreiber, C., 1974a, The role of folic acid binding proteins (FABP) in the cellular uptake of folates, *Proc. Soc. Exp. Biol. Med.* **147**:760.

Waxman, S., and Schreiber, C., 1974b, The isolation of the folate binding protein from commercially purified bovine beta lactoglobulin, *FEBS Lett.* **55**:128.

Waxman, S., and Schreiber, C., 1975a, Isolation and characterization of human milk folate binding protein using affinity chromatography, *Biochemistry* **14**:5422.

Waxman, S., and Schreiber, C., 1975b, Folic acid binding protein (FABP) membrane function and role in intracellular folyl polyglutamate distribution, *Clin. Res.* **23**:284A.

Waxman, S., and Schreiber, C., 1976, Unpublished observations.

Waxman, S., Rubinoff, M., and Schreiber, C., 1975, Immunologic studies on folate binding protein, *Prog. Am. Soc. Hematol.* Abst 114.

Waxman, S., Schreiber, C., and Liu, C. K., 1976, Characteristics and possible clinical significance of the large molecular weight folate binding protein (L-FABP), *Clin. Res.* **24**:223.

Zettner, A., and Duly, P. E., 1974, New evidence for a binding principle specific for folates as a normal constituent of human serum, *Clin. Chem.* **20**:1313.

Zamierowski, M., and Wagner, C., 1974, High molecular weight complexes of folic acid in mammalian tissues, *Biochim. Biophys. Res. Commun.* **60**:81.

Chapter 4

Folate Deficiency in Humans

Neville Colman

1. History

The history of folate is closely linked to that of vitamin B_{12} and can be traced through different scientific disciplines to the convergent purification and identification of a vitamin essential to cellular metabolism. The first clinical report of human deficiency disease caused by a lack of folate was probably that by Channing (1842), which drew attention to a severe form of anemia in pregnancy and the puerperium which was rapidly progressive and sometimes fatal. Osler (1919) described the condition as being clinically and morphologically indistinguishable from pernicious anemia as described by Addison (1855), but postulated that it "is caused by an agent which differs . . . from that which causes the anemia of Addison." Evidence for this hypothesis emerged in the 1930s, when Wills and Mehta (1930) described a macrocytic anemia in pregnant women in Bombay that responded to therapy with Marmite, a commercial yeast extract, and crude liver extract (Wills, 1931), but not to the recently developed purified liver extract, which is now known to be a fairly pure solution of vitamin B_{12} (Castle, 1961). At about the same time, megaloblastic anemia in 6-week-old neonates was described by Faber (1928). This syndrome, known in the German literature as "goat's milk anemia," was successfully treated with crude liver and yeast (Rominger et al., 1933; György, 1934).

Factors subsequently shown to be folic acid that stimulated growth responses of bacteria, chicks, and mammalian bone marrow were identified in yeast, liver extract from which vitamin B_{12} had been removed, alfalfa, and wheat bran, and were called "vitamin M," "vitamin B $_c$," "L. casei factor," "Norite eluate factor," and finally "folic acid" by Mitchell et al. (1941), who

Neville Colman • State University of New York—Downstate Medical Center, and Veterans Administration Hospital, Hematology and Nutrition Laboratory, Bronx, New York.

isolated it from spinach leaves. Folic acid was purified by Stokstad (1943) and isolated in crystalline form from liver (Pfiffner *et al.*, 1943), and then synthesized by Angier *et al.* (1946), who named it pteroylglutamic acid on the basis of its chemical structure.

2. General

Previous reviews of individual aspects of folate metabolism are cited at appropriate places in the text. Most of the leading authorities in the field have contributed to an issue of *Clinics in Haematology* devoted to megaloblastic anemia (Hoffbrand, 1976) and to the proceedings of a National Academy of Sciences workshop on human folate requirements (Food and Nutrition Board, 1977), which together provide an outstanding comprehensive account of current knowledge in the field.

2.1. Frequency of Folate Deficiency

Megaloblastic anemia caused by folate deficiency and its relationship to various conditions has been reported by numerous workers all over the world, which suggests that this disease constitutes a major problem in world health (Herbert, 1968b). On the basis of those reports, it is evident that a majority of cases in all populations are associated with pregnancy, alcoholism, or malabsorption. As discussed above, pregnancy was the earliest and most commonly recognized cause; a review of 27 surveys reported from 1951 to 1968 showed an increasing incidence with time, presumably as a result of increasing recognition, with the most recent studies suggesting a frequency ranging from 13% in London to more than 33% in pregnant women from poorly nourished populations (Rothman, 1970).

The incidence of folate deficiency, as opposed to megaloblastic anemia, has been studied in many groups of pregnant women, in conditions known to cause folate deficiency, and more recently in whole populations. The importance of these surveys lies in the identification of target groups susceptible to the rapid development of megaloblastic anemia and to other possible deleterious effects. Previous evidence that folate levels were low in serum of two-thirds and in erythrocytes of one-third of all pregnant women (Rothman, 1970) has received continuing support in subsequent surveys, including those conducted among black South Africans in rural areas, as well as among Americans of black and Puerto Rican descent in New York and of Mexican descent in Los Angeles (Colman *et al.*, 1975c; Herbert *et al.*, 1975; Jacob *et al.*, 1976). Serum folate was low in 80–90% of alcoholics, with about half that number having low erythrocyte folate and megaloblastic bone marrow (Herbert *et al.*, 1963; Eichner and Hillman, 1971; Herbert and Tisman, 1975; Hines, 1975; Wu *et al.*, 1975).

The incidence of folate deficiency in tropical sprue varies widely in different areas, ranging from 22 to 84% (Klipstein, 1968).

Data on the proportion of the general population who are folate deficient have recently been reported. Although serum and erythrocyte folates were measured in the Ten-State Nutrition Survey of 1968–1970 (U.S., D.H.E.W., 1972), only mean values, and not the incidence of deficiency, were reported. The conclusion drawn was that, "recognizing the limitation of the interpreting data based on mean values alone, folate intakes appear to be adequate in a large percentage of the population." A diametrically opposed conclusion was drawn when the portion of that survey that was conducted in Massachusetts, which did analyze incidence, was reported to the State Commissioner for Public Health (Edozien, 1972). In a sample drawn from low-income households in that state, erythrocyte folate was less than 160 ng/ml in 26% of 1979 subjects, of whom 55% were women and 6% pregnant. A similar survey of 325 rural South African black adults, of whom 57% were women and none pregnant, showed similar results— the overall frequency of erythrocyte folates less than 160 ng/ml was 27% (Colman *et al.*, 1975c). A breakdown of these subjects into sex and age groups, shown in Fig. 1, shows that most of the incidence occurred among women of child-bearing age and subjects of both sexes over 60. Since most affected women of child-bearing age had been pregnant or lactating within 2–3 years of the study, as shown in Table II (page 96), these results indicated that although as many as one-quarter of all adults were folate-deficient, the major target groups were the elderly and women who were pregnant or lactating or who had been during the previous 3 years (Colman *et al.*, 1975c).

There is evidence that folate deficiency may be quite rare in healthy children, even in populations in which it is common in adults (Margo *et al.*, 1977), but this exception does not apply to infants of low birth weight. The necessity for careful monitoring of these infants is clear, and recommendations for prophylactic supplementation are discussed in Section 5.

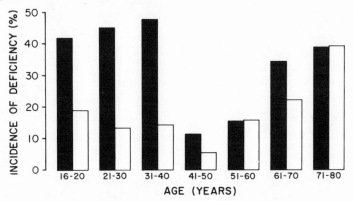

Fig. 1. The incidence of folate deficiency (erythrocyte folate < 160 ng/ml) in adult rural South African blacks grouped according to sex and age (Colman *et al.*, 1975c). (■) Women; (□) men.

2.2.   Nutritional Requirements and the Recommended Daily Allowance

The minimal daily adult requirement for folic acid has been estimated at 50 μg (Herbert, 1968a). Serum folate levels remained constant on tablets containing this dose but rose with a higher dose and fell with 25 μg daily in healthy adult women on a folate-deficient diet (Herbert, 1962b). This dose corrected megaloblastosis (Zalusky and Herbert, 1961) and prevented increased urinary formiminoglutamate excretion in a subject on a folate-deficient diet (Fleming et al., 1963). Herbert (1968a) also calculated that the rate of folate depletion in a subject on a folate-deficient diet (Herbert, 1962a) represented utilization of 50 μg folate daily.

The major source of information regarding folate requirements during pregnancy has been study of the effects of prophylactic supplementation. Willoughby and Jewell (1966) reported that 300 μg folic acid daily protected against falling serum folate and megaloblastic anemia in pregnancy, whereas 100 μg daily did not. Primarily because Alperin et al. (1966) reported that two pregnant subjects had suboptimal responses to treatment of megaloblastic anemia with 200 μg folic acid daily, at least 300 μg has been accepted as the minimum daily requirement. In recent studies, however, increasing the effective daily supplementation dose from 170 to 280, 300, and 560 μg did not cause a significant decrease in the number of women in whom either erythrocyte or serum folate levels fell (Colman et al., 1975b). These studies suggested that depletion of erythrocyte folate progressed in about 10% of pregnant women even when the highest dose was supplemented, and that depletion in the remainder would be prevented by oral administration of less than 200 μg folic acid daily. Higher doses may be necessary for effective treatment of megaloblastic anemia.

The minimal requirement in infants and children cannot be extrapolated on a weight-for-weight basis. In newborn infants of low birth weight, a fall in serum folate occurs and can be prevented by daily administration of 50 μg folic acid (Samuel et al., 1973). Although no beneficial hematological effects of supplementation were observed in a controlled trial (Hibbard and Kenna, 1974), supplementation of the diet has been recommended for low birth-weight infants, particularly if they weigh less than 1500–1700 g (Hoffbrand, 1970; Dallman, 1974). The minimal daily requirement is increased in a number of conditions that may cause folate deficiency, which are discussed below. The daily folate requirement may vary from as little as 3.6 μg/kg body weight in healthy children under the age of 2 (Waslien, 1977) to as much as 13 μg/kg in children recovering from kwashiorkor (Kamel et al., 1972).

The Recommended Dietary Allowances are the levels of intake of essential nutrients considered "to be adequate to meet the known nutritional needs of practically all healthy persons" and, where possible, are geared to include the needs of 97.5% of a normally distributed population (Food and Nutrition Board/National Research Council, 1974). The Recommended Daily Allowance for total folate, as assayed by *L. casei* from dietary sources after conjugase treatment, ranges from 50 μg for infants (under 1 year) through 100 μg (1–3 years), 200 μg (4–6 years), 300 μg (7–10 years) and 400 μg (over 10 years) to 600 μg for lactating women and 800 μg for pregnant women. Unfortunately, the *L. casei* assay of total food folate content has yielded disparate results in different studies, and it is therefore not yet possible to calculate dietary intake from reliable food tables.

3. Causative Factors

There are some 50 known causes of folate deficiency, not all of which are discussed here. They have been listed by Herbert (1975a), who has classified these causes into six groups according to their mechanism—three on the basis of inadequacies (of ingestion, absorption, and utilization), and three caused by increases (of requirement, excretion, and destruction). This classification model is applicable to all nutrient deficiencies and has proved useful in clinical evaluation of a wide variety of conditions. Occasional difficulties may be encountered when using the above-mentioned system because some conditions, particularly those recently described, cause folate deficiency by incompletely elucidated mechanisms. The reader is referred to Herbert (1975a) for an overview of the conditions associated with each mechanism.

3.1. Inadequate Ingestion

It is easy to identify inadequate ingestion as the cause of folate deficiency in some groups because their quantitative and qualitative dietary inadequacies are gross. One such example are the blacks of South Africa, whose folate-poor diet (Edelstein *et al.*, 1966) is largely maize meal, which contains less than 40 μg folate/kg (Metz *et al.*, 1970), and who have a high incidence of folate deficiency (Colman *et al.*, 1975c).

When food supplies are not grossly inadequate, however, the definition of dietary insufficiency may be much more difficult. The major factor contributing to this difficulty has been disagreement among different workers concerning the folate content of average diets and of individual foods. It is universally agreed that the food folate levels reported in the most comprehensive tables available for the past 25 years (Toepfer *et al.*, 1951) are too low because ascorbate was not used to prevent oxidative destruction. After Herbert (1963a) described the necessity of using ascorbate, several reports, which are listed below, indicated that average daily diets in the United States, Canada, Britain, and Sweden all contained about 100–200 μg "free" folate. The results differed when conjugase was added for the assessment of total folate content, with some groups reporting that this procedure increased assayable folate up to fourfold, indicating that average daily diets contained about 680 μg total folate (Butterworth *et al.*, 1963; Chanarin *et al.*, 1968; Hurdle *et al.*, 1968), whereas others found that conjugase treatment only increased *L. casei*-active folate by 50–100%, indicating an average total dietary folate of 200–300 μg daily (Herbert, 1963a; Hoppner *et al.*, 1972; Moscovitch and Cooper, 1973; Jagerstad *et al.*, 1975).

It is possible that these differences are attributable to methodologic errors encountered by those workers who recorded the higher values after conjugase treatment. Jagerstad *et al.* (1975) reported that when incubation with conjugase was allowed to proceed for 21 hr, there was frequent significant contamination with folate-producing, Lancefield group B, β-hemolytic streptococci. These increased folate levels above those seen when bacterial contamination was prevented either by shortening the incubation step, a method used by Moscovitch and Cooper (1973), or by incubating under toluene, as recommended by Herbert (1963a). In this regard, it is significant that one of the two reports that average diets contain about 680 μg folate daily has been retracted, with the authors indicating that their current studies reflect a total folate content of approximately 250 μg daily (Chanarin, 1977a).

For as long as food folate has been studied, it has been divided into "free" and "total" categories according to the growth response of *L. casei* before and after conjugase treatment. But only as a result of recent studies has the precise meaning of these terms emerged. It had been thought that this organism responded equally to mono-, di-, and triglutamates, and not at all to folates containing four or more glutamates. However, Tamura *et al.* (1972) have observed that compared to monoglutamate, there is 66% response to tetraglutamate, 20% to pentaglutamate, and 2–4% to hexaglutamate and heptaglutamate, which indicates that estimation of "free" folate as defined earlier includes folates that contain four or more glutamates. Glutamate chain length is altered by the food preparation method. When raw meat is stored frozen, endogenous conjugases break down polyglutamates to tri- and diglutamates within 3 days and to monoglutamate after prolonged storage (Reed *et al.*, 1976). Such breakdown is not apparent when green vegetables are stored frozen, but it does occur when they are extracted (Shinton and Wilson, 1975). Polyglutamates remain intact during storage and extraction after endogenous conjugases have been inactivated by cooking (Shinton and Wilson, 1975; Reed *et al.*, 1976).

Because conflict concerning the folate content of food persists, it remains difficult to provide nutritionists with reliable information for planning diets that provide the Recommended Daily Allowances. The U.S. Department of Agriculture has attempted to rectify this problem by releasing a provisional table, compiled from various studies, which lists the reported "free" and total folate content of 299 foods (Perloff and Butrum, 1977), but it is not yet clear whether this compilation addresses the conflict among different groups. As an alternative, Thenen (1975) has suggested that the values reported by Toepfer *et al.* (1951) be used after multiplication by a factor of 4.4; the latter is derived from the report by

Moscovitch and Cooper (1973) that average diets contain 4.7 and 4.1 times as much total folate as predicted by the tables of Toepfer *et al.* (1951), which correctly predicted the rank order of folate content in individual foods. An alternative direct source for folate content of foods is the work of Hoppner (1971) and Hoppner *et al.* (1972, 1973).

3.2. Inadequate Absorption

Butterworth and Krumdieck (1975) and Rosenberg (1976) have comprehensively reviewed the mechanisms of folate absorption and their importance in the prevention of deficiency. Among the technologic advances that have been important in the study of these mechanisms have been the use of the triple lumen tube to measure intestinal luminal disappearance of water and solutes (Cooper *et al.*, 1966), *in vitro* study of folate uptake by everted rat gut and individual intestinal cells (Herbert and Shapiro, 1962; Herbert, 1967; Momtazi and Herbert, 1973), and the solid-phase synthesis of folate polyglutamates, which allows radiolabeling at a preselected site on the molecule (Krumdieck and Baugh, 1969). Two separate mechanisms have been identified as significant for folate absorption—hydrolysis of polyglutamates and transport of folate across the mucosal cell and into the portal blood.

3.2.1. Hydrolysis of Polyglutamate

Hydrolysis of polyglutamates to monoglutamate is now accepted as an obligatory step in their absorption (Butterworth and Krumdieck, 1975) despite suggestions that they may be absorbed intact (Cooperman and Luhby, 1965) or even formed during absorption (Baker *et al.*, 1965); a minor exception to this rule is that diglutamate, rarely present in foods, may be absorbed in an unaltered state (Baugh *et al.*, 1971). The conclusive evidence for acceptance of this concept came from the observation that plasma radioactivity and microbiologic folate activity rose in parallel after ingestion of ^{14}C-folate polyglutamates labeled in the first glutamate, but that these were completely dissociated when the ingested folate was labeled in the second glutamate (Butterworth *et al.*, 1969).

The suggestion that this obligatory step was rate-limiting arose from numerous reports, which date back to that of Swendseid *et al.* (1947), that yeast polyglutamate is relatively poorly absorbed except when administered in doses lower than 200 μg (Perry and Chanarin, 1968; Rosenberg *et al.*, 1969; Hoffbrand and Peters, 1970). Studies of luminal disappearance from a 30-cm jejunal segment after perfusion of folates in humans indicated that polyglutamates are absorbed only two-thirds as rapidly as monoglutamate (Halsted *et al.*, 1975), whereas *in vitro* studies using rat intestine suggested that both had the same rate of flux and that hydrolysis was not a rate-limiting step (Wagonfeld *et al.*, 1975).

When absorption was monitored by measurement of urinary excretion, pure polyglutamates were absorbed 70–90% as efficiently as was monoglutamate (Tamura and Stokstad, 1973; Godwin and Rosenberg, 1975; Halsted et al., 1975). In the one reported case studied simultaneously by measurements of luminal disappearance and urinary excretion, the data suggested that differences in absorption rate over a 30-cm jejunal segment became less significant when the remaining unabsorbed poly- and monoglutamates were exposed to the rest of the intestine (Halsted et al., 1975).

Current evidence supports the concept that conjugase action in mammals occurs within the mucosal cell of the intestine rather than at the brush border or within the intestinal lumen. This evidence includes data that folate conjugase activity is minimal in intestinal contents and brush-border subcellular fractions of mucosal cells and maximal in lysosome subcellular fractions, and that the products of hydrolysis are concentrated in isolated mucosal cells in vitro earlier than in the extracellular medium (Hoffbrand and Peters, 1969; Rosenberg and Godwin, 1971; Baugh et al., 1975; Halsted et al., 1975, 1976b). Despite this evidence, aspiration of intestinal contents after perfusion of radiolabeled polyglutamates indicates the accumulation in the intestinal lumen of the products of hydrolysis (Halsted et al., 1975), which suggests that there may be back-diffusion of monoglutamates after intracellular hydrolysis. The suggestion that conjugase may act at the brush border in a manner similar to disaccharidases (Halsted et al., 1975), consistent with the presence of a brush-border binder for folates (Leslie and Rowe, 1972), has been largely, but not totally, excluded by the above-mentioned data and by the finding that mucosal activity of folate conjugase is not affected by tropical sprue in parallel to decreased disaccharidase enzyme activity at the surface of the cell (Corcino et al., 1976).

As most food folates are polyglutamates (Scott and Weir, 1976), it is theoretically feasible that congenital or acquired disease, or both, may cause conjugase deficiency and thereby folate deficiency through impairment of this obligatory route of polyglutamate absorption. However, no case of this enzyme deficiency has been documented. In inborn errors of metabolism attributed to congenital folate malabsorption, the defect appeared to affect monoglutamates and polyglutamates equally (Lanzkowsky et al., 1969; Santiago-Borrero et al., 1973), and in acquired malabsorption caused by tropical sprue, as well as in celiac disease, folate conjugase activity of jejunal biopsies is normal or increased, and polyglutamate malabsorption is no greater than that of monoglutamate (Hoffbrand et al., 1969; Corcino et al., 1976; Halsted et al., 1976a). (See note added in proof, p. 112.)

Impairment of normal conjugase activity may, however, occur as a result of a variety of food factors, some nonspecific and some apparently specific. Conjugase inhibitors were first identified in yeast (Bird et al., 1946) and probably account for the above-mentioned poor absorption of polyglutamates from that food. They have subsequently been identified in beans, peas, and other pulses (Krumdieck *et al.*, 1973) where they are activated by heat, impair conjugase activity *in vivo,* and may be important causes of folate deficiency in populations that consume diets rich in those foods (Butterworth *et al.*, 1974). Apart from their effects *in vivo,* recognition of their presence and control for their effects is important in the assessment of food folate content, since hydrolysis of poly-glutamate to achieve uniform activity for *L. casei* is an integral part of such studies.

Nonspecific inhibition of conjugase action forms the basis of a report that simultaneously explained conflicting data concerning the bioavailability of folate from a particular food, namely orange juice. Orange juice folate activity for *L. casei* before conjugase treatment was 80–90% of total folate, which suggests that almost all was monoglutamate (Streiff, 1971; Tamura and Stokstad, 1973), and appeared to be mainly methyltetrahydrofolate (Dong and Oace, 1973). Although these data suggested that orange juice should be an excellent source of dietary folate, Tamura and Stokstad (1973) observed poor absorption of orange juice folate and of polyglutamate added to orange juice. Subsequently, other workers compared the absorption of orange juice folate and synthetic folic acid and found the former to be 100% of the latter (Nelson *et al.*, 1975). The report that apparently resolved these differences by using more sophisticated techniques than the earlier studies, found only 30–40% of orange juice folate to be monoglutamate and indicated that the required hydrolysis of most of the folate was inhibited *in vivo* by the acid pH of orange juice (Tamura *et al.*, 1976). The inhibitory effect of low pH on polyglutamate absorption could be reproduced using citric acid, but neither orange juice nor citric acid decreased the availability of added monoglutamates (Tamura *et al.*, 1976). The discrepancy between the findings of Nelson *et al.* (1975) and Tamura *et al.* (1976) is probably attributable to differences in experimental design; the former used an orange juice solution with pH 7.0, whereas the latter used one with pH 3.7, and the inhibitory effect of acid is abolished by neutralization to pH 6.4 (Tamura *et al.*, 1976).

3.2.2. Mucosal Transport

As discussed above, it has not yet been established whether monoglutamates and polyglutamates are transported by the same mechanism, presuming that the latter are transported into the cell prior to hydrolysis. The conclusion that monoglutamate absorption occurs by a saturable concentrative mechanism, rather than by diffusion, was suggested in 12 of 23 reports addressing this subject (Rosenberg, 1976), including that of Weir *et al.* (1973), which reported preferential absorption of the naturally occurring isomer of methyltetrahydrofolate compared with the inactive stereoisomer. Although methylation and formylation of physiologic amounts of ingested folates can, and usually does, occur in the intestine (Whitehead *et al.*, 1972; Perry and Chanarin, 1973), this is not an obligatory step in folate absorption (Whitehead and Cooper, 1967; Baugh *et al.*, 1971).

The enhancement of folate absorption by glucose, reported in luminal disappearance studies by Gerson *et al.* (1971), has proved important in designing studies of folate malabsorption. Although folate deficiency in tropical sprue has long been attributed to malabsorption, the latter could not be consistently demonstrated with physiologic doses until independent studies demonstrated that malabsorption of these doses was only detectable in the presence of glucose or galactose (Gerson *et al.*, 1974; Corcino *et al.*, 1975, 1976). The extensive list of gastrointestinal and systemic diseases previously known to be associated with folate malabsorption has been enlarged by evidence that this may occur in cardiac failure (Hyde and Loehry, 1968), dermatitis herpetiformis (Hoffbrand *et al.*, 1970), and systemic bacterial infections (Cook *et al.*, 1974). Alcohol appears to impair absorption of the vitamin only when it is superimposed on a poor diet (Halsted *et al.*, 1971, 1973). In contrast, celiac disease so uniformly causes folate malabsorption that serum and red cell folate assays have been advocated as screening tests for the disease, which is said to be virtually excluded if these are normal (Hoffbrand, 1974).

Diphenylhydantoin, which may cause folate deficiency in other ways, was shown to inhibit monoglutamate absorption (Gerson *et al.*, 1972) after conflicting reports were published concerning the effect of this drug on conjugase *in vivo* and *in vitro*. Although Benn *et al.* (1971) attributed this phenomenon to elevated intestinal pH observed in patients on long-term Dilantin therapy, this elevation was not observed after administering the drug orally to normal subjects (Doe *et al.*, 1971) and could not have affected the studies of Gerson *et al.* (1972), because the perfusion containing 2 mg% Dilantin used in that study was no more alkaline than that without Dilantin (pH 6.5–6.9). The structural similarities among various anticonvulsants, including Dilantin, and folates (Girdwood, 1960; Herbert and Zalusky, 1962; Spector, 1972) support the view that these drugs impair folate absorption by a structure-specific mechanism.

Studies on foods fortified with synthetic folic acid (monoglutamate) indicate that the variable absorption of food folate is attributable not only to effects on hydrolysis, as mentioned above, but also to factors in foods that alter the mucosal transport phase. Figure 2 illustrates an experiment monitored by elevation of serum folate in which the mean absorption of 1 mg folic acid by seven healthy presaturated volunteers was 100% (by definition) when taken in water, 56% when added to corn (maize meal) porridge, 54% when added to rice, and 29% when added to bread (Colman *et al.*, 1975a). Similar comparative data were obtained by monitoring the effect of three of these four (folic acid alone, in corn porridge, and in bread) on the red cell folate levels of 137 women, not preloaded with folate, who took daily doses of the supplement during the last month of pregnancy (Colman *et al.*, 1975b, Margo *et al.*, 1975). Data on a large number of foods ingested at the same time as added folic acid suggest that almost all

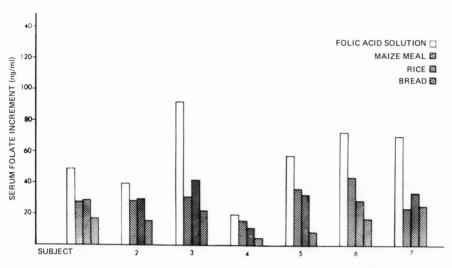

Fig. 2. The absorption of 1 mg folic acid from fortified stable foods, monitored by effect on serum folate levels (Colman *et al.*, 1975a).

foods impair mucosal transport—many to a significant extent (Tamura and Stokstad, 1973).

Identification of the food factors that affect mucosal transport awaits elucidation. Alkaline foods may raise intestinal pH and thereby impair transport (Benn *et al.*, 1971), just as acidic foods may cause impairment of hydrolysis. It has been suggested by various authors that protein binding of food folate may limit its absorption (Tamura and Stokstad, 1973; Butterworth and Krumdieck, 1975; Rosenberg, 1976), but no evidence for this hypothesis has been documented. Although it is possible that excess dietary glycine and methionine may interfere with folate absorption, their megaloblastic effects are probably attributable to impairment of its utilization (Waxman *et al.*, 1970).

3.3. Inadequate Utilization

Inadequate utilization of a nutrient is obviously not always synonymous with deficiency, and it is perhaps more common for total body stores to be increased, as is the case with iron in sideroblastic anemia. In the case of folate, however, there are two principal reasons why it is almost always correct to equate inadequate utilization with deficiency. First, there are seven different folate coenzymes, which vary in one-carbon substitution and state of reduction and which are involved in different enzymatic reactions; therefore, impaired utilization at one site may cause increased concentration of the substrate folate but causes true deficiency of the remaining folates, which perform different functions. Second, objective observations indicate that tissue concentrations of total folate are usually low in these conditions, presumably because of interference with carrier-mediated transport or storage mechanisms, or both. Inadequate utilization of folate may be congenital or acquired, primary or secondary.

3.3.1. Inborn Errors of Metabolism

The six sites implicated in inborn errors of folate metabolism are shown in Table I, which indicates that three of these have been fairly well substantiated. Central nervous system abnormalities have been reported in association with all these disorders and with a number of enzyme deficiencies that cause secondary alterations in folate metabolism. Although this relationship has not always been consistent, it has provided most of the evidence indicating that early folate deficiency may have neurologic effects. An autosomal recessive mode of inheritance was suggested in all three well-documented deficiencies on the basis of occurrence of the disorders in siblings, consanguinity of the parents, genetic studies, and/or half-normal enzyme levels in parents of affected subjects (Luhby and Cooperman, 1967; Rosenblatt and Erbe, 1972; Santiago-Borrero et al., 1973; Tauro et al., 1976). It should be noted that deficiency of two of the enzymes listed in Table I, cyclohydrase and formiminotransferase, would cause more profound effects than is at first apparent because, in mammals, they catalyze not only the reactions for which they are named but also, as shown in copurification studies, the adjacent folate interconversions (Drury et al., 1975; Paukert et al., 1976). An example of a congenital abnormality that secondarily causes folate deficiency is homocystinuria caused by cystathionine synthase deficiency (Waxman et al., 1970). Further details of these congenital abnormalities may be obtained from two excellent reviews of the subject (Erbe, 1975; Cooper, 1976).

3.3.2. Drugs

The drugs that are known to inhibit folate utilization specifically at a single enzyme site are all dihydrofolate reductase inhibitors and thus impair the regen-

Table I. Inborn Errors of Folate Metabolism[a]

Pathway affected	No. of cases reported
Folate malabsorption (? mucosal transport defect)	4
Dihydrofolate reductase deficiency	3[d]
Methenyl-THF[b] cyclohydrase deficiency[c]	3[d]
Methylene-THF reductase deficiency	6
Methyl-THF : homocysteine methyltransferase deficiency	1[d]
Glutamate formiminotransferase deficiency[c]	9

[a] Adapted from Erbe (1975), with permission.
[b] THF, tetrahydrofolate.
[c] In mammals, these enzymes catalyze more than one reaction involving folate (see text).
[d] The evidence for existence of the disorder in these subjects has been challenged.

eration of tetrahydrofolate after thymidylate synthesis and its generation from ingested unreduced folate. They fall into two classes: the 4-amino folate analogues and the 2,4-diaminopyrimidines. Those in the first group include methotrexate (4-amino-10-methyl folic acid), aminopterin (4-amino folic acid), and the diuretic triamterene (a triaminopteridine); those in the second class are the antimicrobials pyrimethamine and trimethoprim. Megaloblastic anemia has been reported in association with all these drugs but is uncommon with triam- terene and trimethoprim. Methotrexate, particularly when used in high doses, also acts by competitive inhibition for the carrier-mediated transport of reduced folate into cells and by a direct effect on purine synthesis (Goldman, 1971; Hryniuk, 1972).

Although it has been unambiguously demonstrated that diphenylhydantoin inhibits folate transport across the intestinal mucosa (Gerson *et al.*, 1972), it has not been established that anticonvulsants impair folate utilization, although there is evidence that suggests that they do. The structural similarity between several anticonvulsants and folic acid was noted by Girdwood (1960) and elaborated upon by Spector (1972). Herbert and Zalusky (1962) pointed out the close resemblance between Dilantin (diphenylhydantoin) and the 5-membered ring of 5,10-methylenetetrahydrofolic acid and speculated that the former may compet- itively inhibit reduction of the latter. In light of the psychiatric manifestations of anticonvulsant-induced folate deficiency (Reynolds, 1976) and of congenital deficiency of methylenetetrahydrofolate reductase (Freeman *et al.*, 1975), as well as the evidence that methylenetetrahydrofolate may be the source of a 1-carbon unit that condenses with biogenic amines (Pearson and Turner, 1975) to form psychotropic compounds (Baldessarini, 1975), we have recently modified the hypothesis of Laduron (1974) by suggesting the possibility that Dilantin therapy, the above-mentioned congenital defect, and some other organic psycho- ses may all exhibit psychiatric manifestations because of increased availability of methylenetetrahydrofolate (Colman and Herbert, 1977a). This suggestion is, of course, still highly tentative.

3.3.3. Alcohol

Herbert *et al.* (1963) studied 70 alcoholic patients and reported that megaloblastic anemia and macrocytosis were common, and that only 7% had normal serum folate levels. On the basis of the high folate content of beer (Herbert, 1963a), they correctly predicted the subsequent observation by Wu *et al.* (1975) that folate deficiency was not as common in beer-drinking alcoholics. In contrast to observations in nonalcoholic subjects, serum folate in the al- coholics studied by Herbert *et al.* (1963) correlated poorly with macrocytosis, anemia, and diet, but there was a strong inverse correlation between serum folate levels and recent alcohol ingestion. The temporal aspect of this relationship was defined by Eichner and Hillman (1973), who reported that serum folate fell rapidly within hours of moderate alcohol ingestion by humans. Their group subsequently confirmed this observation in rats (McGuffin et al., 1975).

Sullivan and Herbert (1964) reported that alcohol suppressed the hematopoietic response of folate-deficient patients to folic acid and suggested that the mechanism whereby alcohol caused folate deficiency was at least partly dependent on interference with folate utilization. McGuffin et al., (1975) were unable to demonstrate an effect of alcohol on the rate of liver folate depletion in rats receiving a folate-deficient diet, and concluded that the effect of alcohol on serum levels was mediated at a point other than the delivery of folate from stores to plasma. In subsequent human studies, however, they were able to demonstrate an effect of alcohol when folate metabolism was monitored with radiolabeled methyltetrahydrofolate, but not when folic acid was used. After allowing 5 days for equilibration of injected tracer, the amount flushed into urine by large doses of nonradioactive folate was significantly increased in folate-replete but not folate-deficient subjects if alcohol was administered during the equilibration period (Lane et al., 1976). These authors suggested that alcohol acted by impairing normal release of tissue folate to plasma, resulting in low serum folate and an increased amount of tissue tracer removable by the flushing technique (Lane et al., 1976). Further studies in an animal model by this group suggest that alcohol impairs normal release of folate into bile, and that the resultant decrease in enterohepatic circulation is responsible for depression of serum folate levels (Hillman et al., 1977); in that study, alcohol appeared to shunt folate from a pathway involving methylation and transfer to bile toward one concerned with formation of polyglutamate. In contrast, an earlier report suggested that alcohol interfered with formation of polyglutamate in liver (Brown et al., 1973).

As mentioned above, alcohol impairs folic acid absorption in folate-deficient subjects (Halsted et al., 1973), but current information supports the view of Sullivan and Herbert (1964) that its effect on folate metabolism is mainly through impairment of utilization.

3.3.4. Vitamin B_{12} Deficiency

No aspect of folate metabolism has spurred more investigation than that related to vitamin B_{12} deficiency, comprehensively reviewed by Das and Herbert (1976a). The working hypothesis that has been the pivot of this field for 15 years was proposed by Herbert and Zalusky (1962) and Noronha and Silverman (1962) and is known as the "methyl folate trap hypothesis." The original hypothesis was formulated shortly after documentation that the form of folate transported in serum was methyltetrahydrofolate (Herbert et al., 1962), which was the form involved in the vitamin B_{12}-dependent conversion of homocysteine to methionine (Larrabee et al., 1961), and observations that serum folate was elevated in 26 of 100 subjects with untreated pernicious anemia, that erythrocyte folate was decreased, and that vitamin B_{12} therapy caused the serum folate level to fall (Herbert and Zalusky, 1962). Stated in simple terms, the hypothesis was that vitamin B_{12} deficiency resulted in "trapping" of folate in the methyl form, thereby depleting the tetrahydrofolate pool of coenzymes necessary for other reactions.

This hypothesis has been supported by almost all, expanded by many, and contradicted by almost none of the data and interpretation generated by workers in this area, although the persistence of skepticism in some minds (Chanarin, 1977b) must be noted. Numerous clinical studies indicate that total cell folate is frequently depleted in B_{12} deficiency and that megaloblastosis caused by vitamin B_{12} and folate deficiency is indistinguishable except by direct measurement of the respective vitamins or by *in vivo* or *in vitro* therapeutic trial (Herbert, 1975b). These support the unchallenged concept that the final common pathway is depletion of 5,10-methylenetetrahydrofolate required for thymidylate synthesis. The proportion of folate in the methyl form is increased in experimental vitamin B_{12} deficiency and normalizes after B_{12} administration (Smith and Osborne-White, 1973; Thenen and Stokstad, 1973). Stokstad (1976), whose group has probably been the most active to investigate potential routes for demethylation of folate in vitamin B_{12} deficiency, recently reviewed the large body of data indicating that direct transmethylation is wholly B_{12}-dependent and that a potential indirect route via oxidation to 5,10-methylenetetrahydrofolate is made even more unlikely in B_{12} deficiency. This last view has received further support from Krebs *et al.* (1976).

One of the most interesting expansions of the methylfolate trap hypothesis is that relating to cellular transport of folate. Das and Hoffbrand (1970a) reported that stimulated lymphocytes from pernicious anemia patients had defective uptake of methyltetrahydrofolate. Tisman and Herbert (1973) confirmed this observation in human bone marrow and showed that it could be corrected by addition of B_{12} to the incubation medium. Lavoie *et al.* (1974) observed that decreased uptake by B_{12}-deficient lymphocytes was associated with decreased transfer of labeled methyl groups from methyltetrahydrofolate which was taken up. In all of the above studies uptake of folic acid was unaffected. In the light of current concepts of cellular folate transport (Goldman, 1971) these observations can be fully accounted for by a methyl folate trap. In the B_{12}-deficient cell, failure to remove methyl groups from transported folate would raise the concentration of methyltetrahydrofolate and inhibit dissociation of the carrier–folate complex at the inner cell wall, increasing countertransport. An untested hypothesis is that transmethylation may occur at the cell membrane or otherwise directly participate in cell transport (Tisman and Herbert, 1973).

3.3.5. Uremia

It has been convincingly demonstrated that the cellular transport of folates is impaired by uremic serum, although it is not established that this is a significant cause of deficiency. Transport of methotrexate, which occurs by the same mechanism as that of methyltetrahydrofolate, is impaired in the presence of extracellular organic and inorganic anions, probably via a requirement for countertransport of intracellular anions (Goldman, 1971). Inhibition of methotrexate transport into rapidly proliferating cells by uremic sera was demonstrated by Jennette and Goldman (1975), who showed that this correlated well with the anion gap. The degree of inhibition by ultrafiltrates of eight uremic sera averaged about 30%, and was halved by dialysis (Jennette and Goldman, 1975).

A different postulate for cell folate depletion despite normal serum folate levels emerged with the suggestion that uremic subjects have abnormal folate binding proteins (Hines *et al.*, 1973). These workers reported wide differences between folate radioassay of uremic serum before and after heat extraction, attributing these to binding of most of the serum folate by an abnormal factor, and suggested a mechanism based thereon as an alternative to that of Jennette and Goldman. Although it has been documented that folate bound at high affinity is not available for cell uptake (Tisman and Herbert, 1971; Waxman and Schreiber, 1974), this hypothesis would not account for inhibition of folate transport by ultrafiltrates (Jennette and Goldman, 1975), and recent studies indicate that the effect of heat extraction on folate radioassay is attributable mainly to removal of unsaturated binders which invalidate radioassay rather than to release of bound folate (Colman and Herbert, 1977b). Furthermore, it is now clear that all mammalian sera contain folate binders (Colman and Herbert, 1974, 1976). Although levels of these are doubled in uremia, their affinity for serum folate is apparently unaltered, and the amount of bound folate calculated from labeling of dissociated binding sites was about one-twentieth of that suggested by Hines and his coworkers (Colman and Herbert, 1976).

3.3.6. Female Sex Hormones and Oral Contraceptives

Although increased requirements due to increase in the number of rapidly proliferating cells is the basis of folate deficiency in pregnancy, recent evidence indicates that folate metabolism may be directly altered in other conditions associated with either increase or decrease of female sex hormones. In rat studies, oophorectomy produced significant decreases in uterine 10-formyltetrahydrofolate which could be prevented by administration of an estrogen compound (Laffi *et al.*, 1972); similar effects on folate-dependent enzymes were noted by Tolomelli *et al.* (1972). Krumdieck *et al.* (1976) reported that conjugase activity was twice as great during proestrus as at any other stage of the rat menstrual cycle, with corresponding differences in the proportion of substrate to product of this enzyme, measured crudely using *L. casei*. In addition, these workers have shown an increase in conjugase levels and similar increase in the proportion of folate containing three or less glutamate moieties following administration of estrogens to oophorectomised rats (Krumdieck *et al.*, 1975). Finally, measurements of serum folate in the baboon indicated not only diurnal variation but also menstrual cycle variation, with a trough in the first ten days of the cycle being followed immediately by a high peak (Boots *et al.*, 1975). It has been suggested that changes in conjugase activity and polyglutamate chain length, such as those mentioned here, are important regulators of folate utilization, and that the abnormalities of folate metabolism confirmed by some but not all studies of oral contraceptive users represent exaggerated responses to physiological events (Krumdieck *et al.*, 1975).

3.4. Increased Requirements

Body folate stores, normally of the order of 5–10 mg, last approximately 4 months without dietary replenishment in the normal adult (Herbert, 1962a). Depletion is accelerated when requirements increase, most commonly as a result of increased DNA synthesis, but occasionally as a result of increased demands of other pathways.

3.4.1. Pregnancy

The importance of pregnancy as a cause of folate deficiency and consequent megaloblastic anemia is attested to by the fact that it was in relation to this condition that the first recognized case reports appeared (Channing, 1842; Osler, 1919), that a differentiation was made between folate and B_{12} deficiency in terms of pathogenesis (Osler, 1919; Wills, 1931), and that the therapeutic effect of folic acid was first demonstrated (Moore *et al.*, 1945). Most of our current information was gained in early studies and published in previous reviews by Rothman (1970) and Cooper (1973).

There is, however, a need for continuing emphasis on two points discussed elsewhere in this review, namely the fact that about half the reported cases of megaloblastic anemia associated with folate deficiency in pregnancy presented during the postpartum period, and that recognition of the necessity and efficacy of providing folate supplements during pregnancy has not resulted in impressive reduction in the incidence of disease in many populations.

The assessment that about one-third of the pregnant women in the world have folate deficiency (Herbert, 1972b) has received continuing support from nutritional surveys (Colman *et al.*, 1975c, Herbert *et al.*, 1975; Jacob *et al.*, 1976). Weekly studies of 18 healthy third-trimester pregnant women with hemoglobin greater than 11 g%, receiving a hospital diet relatively low in folate and no additional supplement, showed no significant change in serum folate, which reflects a prior fall caused by negative folate balance, whereas erythrocyte folate fell by an average of 7 ng/ml per week (Colman *et al.*, 1974a).

An aspect of folate deficiency in pregnancy that has received recent attention is the neutrophil lobe average. Although it had been suggested that this test was "a specific, accurate, and reasonably sensitive method to detect and define folic acid deficiency of pregnancy prior to overt manifestation of the disease state" (Kitay *et al.*, 1969), prior studies indicated that it correlated poorly with folate deficiency in pregnancy (Chanarin *et al.*, 1965; Lowenstein *et al.*, 1966). A recent report demonstrated no correlation between folate deficiency and either the lobe average or other tests of neutrophil hypersegmentation, and after reanalysis and review of the data from nine reports in this area it was concluded that hypersegmentation was a poor predictor of folate deficiency in pregnancy (Herbert *et al.*, 1975). A factor that may be responsible for making this otherwise useful test questionable in pregnancy is the decreased lobe average reported in pregnant subjects (Lowenstein *et al.*, 1966; Herbert *et al.*, 1975).

Although the urinary excretion of folate is increased in pregnancy (Fleming, 1972), the amounts involved are infinitesimal compared with the observed depletion in folate stores. Despite the possibility of alternative causative mechanisms (Cooper, 1973) there is little if any doubt that folate deficiency in pregnancy is attributable to increased requirement.

3.4.2. Hyperalimentation

It has long been recognized that increased metabolic activity, such as occurs in hyperthyroidism or chronic pyrexias, can increase folate requirements sufficiently to cause megaloblastic anemia (Herbert, 1975a). This mechanism appeared significant in a number of recently described acutely ill patients in whom megaloblastic anemia developed rapidly while they were receiving total parenteral nutrition. Ballard and Lindenbaum (1974) first drew attention to this syndrome, and a number of subsequent reports and letters in the British literature suggested that the anemia developed too rapidly to be wholly attributable to inadequate ingestion, that alcohol in the infusion fluid was not essential to the genesis of the syndrome, and that some of these patients had additional bizarre morphologic features in the bone marrow (Ibbotson *et al.*, 1975a,b; Saary *et al.*, 1975; Wardrop *et al.*, 1975a,b; Saary and Hoffbrand, 1976).

3.4.3. Miscellaneous

Folate deficiency caused by increased utilization has been implicated in an extremely large number of diseases that can be approached systematically by considering each step in folate metabolism. With the exception of the above-mentioned conditions and prematurity in infants, the increased demand in most conditions originates in the bone marrow.

Many conditions are known to cause both iron deficiency and folate deficiency, such as chronic malnutrition, bleeding, and pregnancy. However, experimental studies in rats on an iron-deficient folate-replete diet suggest that iron deficiency may itself cause folate deficiency through ineffective erythropoiesis and hemolysis (Toskes *et al.*, 1974). A similar conclusion was reached in human studies in which no data were found to support the view that the folate-dependent enzyme, formiminotransferase, was iron-dependent (Roberts *et al.*, 1971). A similar mechanism was implicated in reports that one-half of 32 patients with multiple myeloma and two-thirds of 48 cases with agnogenic myeloid metaplasia (myelosclerosis) developed folate-deficient megaloblastic change during follow up (Hoffbrand *et al.*, 1967, 1968).

Other conditions in which an increased demand for folate was suggested were sickle-cell anemia in children (Liu, 1974), in which it was not associated with growth retardation as suggested earlier (Watson-Williams, 1962), hemolytic anemia (Jandl and Greenberg, 1959), the Lesch-Nyhan syndrome, in which increased folate may be required to offset failure of hypoxanthine salvage as has been objectively observed in fibroblast cultures (Felix and DeMars, 1969), and in premature infants. Experimental studies in rats, not yet confirmed in humans, suggest that patients on L-dopa therapy may have increased folate requirements (Ordonez and Wurtman, 1974).

3.5. Increased Excretion, Secretion, or Loss

3.5.1. Lactation

The importance of lactation in preventing the restoration of normal folate status after pregnancy has been emphasized (Metz, 1970). Review of the literature indicated that about one-half the reported cases of megaloblastic anemia "of pregnancy" were diagnosed after delivery (Rothman, 1970). Although many were detected in the early puerperium before lactation could have had a significant effect, in some populations more than 20% present later than 6 months after delivery, probably because of loss of folate in milk (Metz, 1970). In a study of folate binders, Metz *et al.* (1968b) showed that small doses of folate administered to lactating folate-deficient women were preferentially taken up in milk, while serum folate and reticulocyte counts remained low.

A recent survey indicates that the effects of pregnancy and lactation on folate status are not confined to those periods but profoundly affect the incidence of folate deficiency in nonpregnant women of child-bearing age. The chronologic relationship between the incidence of folate deficiency, i.e., low erythrocyte folate, and the last pregnancy or lactation of nonpregnant women in a population with a high incidence of folate deficiency is shown in Table II; only subjects aged 16–65 who had stopped lactating more than 6 months prior to study are shown. The incidence in nulliparous women was similar to that in men of the same age group, whereas it was significantly higher in parous women, particularly if they had been pregnant or lactating within 2 years of the study (Colman *et al.*, 1975c). The increased incidence of folate deficiency in young adult women compared with men of similar age, shown in Fig. 1, is therefore wholly attributable to the long-term effects of pregnancy and lactation.

Table II. The Effect of Previous Pregnancy and Lactation
on the Incidence of Folate Deficiency in Nonpregnant
Women of Child-Bearing Age (16–45 Years)[a]

Years since pregnancy or lactation	No. of subjects	No. deficient	Incidence of deficiency (%)
$1/2$–1	43	21	48.8
1–2	16	9	56.3
> 2	26	7	26.9
Nulliparous	16	3	18.8

[a]From Colman et al. (1975c).

3.5.2. Dialysis

An additional causative factor for folate deficiency in uremic patients and others subjected to dialysis is loss of folate in the dialysis fluid, reported to be uniformly significant in 17 patients by Sevitt and Hoffbrand (1969). The fact that serum folate fell by at least 50% during dialysis of 9 of 10 subjects with chronic renal disease (Sevitt and Hoffbrand, 1969), and by an average of greater than 50% in an additional 15 subjects not reported individually (Whitehead et al., 1968), provides further evidence against the suggestion of Hines et al. (1973) that almost all serum folate in these subjects is firmly bound to a high-affinity binder.

3.5.3. Miscellaneous

There is indirect evidence that the decline in total body folate observed in many patients with vitamin B_{12} deficiency may be caused by increased excretion or a failure to decrease excretion in response to decreased tissue stores. The biliary concentration of L. casei active folate in a subject with pernicious anemia was eight times higher than that in serum and amounted to twice the minimal daily requirement (Herbert, 1965). Although methyltetrahydrofolate is well absorbed in pernicious anemia (Colman et al., 1975d), which suggests that the enterohepatic circulation of natural folates is unimpaired, the efficiency of this circulation has not been carefully studied. Lavoie and Cooper (1974) reported the rapid transfer of injected radioactive folic acid from blood to bile in humans. A significant proportion of radioactivity was associated with a previously unidentified compound, inactive for L. casei, which contained both the pteridine and p-aminobenzoic acid portions of the molecule and was therefore not the common oxidative breakdown product of folic acid. It was also not a normal polyglutamate. The possibility that this compound represents an unsalvageable folate excretion product awaits resolution in further studies in which the nature of the entero-hepatic circulation of folate must also be defined. A preliminary report of studies in rats suggests that this circulation is important in maintaining normal serum folate levels (Hillman *et al.*, 1977).

Although there have been numerous short-term studies of folic acid excretion in animals and humans, Murphy *et al.* (1976) have reported data to suggest that our knowledge of folate catabolism and excretion is still in its infancy. These workers reported that the principal catabolic product of injected tritiated folic acid that appears in the urine was an acetylated derivative formed after cleavage of the 9–10 bond, i.e., acetamidobenzoylglutamate. There was evidence that the corresponding pteridine cleavage products were also excreted in urine, but the authors suggested that these products may also appear in bile. They were not the same as the biologically inactive folate derivative observed in bile by Lavoie and Cooper (1974), which was not a product of 9–10 cleavage. Murphy *et al.* (1976) suggest that the failure of previous authors to identify 9–10 cleavage as the principal route of folate catabolism in the rat was attributable to the use of too short a study to allow equilibration of the tracer with endogenous folate pools, which apparently only occurs after 13 days, or to incorrect identification of the pteridine catabolite as an intact folate derivative. The more recent study showed that rat urine 13 days after injection of tracer folic acid contained five catabolites, i.e., acetamidobenzoylglutamate and the cleavage produce from which it was derived and three other compounds, probably pteridines (Murphy *et al.*, 1976). These studies illustrate the necessity for reappraisal of folate catabolism and excretion in humans.

3.6. Increased Destruction

Although oxidative destruction of food folate has been clearly demonstrated (Herbert, 1963a; O'Brien *et al.*, 1975), there is no firm evidence that folate deficiency may result from increased destruction of the vitamin in the body. However, recent evidence that vitamin B_{12} deficiency may be caused by excessive destruction of cobalamins in the body (Jacob *et al.*, 1973) has again raised the possibility that this mechanism may be relevant to other vitamins.

The primary data concerning this mechanism as a cause of folate deficiency are based on reports that megaloblastic anemia that occurs in scurvy may sometimes respond to ascorbic acid alone (Brown, 1955; Asquith *et al.*, 1967), although lack of response to ascorbate, with response to folic acid alone, is the rule (Zalusky and Herbert, 1961). It has been pointed out that because the distribution of folate and ascorbate in foods is similar, a scorbutigenic diet contains very little folate (Zalusky and Herbert, 1961), and it is therefore necessary to exclude increased folate ingestion in hospital before a therapeutic trial of ascorbic acid can be considered positive.

Stokes *et al.* (1975) studied the effect of ascorbate ingestion on urinary folates in a scorbutic patient with megaloblastic anemia treated by oral administration of folinic acid. They reported that the major urinary folate prior to ascorbate or folinic acid ingestion was 10-formylfolic acid, whereas that after therapy with both was 5-methyltetrahydrofolic acid, and that the latter pattern corresponded with that in 20 normal subjects. They suggested that 10-formyl-tetrahydrofolate, a natural coenzyme, was irreversibly oxidized to 10-formyl folic acid, which was metabolically inactive, and that ascorbate may prevent such oxidation in normal subjects. The significance of these data, along with other information regarding folate excretion, may require reappraisal in light of the previously discussed report concerning folate excretion (Murphy et al., 1976).

4.   Effects of Folate Deficiency

In discussing the effects of folate deficiency, it is convenient to outline simultaneously those characteristics that are used in clinical diagnosis. As previously mentioned, all the effects of folate deficiency may be seen in B_{12} deficiency except those related to serum-folate levels and the effect of respective vitamins in therapeutic trials. This is consistent with the methylfolate trap concept, which indicates that deficiency of methylcobalamin, the B_{12} coenzyme involved in methyltetrahydrofolate: homocysteine transmethylation, results in depletion of the tetrahydrofolate pool. The reverse is not true, i.e., not all of the effects of vitamin B_{12} deficiency are seen in folate deficiency, at least partly because the other vitamin B_{12} coenzyme, adenosylcobalamin, is involved in a folate-independent pathway.

4.1.   Biochemical and Morphological Effects on the Blood and Bone Marrow and Their Use in Clinical Diagnosis

The sequence in which effects of folate deficiency appear was described in a report of a healthy 77-kg male, initially folate replete, fed a folate-deficient diet until megaloblastic anemia developed (Herbert, 1962a). All the tests conventionally used to diagnose folate deficiency are based on effects evident in the blood and bone marrow. Discussion of urinary excretion of formimino-glutamic acid (FiGlu) and other metabolites, previously used for diagnostic purposes, is included here for convenience, with the recognition that such excretion is secondary to elevated concentration of these metabolites in blood. These effects are considered in the chronologic sequence in which they appeared in Herbert's study (1962a); the duration of folate deprivation at the time they were first observed is shown in parentheses.

4.1.1.   Low Serum Folate (Three Weeks)

Assay of serum-folate levels remains the most common test used in diagnosing folate deficiency. As indicated here, however, it is also the most sensitive test and therefore most unreliable as a predictor of true folate deficiency. The chronologic data indicate that a low serum folate indicates nothing more than negative folate balance, attributable to any cause, for the preceding 3 weeks. Serum folate is useful in differentiating folate deficiency from vitamin B_{12} deficiency, because it is usually normal or elevated in the latter, and as a screening test, which can suggest but not prove folate deficiency.

Microbiologic folate assay using *L. casei* (Herbert, 1961) has been simplified by elimination of the necessity for protein extraction (Herbert, 1966) and use of a chloramphenicol-resistant organism (Davis *et al.*, 1970) and can now be conducted by a nonsterile procedure in disposable glassware by a method requiring no special equipment or skills (Scott *et al.*, 1973). Folate radioassay has become a widely used technique, most procedures being based on the original report of Waxman *et al.* (1971). Elimination of the use of unstable methyltetrahydrofolate standards by assay at pH 9.3 and of β-emitting tracer has substantially improved the assay (Longo and Herbert, 1976). Since serum folate and B_{12} assays are both indicated when deficiency of either is suspected, a considerable gain in time and expense can be achieved by the simultaneous assay developed by Gutcho and co-workers at Schwartz/Mann Laboratories, which measures both vitamins in the same test tube using $^{57}Co\text{-}B_{12}$ and ^{125}I-folate as tracers counted simultaneously on different channels. This assay was found to compare well with individual radioassays for the two vitamins (Jacob *et al.*, 1977).

4.1.2. Neutrophil Hypersegmentation (Seven Weeks)

Hypersegmentation, a feature observed in the peripheral blood, is defined either as a neutrophil lobe average exceeding 3.5, or as the presence of five or more lobes in at least 5% of neutrophils (Herbert, 1959). It remains a useful test in the diagnosis of folate and B_{12} deficiency, but it is particularly important when there is concomitant iron deficiency. In iron deficiency, the morphologic and biochemical effects of megaloblastosis are often masked in erythroid cells, and the only morphologic features of megaloblastic anemia may be those seen in the granulocyte series, i.e., neutrophil hypersegmentation in the peripheral blood and giant myeloid cells in the bone marrow (Herbert, 1975a). Although it has been reported that iron deficiency alone may cause hypersegmentation in some subjects (Dine and Snyder, 1970), the possibility of prior folate deficiency was not excluded (Herbert, 1970). It is important to recognize the difference between hypersegmented neutrophils and "twinning deformities," and to be aware that congenital hypersegmentation has been reported in approximately 1% of normal adults (Herbert, 1964).

4.1.3. Increased Urinary FiGlu Excretion (Thirteen-and-One-Half Weeks)

Among the well-recognized biochemical effects of folate deficiency are elevated levels of substrates for various folate-dependent enzymes, especially after oral loading. Urinary excretion of FiGlu, aminoimidazolecarboxamide, and formate has been measured for diagnostic purposes but did not prove useful in the differentiation of folate and B_{12} deficiency and is seldom performed in current practice.

4.1.4. Low Erythrocyte Folate (Seventeen-and-One-Half Weeks)

Low erythrocyte folate is generally accepted as the best clinical index of depleted tissue stores. It correlates extremely well with folate concentration in the liver (Wu *et al.*, 1975). The basis for this measure being less labile than serum folate rests in the fact that erythrocytes that have matured beyond the reticulocyte stage do not take up folate (Herbert and Zalusky, 1962); thus, the folate content of a mixed red cell population in the absence of hemolysis reflects folate availability during the previous 3 months. As previously mentioned, erythrocyte folate may be low in B_{12} deficiency. It may be normal in folate-deficient subjects if they have received a transfusion of folate-replete erythrocytes, and Chanarin (1976) has noted that it may also be normal in megaloblastic anemia with marked hemolysis, because the folate content of reticulocytes is high, or when there is rapid progression of folate deficiency while the blood still contains many erythrocytes that matured before deficiency ensued.

4.1.5. Macroovalocytosis (Eighteen Weeks) and Macrocytosis

The presence of large oval erythrocytes on a peripheral blood smear is an extremely specific sign of megaloblastic anemia (Herbert, 1975). The recognition of macrocytosis, rather than macroovalocytosis, has been dramatically improved by the use of automated blood counters that measure mean corpuscular volume (MCV), as outlined by Chanarin (1976). The advisability of investigating subjects with persistently raised MCV is apparent from studies reflecting a narrow range in normals (Silver and Frankel, 1971; England *et al.*, 1972; McPhedran *et al.*, 1973; Croft *et al.*, 1974). Macrocytosis, but not macroovalocytosis, may occur in the absence of megaloblastic anemia in liver disease, aplastic and sideroblastic anemias, hypothyroidism, myeloma, neoplasia, antimetabolite drug therapy, and in any condition associated with reticulocytosis (Chanarin *et al.*, 1973; McPhedran *et al.*, 1973). Macrocytosis in alcoholics and anticonvulsant-treated epileptics may be caused either by folate deficiency or by what appears to be a direct effect of alcohol or anticonvulsant drug in the absence of folate deficiency (Wickramasinghe *et al.*, 1975; Wu *et al.*, 1975).

Clarkson and Moore (1976) reasoned that the several-month delay between appearance of the first signs of folate or iron deficiency and the detection of consequent macrocytosis or microcytosis (Herbert, 1962a; Conrad and Crosby, 1962) was partly caused by nonselective screening of the whole erythrocyte population, including cells surviving from the period preceding deficiency. They developed a sensitive test for macrocytosis and microcytosis based on supravital staining, photography, and planimetry of reticulocytes in wet preparations. Using this technique, Clarkson and Moore (1976) detected reticulocyte macrocytosis within 5 days of methotrexate injection and microcytosis within 2 days of the fall in transferrin saturation in a phlebotomized volunteer.

Although this test is currently rather laborious, it is evident that the mode of action of recently developed automated differential counters gives them the potential to screen for changes in reticulocyte MCV, and even for macroovalocytosis, just as cell counters currently screen for the relatively insensitive and nonspecific changes in erythrocyte MCV.

4.1.6. Megaloblastic Bone Marrow (Nineteen Weeks)

Morphologic megaloblastosis is characterized by changes that affect the erythroid and myeloid series; it may also affect platelets. The affected erythroid cell is known as a megaloblast and is characterized by an enlarged nucleus with immature features, primarily finely particulate chromatin, and cytoplasm that is even more enlarged but shows mature features. This phenomenon is known as nucleocytoplasmic dissociation—the major characteristic of megaloblastosis— and is easily explained by the concept that megaloblastosis is caused by impaired DNA synthesis (Herbert, 1975a). Affected myeloid cells show similar features and are primarily giant metamyelocytes and staff (or band) cells. Megaloblastosis may affect any rapidly dividing cells, but in uncomplicated folate deficiency· morphologic changes seem largely confined to hematopoietic cells and the epithelium of the female genital tract.

Biochemical megaloblastosis is routinely defined by the deoxyuridine (dU) suppression test (Metz et al., 1968; Herbert et al., 1973), which measures the "suppression" by added nonradioactive dU of radioactive thymidine incorporation into DNA. Normal folate coenzyme function is essential for the methylation of dU to thymidine, which in this test decreases the specific activity of the thymidine pool and is responsible for such "suppression." The test is therefore abnormal in both folate and B_{12} deficiency, but the difference can be defined by correction after adding the specific nutrient. Normal dU-suppression tests in the presence of morphologic megaloblastosis are seen during therapy with drugs that impair DNA synthesis at sites other than thymidylate synthetase and have been reported in patients taking anticonvulsants (Wickramasinghe et al., 1975). A preliminary report indicates that dU-suppression tests in stimulated lymphocytes allow retrospective diagnosis of folate deficiency despite prior "shotgun" therapy, a persistent clinical problem, presumably because the lymphocyte does not alter its internal environment after therapy unless stimulated to go into the S-phase of cell division (Das and Herbert, 1976b).

Hoffbrand and co-workers (1976) have marshaled evidence from their own and other laboratories to propose a hypothesis of the underlying defect in DNA synthesis in megaloblastic anemia. This hypothesis is based on evidence that the extremely large DNA molecules on individual chromosomes do not replicate continuously from one end to the other. After partial unwinding of the parental strands, a large number of initiation points are activated on both strands and at each a primer RNA is formed. Fragments of new DNA, primed by RNA, are then synthesized on the parental template under the action of a DNA polymerase. When the primer RNA is removed, the gap left is filled by further DNA synthesis through the action of a different polymerase. The fragments of new DNA are then joined together to form "replicons," which are ligated together to form bulk DNA. Hoffbrand *et al.* (1976) reported that DNA from stimulated lymphocytes in megaloblastic anemia broke up into smaller fragments than that from controls, and that methotrexate induced the formation of increased amounts of low-molecular-weight DNA in normal cells. Their suggestion was that megaloblastosis was caused by a concentration of one or other deoxyribonucleoside triphosphate falling below that required for the action of the "gap-filling" DNA polymerase but not below that for the polymerase involved in initiation of DNA synthesis. This would result in initiation of DNA synthesis in excess of DNA chain elongation, the basis they postulated for the defect in megaloblastic anemia. They point out that this mechanism would require active protein synthesis in order to enter the S-phase of the cell cycle and relatively normal RNA synthesis in order to prime new DNA synthesis, and that these two functions have appeared to be normal in clinical, laboratory, and biochemical observations of megaloblastic anemia.

4.1.7. Anemia (Nineteen-and-One-Half Weeks)

Anemia became evident 3 days after the appearance of megaloblastic marrow changes in Herbert's (1962a) study, thereby indicating a chronologic correlation between these two phenomena. Megaloblastic anemia is functionally a state of ineffective erythropoiesis, and is thus characterized by a high serum lactic dehydrogenase, moderately elevated serum bilirubin, and increased saturation of serum iron-binding capacity and iron stores caused by ineffective utilization of iron. The latter may mask underlying iron deficiency, which may become evident only after treatment (Herbert, 1975b). It was suggested that the moderate hemolysis of megaloblastic anemia was attributable to increased splenic entrapment caused by decreased erythrocyte deformability (Ballas *et al.*, 1976).

4.1.8. Miscellaneous

The activity of a number of enzymes involved in pyrimidine metabolism is reportedly elevated in megaloblastic anemia. Sakamoto and Takaku (1973) reported that the thymidylate synthetase apoenzyme had increased activity in the bone marrow of folate-deficient rats when the coenzyme, 5,10-methylenetetrahydrofolate, was added to cultures from these and normal controls; relevant studies suggested that the mode of enzyme induction was dependent on accumulation of the substrate, rather than release of enzyme from end-product inhibition. Haurani (1973) reported decreased activity of this enzyme in stimulated lymphocytes in pernicious anemia, but Sakamoto *et al.* (1975) reported that it was increased by ninefold in the bone marrow of these subjects and suggested that Haurani's findings might have been attributable to the use of nonhemopoietic cells or to the fact that he used a folate compound that required interconversion before it could serve as coenzyme in the enzyme assay. A closely related, folate-independent enzyme, i.e., thymidine kinase, was also elevated in pernicious anemia bone marrow (Nakao and Fujioka, 1968), but unlike thymidylate synthetase (Haurani, 1973) this enzyme also had increased activity in stimulated lymphocytes which could be reversed *in vitro* by addition of folate and induced in normal lymphocytes by addition of methotrexate (Hooton and Hoffbrand, 1976). In pernicious anemia, it has been reported that there is increased activity of aspartate carbamyltransferase and dehydroorotase (Smith and Baker, 1960) and of ribonucleotide reductase (Fujioka and Silver, 1971), all of which are involved in pyrimidine metabolism.

Fox *et al.* (1976) studied the activity of three purine synthetic enzymes in seven cases of megaloblastic anemia, of whom three had folate deficiency and one combined deficiency of B_{12} and folate. They reported increased specific activity of erythrocyte adenine phosphoribosyltransferase in megaloblastic patients compared with controls, and similar increases in two subjects with hypoxanthine-guanine phosphoribosyltransferase deficiency, which causes folate deficiency in affected subjects and is associated with increased folate requirements of cultured fibroblasts (Felix and DeMars, 1969). During the reticulocyte phase of therapy for megaloblastic anemia, there was a further rise in this enzyme and in two other purine synthetic enzymes (Fox *et al.*, 1976).

The sum of serum uracil and uridine concentrations was one-third lower in 21 patients with pernicious anemia than in controls (Parry and Blackmore, 1976a,b), in contrast to what might have been predicted from the known inhibition of thymidylate synthetase. The explanation for this observation and its relevance to folate deficiency are still unknown.

Bazzano (1969) reported lowered serum cholesterol in megaloblastic anemia, which increased after folic acid therapy, and suggested the possibility of a defect in lipid transport induced by folate deficiency. This observation was pursued in controlled rat studies, which indicated that the addition of folic acid to a folate-deficient diet caused a significant increase in serum cholesterol and a significant decrease in liver cholesterol, and that an opposite effect of dietary casein supplementation was only detectable in rats that received concomitant folate supplements (Hatwaine, 1975). A caution concerning this study is appropriate, as the diet was fed for only 6 weeks and there was no documentation of either the folate content of the diet or the extent of folate depletion in the rats.

Two recent reports have suggested that folate deficiency may be associated with impaired cell-mediated immunity and with impairment of mental development in children of affected mothers (Gross *et al.*, 1974, 1975). In both groups, taken from a population in which folate deficiency is common and B_{12} deficiency rare, the diagnosis was supported by low serum folates in the presence of a megaloblastic bone marrow, but erythrocyte folate and dU suppression were not measured. The association with impaired infant development was not conclusive, and the authors suggested that further study be undertaken (Gross *et al.*, 1974). In the other report (Gross *et al.*, 1975), there was an impaired cutaneous response to 2,4-dinitrochlorobenzene in 22 of 23 folate-deficient subjects and only 1 of 18 controls, and significantly decreased lymphocyte transformation to PHA in all folate-deficient subjects compared with controls. These results contrasted with those of Das and Hoffbrand (1970b), who reported increased incorporation of tritiated thymidine in lymphocytes of six subjects with pernicious anemia. This increase was probably attributable to the requirement for preformed thymidine in megaloblastic cells owing to a block in the salvage pathway (Metz *et al.*, 1968a) rather than to a net increase in new DNA synthesis. However, the incorporation of tritiated thymidine in the single folate-deficient subject studied by Das and Hoffbrand (1970b) was 45% of control values, similar to the 30% average in 23 folate-deficient subjects studied by Gross *et al.* (1975). The reported defect in cell-mediated immunity in folate deficiency awaits confirmation.

4.2. The Nervous System

In contrast to vitamin B_{12} deficiency, folate deficiency is not generally accepted as a cause of neurologic disease in adults, but a number of recent reports have again raised the possibility that such syndromes may occur. The data on the subject have been reviewed by Reynolds (1976) and Colman and Herbert (1977a), the former espousing the view that such syndromes have been convincingly documented and the latter that they have not. The brains of experimental animals have levels of folate and some of its interconverting enzymes several orders of magnitude greater *in utero* and during the neonatal period than in adults, corresponding to the demands of growth and cell division; the effects of folate deficiency must therefore be considered separately in these two age groups.

4.2.1. During Gestation and the Neonatal Period

Folate deprivation in pregnant rats results in up to 95% having litters with congenital abnormalities of various organs, including the nervous system (Nelson *et al.*, 1952). A report that anencephaly was four times as common in human fetuses if mothers were folate-deficient (Hibbard and Smithells, 1965) was not substantiated by subsequent studies (Pritchard *et al.*, 1970; Scott *et al.*, 1970).

The problem with human studies is that it is impossible to assess whether or not folate deficiency was present early in pregnancy when the fetus was subject to teratogenic effects.

Although central nervous system disorders have been associated with each of the reported inborn errors of folate metabolism, the nature and severity has varied and several cases had enzyme deficiency without manifest neurologic changes. It is therefore unclear whether or not these subjects had multiple congenital abnormalities, although it is noteworthy that some were reported to have striking responses to folate treatment. It has been reported that the total RNA content of Purkinje cells was decreased in chicks fed a folate-deficient diet from birth (Haltia, 1970).

The supply of folate to the brain appears to be mediated by the active transport mechanism of cerebrospinal fluid (CSF) (Spector and Lorenzo, 1975b), as there is no evidence for active uptake of folate by brain (Colman and Herbert, 1977a). This system, which may be controlled by the choroid plexus (Spector and Lorenzo, 1975a; Chen and Wagner, 1975), has a higher affinity for folate than do non-neurologic cell-uptake mechanisms (Goldman, 1971) and could thus be the basis for the preliminary observation that in weanling rats fed a folate-deficient diet, folate concentration falls less rapidly in brain than in other tissues (Allen and Klipstein, 1970). This view is supported by evidence that folate was conserved as effectively in CSF as in rat brain (Fehling *et al.*, 1975).

4.2.2. In Adults

Human CSF folate concentration remains at least three times greater than that in serum in the presence of tissue deficiency (Herbert, 1964). Adult rats deprived of folate for a year remained neurologically normal in the face of decreased brain folate, despite tissue folate deficiency more severe than that in a second group with decreased nerve conduction velocity after administration of succinylsulfathiazol and a folate-deficient diet (Fehling *et al.*, 1975). In that study, peripheral nerve folate, but not brain or CSF folate, fell at a rate similar to that in liver, thereby indicating that it was only the central and not the peripheral nervous system which was protected against deficiency.

Case reports of neurologic disease attributed to folate deficiency in adults include 16 patients reported in 1976 (Manzoor and Runcie, 1976; Botez *et al.*, 1976). The criticisms of these and previous reports are extremely fundamental, i.e., that folate deficiency was complicated by the presence of other disorders, such as alcoholism or perhaps even vitamin B_{12} deficiency (Herbert, 1972a), and that those reports claiming a therapeutic response were not controlled (Bateson, 1976). In addition to neurologic disorders, psychiatric disorders including dementia have been attributed to folate deficiency (Carney, 1967; Strachan and Henderson, 1967; Melamed *et al.*, 1975; Sapira *et al.*, 1975). None of these reports presents evidence to contradict the conclusion of Grant *et al.* (1965) who reported seven patients with neurologic disease and folate deficiency, that the deficiency probably followed inadequate ingestion secondary to neurologic disease, and that "it is unjustifiable to conclude that the neurological syndromes . . . were the direct result of folate deficiency."

In experimental folate deficiency in humans, insomnia and amnesia developed during the 4th month and were followed by progressive irritability for the following month; all these symptoms disappeared 2 days after folate replenishment (Herbert, 1962a). It could be suggested that further neurologic damage would have occurred in this case if folate deficiency had progressed, since the above-mentioned evidence indicates that depletion of brain folate requires deficiency far more severe than that required for depletion of, for example, marrow folate. The same argument may be used, however, to suggest that organic central nervous system (CNS) damage could never be caused by folate deficiency, because megaloblastic anemia would be fatal before the degree of deficiency was sufficient to cause damage to the brain. Indirect methods whereby folate deficiency might cause neurologic damage have been reported by two groups. Frenkel et al. (1973) have suggested that CNS B_{12} deficiency may occur in folate-deficient subjects on anticonvulsants. Thomson et al. (1972) report that folate deficiency may cause thiamine malabsorption in the presence of adequate dietary thiamine and that secondary neurologic effects may then ensue because of thiamine deficiency.

The hypothesis that methyltetrahydrofolate deficiency might cause schizophrenic symptomatology (Mudd and Freeman, 1974) was based on observations of a subject who lacked the enzyme necessary to reduce methylenetetrahydrofolate to methyltetrahydrofolate and on the earlier belief that methyltetrahydrofolate was the 1-carbon donor involved in methylation of biogenic amines. Subsequently it appeared that this "methylation" was in fact a nonenzymatic condensation with formaldehyde released by methylenetetrahydrofolate. Although some workers consider these condensation reactions in vitro artifacts, others believe they may occur in vivo and generate psychotropic compounds. If they do, any association between them and the patient discussed by Mudd and Freeman (1974) would relate to excess of methylenetetrahydrofolate rather than to deficiency of methyltetrahydrofolate (Colman and Herbert, 1977a).

4.3. Obstetric Complications and the Female Genital Tract

At various times in the past, folate deficiency has been regarded as a causative factor in abruptio placentae, abortion, eclampsia, perinatal mortality, fetal abnormalities, premature labor, and low birth weight. With the exception of the last of these, all were subsequently found to be unassociated with folate deficiency in human studies. A detailed review of the positive findings and their refutations appears elsewhere (Rothman, 1970). Although there did not appear to be any clear relationship between low birth weight and folate deficiency at the time Rothman (1970) wrote her review, subsequent work has suggested a rather unusual but significant relationship. Baumslag et al. (1970) suggested that the relationship might be more significant in populations with suboptimal folate nutrition. They conducted controlled clinical trials of folic acid supplements in 183 pregnant South African blacks, in whom folate deficiency is common, and 172 South African whites, in whom folate deficiency is rare. No effect of folate on birth weight was noted in white infants, but 20% of black infants whose mothers were receiving no folate weighed less than 4 lb, whereas all 110 black infants whose mothers received folate weighed more than 4 lb. This highly significant relationship, which appears to be detectable only in populations in whom folate deficiency is common, has been supported by some subsequent reports from such populations in Nigeria (Harrison and Ibeziako, 1973) and India (Iyengar and Rajalakshmi, 1975).

Megaloblastosis in the female genital tract is a well-known association of tissue-folate deficiency (Van Niekerk, 1962, 1966; Klaus, 1971). Similar changes, also reversible with folic acid therapy, have subsequently been observed, in the absence of other evidence of folate deficiency, in women taking oral contraceptives (Whitehead *et al.*, 1973). These workers have suggested that the underlying mechanism may lie in the close relationship between female sex hormones and folate coenzymes (Lindenbaum *et al.*, 1975) mentioned in Section 3.3.6., which they review in greater detail.

4.4. The Gastrointestinal Tract

Megaloblastosis affects all proliferating cells, including the entire length of the gastrointestinal tract. In patients with folate deficiency, morphologic changes essentially the same as those seen in the marrow have been demonstrated in the tongue and buccal mucosa (Boddington and Spriggs, 1959) and in the jejunum in association with coexisting sprue (Veeger *et al.*, 1965). Jejunal biopsies were normal, however, in experimental folate deficiency (Herbert, 1962a) and in six patients with megaloblastic anemia, four of whom were alcoholics (Winawer *et al.*, 1965). In a subsequent report of an alcoholic subject with folate-deficient megaloblastic anemia, duodenal biopsies showed megaloblastic changes that remitted when he was treated with folic acid (Bianchi *et al.*, 1970).

It has been suggested that the relative resistance of the jejunal mucosa to folate deficiency may be caused by the retention of sufficient folate therein to supply its own needs even when dietary supplies are very small (Herbert, 1964). Functional rather than morphologic gastrointestinal changes are common in folate deficiency and include acute and chronic glossitis, diarrhea, anorexia, and cachexia. A recent report of a controlled trial suggested that folic acid administration to folate-replete subjects increased the resistance of the gums to local irritants (Vogel *et al.*, 1976); although local morphology was not studied, the authors suggest the possibility of selective folate deficiency in the gums similar to that observed by Whitehead *et al.* (1973) in the cervical epithelium.

Folate deficiency has been suggested in a single report as a cause of angular cheilosis because (1) serum folate levels were in the deficient range in 9 of 50 affected patients and were significantly lower than the mean of controls, and

(2) two patients apparently responded to folic acid therapy (Rose, 1971). Apart from the fact that tissue-folate deficiency was not documented and other more important vitamins were not measured in that study, it has been pointed out that cheilosis may be a nonspecific consequence of exposure to the elements or constant mouth breathing (Sandstead and Pearson, 1973).

Herman and Rosensweig led studies in which jejunal glycolytic enzyme activity was low in folate deficiency and responded to oral folic acid, which was beneficial in treatment of a few cases of congenital enzyme deficiency (Rosensweig, 1975).

5. The Treatment and Prevention of Folate Deficiency

5.1. Treatment of Folate-Deficient Megaloblastic Anemia

Folate-deficient megaloblastic anemia is best treated with 0.1 mg folic acid daily unless a higher dose is indicated by malabsorption or increased requirements. Folinic acid therapy may be necessary to supply tetrahydrofolate coenzymes in some congenital enzyme deficiencies or after inhibition of dihydrofolate reductase with methotrexate. This dosage (0.1 mg folic acid daily) originally came into use because 0.1–0.2 mg daily did not cause reticulocyte responses in vitamin B_{12} deficiency, whereas 0.4 mg daily occasionally did (Herbert, 1963b). Although the dU-suppression test can now be used to conduct therapeutic trials *in vitro* (Herbert *et al.*, 1973), therapeutic trials *in vivo* remain useful and should be monitored by daily reticulocyte counts until an adequate reticulocytosis and rising hemoglobin are observed.

Lawson *et al.* (1972) suggested that severe megaloblastic anemia caused depletion of whole-body potassium and that subsequent therapy resulted in a massive shift of this electrolyte from extracellular to intracellular compartments, with resultant cardiotropic effects. They suggested that this might have explained the high mortality of severe megaloblastic anemia (14%) compared with severe iron deficiency anemia (1.6%) in patients of similar age in their hospital. Hesp *et al.* (1975) confirmed that total-body potassium was lower before than after therapy in severe megaloblastic anemia, but did not detect clinically significant changes in serum potassium concentration. Based on the latter study, it has been recommended that potassium supplements not be given during treatment of megaloblastic anemia (Chanarin, 1976).

A major clinical problem in severe megaloblastic anemia is whether, when, and how transfusions should be administered, as patients with chronic anemia frequently tolerate such transfusions poorly. Herbert (1975b) recommends such treatment only when the patient manifests with dyspnea, congestive cardiac failure, or angina, which usually occur when the hematocrit is lower than 15%, and advises use of exchange transfusion as reported by Duke *et al.* (1964). Chanarin (1976) recommends that patients over 50 years of age should receive one unit of packed cells slowly over 12 or more hours, accompanied by a diuretic and careful clinical monitoring. For treatment to be successful in the long term, it is essential that the underlying cause of folate deficiency be identified and wherever possible corrected. More specific details regarding therapy may be found in the above-mentioned reviews (Herbert, 1975b; Chanarin, 1976).

5.2. The Prevention of Folate Deficiency

As reviewed elsewhere (Herbert *et al.*, 1975), there is a general consensus that folic acid supplements should be administered to pregnant women. In addition, such supplements have been recommended for infants of low birth weight (Hoffbrand, 1970; Dallman, 1974) and in a number of clinical disorders discussed herein as causes of folate deficiency.

The major problem in preventing megaloblastic anemia of pregnancy is not that folic acid is not effective, but that it can only be effective when it is taken. Bonnar *et al.* (1969) found that significant numbers of patients attending antenatal clinics did not take the iron tablets supplied to them; this problem may not be as severe when tablets containing both iron and folate are given, since there has been a report that the adverse gastrointestinal effects of iron ingestion may be decreased when folic acid is simultaneously ingested (Sood *et al.*, 1975). The largest component of the problem, however, is that antenatal care is not available to, or taken advantage of, by large numbers of pregnant women, particularly in those populations in which folate deficiency is most common. This phenomenon is discussed elsewhere (Colman *et al.*, 1974a) and has again been highlighted by the fact that of 110 consecutive women seeking prenatal care at Morrisania Hospital in the South Bronx, 61 were in the second trimester and 26 already in the third trimester of pregnancy (Herbert *et al.*, 1975).

As an alternative approach to alleviating this problem, a series of studies was conducted to determine the feasibility of fortifying maize (corn) meal, the staple food of most South African Blacks, with folic acid (Colman *et al.*, 1974a,b, 1975a,b,c). Folic acid in fortified maize meal was stable under conventional storage conditions for up to 18 months, and the addition of the vitamin did not have undesirable effects on the color, odor, taste, or cooking properties of the food; unlike natural folates, folic acid resisted destruction by boiling for at least 2 hr and by baking at 230°C for 1 hr (Colman *et al.*, 1975a). As previously mentioned, and as shown in Fig. 2, the availability of folic acid from fortified maize and rice was about 55% and that from bread was 30% in comparison with absorption of folic acid from solution. Clinical trials in 122 healthy, nonanemic pregnant women receiving iron, studied in a hospital environment during the last month of pregnancy, compared the effects of three levels of fortification in maize meal, folic acid tablets, and no folate supplementation (Colman *et al.*, 1974a, 1975b). The changes in red cell folate in each group, expressed as linear regressions, are shown in Fig. 3. Using the regression in the control group as a base line and the change in red cell folate as an index of availability, folic acid in

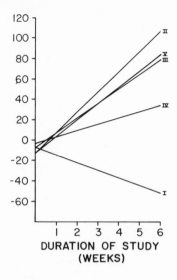

Fig. 3. Changes in erythrocyte folate (regression lines) in five groups of pregnant women receiving folic acid supplements in different forms; (I) no supplement; (II) 1000 μg daily in maize; (III) 500 μg daily in maize; (IV) 300 μg daily in maize; (V) 300 μg daily as tablets (Colman et al., 1975b).

fortified maize was 55–58% as available as that taken alone. This is in close agreement with the previous absorption studies monitored by serum-folate assay, shown in Fig. 2 (Colman et al., 1975a). Similar analysis of the effects on red cell folate of folic acid-fortified bread (Margo et al., 1975) indicated close correlation with the laboratory-based absorption tests.

Although red cell folate concentration fell in a few subjects in every supplemented group, fortified maize meal containing an effective dose of approximately 170 μg folic acid daily was as successful as more strongly fortified maize and 300 μg folic acid tablets in preventing a fall in erythrocyte folate concentration in individual subjects. Serum vitamin B_{12} concentrations did not change significantly. As these patients were taken from a population with a very high incidence of folate deficiency (Colman et al., 1975c), there is apparently little purpose in supplying pregnant women in any population with folic acid supplements greater than 200 μg daily. A dietary survey conducted among 469 families in the surrounding area indicated that if the level of fortification was based on the average monthly consumption of maize, 90% of the population would receive half to twice the intended dose of the supplement, with 5% receiving less than half the dose and 5% receiving about 2½ times the dose (Colman et al., 1975c). The antimegaloblastic effects of folic acid-fortified maize meal was demonstrated in therapeutic trials in megaloblastic anemia during lactation (Colman et al., 1974b). The processing and distribution of maize meal and many staple foods in developing countries is limited to a few centralized institutions. Hence, there are few practical obstacles to the institution of such fortification, which in 1974 would have cost 1–2 cents per adult in the population per year.

It is necessary to consider two theoretical hazards of food fortification with folic acid in populations in which folate deficiency is common. The first of these is the possibility that some anticonvulsant-treated epileptics may respond to folic acid therapy with increased fit frequency (Chanarin *et al.*, 1960). Although no such effect has been observed in numerous controlled studies using an oral dose of 15 mg folic acid daily, there is now a firm experimental and clinical evidence that high concentrations of folic acid can have a convulsant effect (Colman and Herbert, 1977a). It must be emphasized that the concentrations of folic acid that were reported to be convulsant were high. Most of the data were obtained in experimental studies in animals by Hommes and Obbens and indicated that the convulsant dose of folic acid in normal rats was 45–125 mg if administered intravenously and 15–30 mg if preceded by induction of a focal cortical lesion (Hommes and Obbens, 1972; Obbens, 1973; Obbens and Hommes, 1973; Hommes *et al.*, 1973). The clinical case report in an epileptic concerned a convulsion that occurred after the rapid intravenous infusion of 7.2 mg folic acid (Ch'ien *et al.*, 1975). There have been no suggestions of, and no evidence for, a convulsant effect of folic acid when taken in the low doses suggested for food fortification.

In contrast, there is little evidence that even megadoses of folic acid may cause the second suggested hazard, i.e., the precipitation of neuropathy in pernicious anemia. The major studies on which this suggestion was based were conducted by Schwartz *et al.* (1950), who treated 85 cases of pernicious anemia with 5 mg folic acid daily for 3½ years and by Will *et al.* (1959), who gave 30 mg folic acid daily for up to 10 years to 36 patients with this disease. The incidence of neuropathy, including parasthesiae, was 37% and 44%, respectively, compared with a frequency of 61% reported in 466 cases of untreated pernicious anemia (Cox, 1962). Only when the 58% of subjects who had hematologic relapses on 30 mg folic acid daily, presumably because of more severe B_{12} deficiency, were given 50–100 mg folic acid daily in the 10-year study (Will *et al.*, 1959) did neuropathy become more common than in the above-mentioned untreated cases. The reasonable interpretation of these data would appear to be that failure to treat pernicious anemia with vitamin B_{12} will eventually result in the development of neuropathy in the most severe cases, and that there is no evidence that this is precipitated by folic acid megadoses.

There have been no studies to test the hypothesis that doses of folic acid as low as 0.2 mg daily "may well produce neurologic damage in some patients" (Chanarin, 1969), or the subsequent firmer assertion that "after prolonged use (years rather than weeks) folate can precipitate B_{12} neuropathy in a patient with subclinical or early evidence of B_{12} deficiency" (Chanarin, 1976). Relevant to the issue of food fortification is that of the patients receiving folic acid megadoses in pernicious anemia, 92% and 86%, respectively, developed symptoms that would have led them to seek medical attention (Schwartz *et al.*, 1950; Will *et al.*, 1959). Although this dispute has little relevance to South African blacks, because they rarely have B_{12} deficiency (Colman *et al.*, 1975c), it is relevant in countries like India where both folate and B_{12} deficiency are common (Sood *et al.*, 1975). It would appear untenable to withhold food fortification because of an opinion unsupported by data, when the only effect of such fortification would be to raise the dietary folate content of large undernourished populations to the level enjoyed by more affluent people.

Folic acid fortification of the staple food of populations with a high incidence of folate deficiency complies with all the recommendations concerning food fortification outlined by a Joint Expert Committee of the Food and Agricultural Organization and World Health Organization (FAO/WHO, 1971). Most of these recommendations relate to findings reviewed in this chapter. As a result of the studies mentioned herein, a joint meeting of the World Health Organization and two interested bodies recommended that "in populations with a high incidence of folate deficiency, attempts should be made to find a suitable vehicle for fortification and appropriate trials carried out to determine the feasibility and effectiveness of fortification" (WHO, 1975).

Note Added in Proof

Three studies published after preparation of this manuscript have changed concepts of polyglutamate hydrolysis (p. 84). Description of brush border folate conjugase distinct from that in cells suggested that hydrolysis occurs at the cell surface (Reisenauer et al., 1977). Kinetic studies supported this concept (Dhar et al., 1977). Efficiency of hydrolysis in celiac sprue, initially considered unimpaired (Halsted et al., 1976a), was recalculated to include removal of products from the luminal site and appeared to be significantly decreased (Halsted et al., 1977).

References

Addison, T., 1855, *On the Constitutional and Local Effects of Disease of the Suprarenal Capsules,* Highley, London.

Allen, C. D., and Klipstein, F. A., 1970, Brain folate concentration in rats receiving diphenylhydantoin, *Neurology (Minneap.)* **4**:403.

Alperin, J. B., Hutchinson, H. T., and Levin, W. C., 1966, Studies of folic acid requirements in megaloblastic anemia of pregnancy, *Arch. Int. Med.* **117**:681.

Angier, R. B., Boothe, J. H., Hutchings, B. L., Mowat, J. H., Semb, J., Stokstad, E. L. R., Subba Row, Y., Waller, C. W., Cosulich, D. B., Fahrenbach, M. J., Hultquist, M. E., Kuh, E., Northey, E. H., Seeger, D. R., Sickels, J. P., and Smith, J. M., 1946, The structure and synthesis of the liver *L. casei* factor, *Science,* **103**:667.

Asquith, P., Oelbaum, M. H., and Dawson, D. W., 1967, Scorbutic megaloblastic anaemia responding to ascorbic acid alone, *Br. Med. J.* **4**:402.

Baker, H., Frank, O., Feingold, S., Ziffer, H., Gellene, R. A., Leevy, C. M., and Sobotka, H., 1965, The fate of orally and parenterally administered folates, *Am. J. Clin. Nutr.* **17**:88.

Baldessarini, R. J., 1975, Metabolic hypotheses in schizophrenia, *New Engl. J. Med.* **292**:527.

Ballard, H. S., and Lindenbaum, J., 1974, Megaloblastic anemia complicating hyperalimentation therapy, *Am. J. Med.* **56**:740.

Ballas, S. K., Saidi, P., and Constantino, M., 1976, Reduced erythrocytic deformability in megaloblastic anemia, *Am. J. Clin. Pathol.* **66**:953.

Bateson, M. C., 1976, Folate-responsive neuropathy, *Br. Med. J.* **1**:1528.

Baugh, C. M., Krumdieck, C. L., Baker, H. J., and Butterworth, C. E., 1971, Studies on the absorption and metabolism of folic acid. I. Folate absorption in the dog after exposure of isolated intestinal segments to synthetic pteroyl-polyglutamates of various chain lengths, *J. Clin. Invest.* **50**:2009.

Baugh, C. M., Krumdieck, C. L., Baker, H. J., and Butterworth, C. E., Jr., 1975, Absorption of folic acid poly-γ-glutamates in dogs, *J. Nutr.* **105**:80.

Baumslag, N., Edelstein, T., and Metz, J., 1970, Reduction of incidence of prematurity by folic acid supplementation in pregnancy, *Br. Med. J.* **1**:16.

Bazzano, G., 1969, Effects of folic acid metabolism on serum cholesterol levels, *Arch. Int. Med.* **124**:710.

Benn, A., Swan, C. H. J., Cooke, W. T., Blair, J. A., Matty, A. J., and Smith, M. E., 1971, Effect of intraluminal pH on the absorption of pteroylmonoglutamic acid, *Br. Med. J.* **i**:148.

Bianchi, A., Chipman, D. W., Dreskin, A., and Rosenzweig, N. S., 1970, Nutritional folic acid deficiency with megaloblastic changes in the small bowel epithelium, *New Engl. J. Med.* **282**:859.

Bird, O. D., Robbins, M., Vanderbelt, J. M., and Pfiffner, J. J., 1946, Observations on vitamin B_c conjugase from hog kidney, *J. Biol. Chem.* **163**:649.

Boddington, M. M., and Spriggs, A. I., 1959, The epithelial cells in megaloblastic anaemias, *J. Clin. Pathol.* **12**:228.

Bonnar, J., Goldberg, A., and Smith, J. A., 1969, Do pregnant women take their iron?, *Lancet* **1**:457.

Boots, L. R., Cornwell, P. E., and Beck, L. R., 1975, Diurnal and menstrual cycle variations in serum vitamin B_{12} and folacin activity in baboons, *J. Nutr.* **105**:571.

Botez, M. I., Cadotte, M., Beaulieu, R., Pichette, L. P., and Pison, C., 1976, Neurologic disorders responsive to folic acid therapy, *Can. Med. Assoc. J.* **115**:217.

Brown, A., 1955, Megaloblastic anaemia associated with adult scurvy. Report of a case which responded to synthetic ascorbic acid alone, *Br. J. Haematol.* **1**:345.

Brown, J. P., Davidson, G. E., Scott, J. M., and Weir, D. G., 1973, Effect of diphenylhydantoin and ethanol feeding on the synthesis of rat liver folates from exogenous pteroylglutamate [^{3}H], *Biochem. Pharmacol.* **22**:3287.

Butterworth, C. E., and Krumdieck, C. L., 1975, Intestinal absorption of folic acid monoglutamates and polyglutamates; A brief review of some recent developments, *Br. J. Haematol.* [Suppl.]**31**: 111.

Butterworth, C. E., Santini, R., and Frommeyer, W. B., 1963, The pteroylglutamate components of American diets as determined by chromatographic fractionation, *J. Clin. Invest.* **42**:1929.

Butterworth, C. E., Jr., Baugh, C. M., and Krumdieck, C., 1969, A study of folate absorption and metabolism in man utilizing carbon-14-labeled polyglutamates synthesized by the solid phase method, *J. Clin. Invest.* **48**:1131.

Butterworth, C. E., Newman, A. J., and Krumdieck, C. L., 1974, Tropical sprue: A consideration of possible etiologic mechanisms with emphasis on pteroylpolyglutamate metabolism, *Trans. Am. Clin. Climatol. Assoc.* **86**:11.

Carney, M. W. P., 1967, Serum folate values in 423 psychiatric patients, *Br. Med. J.* **4**:512.

Castle, W. B., 1961, The Gordon Wilson Lecture. A century of curiosity about pernicious anemia, *Trans. Am. Clin. Climatol. Assoc.* **73**:54.

Chanarin, I, 1969, *The Megaloblastic Anaemias,* pp. 585–588, F. A. Davis, Philadelphia.

Chanarin, I., 1976, Investigation and management of megaloblastic anemia, *Clin. Haematol.* **5**:747.

Chanarin, I., 1977a, cited by V. Herbert, Summary of the workshop, p. 286, in Food and Nutrition Board, 1977.

Chanarin, I., 1977b, Vitamin B_{12}-folate interrelationships in *Proceedings of the 16th International Congress of Hematology, Kyoto, 1976,* Excerpta Medica, Amsterdam (in press).

Chanarin, I., Laidlaw, J., Loughridge, L. W., and Mollin, D. L., 1960, Megaloblastic anemia due to phenobarbitone. The convulsant action of therapeutic doses of folic acid, *Br. Med. J.* **i**:1099.

Chanarin, I., Rothman, D., Ardeman, S. and Berry, V., 1965, Some observations on the changes preceding the development of megaloblastic anaemia in pregnancy with particular reference to the neutrophil leucocytes, *Br. J. Haematol.* **11**:557.

Chanarin, I., Rothman, D., Perry, J. and Stratfull, D., 1968, Normal dietary folate, iron, and protein intake, with particular reference to pregnancy, *Br. Med. J.* **2**:394.

Chanarin, I., England, J. M., and Hoffbrand, A. V., 1973, Significance of large red blood cells, *Br. J. Haematol.* **25**:351.

Channing, W., 1842, Notes on anhaemia, principally in its connections with the puerperal state and with functional disease of the uterus; with cases. *New Engl. QJ Med. Surg.* **1**:157.

Chen, C., and Wagner, C., 1975, Folate transport in the choroid plexus, *Life Sci.* **16**:1571.

Chi'en, L. T., Krumdieck, C. L., Scott, C. W., Jr., and Butterworth, C. E., 1975, Harmful effect of megadoses of vitamins: Electroencephalogram abnormalities and seizures induced by intravenous folate in drug-treated epileptics, *Am. J. Clin. Nutr.* **28**:51.

Clarkson, D. R., and Moore, E. M., 1976, Reticulocyte size in nutritional anemias, *Blood* **48**:669.

Colman, N., and Herbert, V., 1974, Measurement of total folate binding capacity (TFBC) of serum after removal of endogenous folate at acid pH, *Clin. Res.* **22**:700A.

Colman, N., and Herbert, V., 1975, Folate binding in folate deficiency, *Blood* **46**:992.

Colman, N., and Herbert, V., 1976, Total folate binding capacity of normal human plasma and variations in uremia, cirrhosis, and pregnancy, *Blood* **48**:911.

Colman, N., and Herbert, V., 1977a, Folate metabolism in brain, in *Biochemistry of Brain* (S. Kumar, ed.), Pergamon Press, Oxford (in press).

Colman, N., and Herbert, V., 1977b, Folate binders in cerebrospinal fluid, *Clin. Res.* **25**:336A.

Colman, N., Barker, M., Green, R., and Metz, J., 1974a, Prevention of folate deficiency in pregnancy by food fortification, *Am. J. Clin. Nutr.* **27**:339.

Colman, N., Green, R., Stevens, K., and Metz, J., 1974b, Prevention of folate deficiency by food fortification. VI. The anti-megaloblastic effect of folic acid-fortified maize meal, *S. Afr. Med. J.* **48**:1795.

Colman, N., Green, R., and Metz, J., 1975a, Prevention of folate deficiency in food fortification. II. Absorption of folic acid from fortified staple foods, *Am. J. Clin. Nutr.* **28**:459.

Colman, N., Larsen, J. V., Barker, M., Barker, E. A., Green, R., and Metz, J., 1975b, Prevention of folate deficiency by food fortification. III. Effect in pregnant subjects of varying amounts of added folic acid, *Am. J. Clin. Nutr.* **28**:465.

Colman, N., Barker, E. A., Barker, M., Green, R., and Metz, J., 1975c, Prevention of folate deficiency by food fortification. IV. Identification of target groups in addition to pregnant women in an adult rural population, *Am. J. Clin. Nutr.* **28**:471.

Colman, N., Bernstein, L. H., and Herbert, V., 1975d, Preferential absorption of methyltetrahydrofolate over folic acid in pernicious anemia, *Clin. Res.* **23**:576A.

Conrad, M. E., and Crosby, W. H., 1962, The natural history of iron deficiency induced by phlebotomy, *Blood* **20**:173.

Cook, G. C., Morgan, J. O., and Hoffbrand, A. V., 1974, Impairment of folate absorption by systemic bacterial infections, *Lancet* **2**:1416.

Cooper, B. A., 1973, Folate and vitamin B_{12} in pregnancy, *Clin. Haematol.* **2**:461.

Cooper, B. A., 1976, Megaloblastic anaemia and disorders affecting utilisation of vitamin B_{12} and folate in childhood, *Clins. Haematol.* **5**:631.

Cooper, H., Levitan, R., Fordtran, J. W., and Ingelfinger, F. J., 1966, A method for studying absorption of water and solute from the small intestine, *Gastroenterology* **50**:1.

Cooperman, J. M., and Luhby, A., 1965, The physiological fate in man of some naturally occurring polyglutamates of folic acid, *Isr. J. Med. Sci.* **1**:704.

Corcino, J. J., Coll, G., and Klipstein, F. A., 1975, Pteroylglutamic acid malabsorption in tropical sprue, *Blood* **45**:577.

Corcino, J. J., Reisenauer, A. M., and Halsted, C. H., 1976, Jejunal perfusion of simple and conjugated folates in tropical sprue, *J. Clin. Invest.* **58**:298.

Cox, E. V., 1962, The clinical manifestations of vitamin B_{12} deficiency in Addisonian pernicious anemia, in *Vitamin B_{12} und Intrinsic Factor: 2nd European Symposium* (H. C. Heinrich, ed.), Enke, Stuttgart.

Croft, R. F., Streeter, A. M., and O'Neill, B. J., 1974, Red cell indices in megaloblastosis and iron deficiency, *Pathology* **6**:107.

Dallman, P. R., 1974, Iron, vitamin E, and folate in the preterm infant, *J. Pediat.* **85**:742.

Das, K. C., and Hoffbrand, A. V., 1970a, Studies of folate uptake by phytohaemagglutinin stimulated lymphocytes, *Br. J. Haematol.* **19**:203.

Das, K. C., and Hoffbrand, A. V., 1970b, Lymphocyte transformation in megaloblastic anemia. Morphology and DNA synthesis, *Br. J. Haematol.* **19**:459.

Das, K. C., and Herbert, V., 1976a, Vitamin B_{12}-folate interrelations. *Clins. Haematol.* **5**:697.

Das, K. C., and Herbert, V., 1976b, Use of the lymphocyte deoxyuridine (dU) suppression test and lymphocyte chromosome study for retrospective diagnosis of vitamin B_{12} and/or folate deficiency despite "shotgun" treatment, *Clin. Res.* **24**:480A.

Davis, R. E., Nicol, D. J., and Kelly, A., 1970, An automated method for the measurement of folate activity, *J. Clin. Pathol.* **23**:47.

Dhar, G. J., Selhub, J., Gay, C., and Rosenberg, I., 1977, Direct *in vivo* demonstration of the sequence of events in intestinal polyglutamyl folate absorption, *Clin. Res.* **25**:309A.

Dine, M. E., and Snyder, L. M., 1970, Iron deficiency and PMN segmentation, *New Engl. J. Med.* **282**:691.

Doe, W. F., Hoffbrand, A. V., Reed, P. I., and Scott, J. M., 1971, Jejunal pH and folic acid, *Br. Med. J.* **i**:669.

Dong, F. M., and Oace, S. M., 1973, Folate distribution in fruit juices, *J. Am. Diet. Assoc.* **62**: 162.

Drury, E. J., Bazar, L. S., and MacKenzie, R. E., 1975, Formiminotransferase-cyclodeaminase from porcine liver: Purification and physical properties of the enzyme complex, *Arch. Biochem. Biophys.* **169**:662.

Duke, M., Herbert, V., and Abelman, W. H., 1964, Hemodynamic effects of blood transfusion in chronic anemia, *New Engl. J. Med.* **271**:975.

Edelstein, T., Stevens, K., Brandt, V., Baumslag, N., and Metz, J., 1966, Tests of folate and vitamin B_{12} nutrition during pregnancy and the puerperium in a population subsisting on a suboptimal diet, *J. Obstet. Gynaecol. Br. Commonw.* **73**:197.

Edozien, J. C., 1972, National Nutrition Survey, Massachusetts, July 1969–June 1971, Report of the Survey Director to the Commissioner for Public Health, Commonwealth of Massachusetts (supplied to V. Herbert by D. Robinson, Director, Division of Community Operations, Department of Public Health, Commonwealth of Massachusetts).

Eichner, E. R., and Hillman, R. S., 1971, The evolution of anemia in alcoholic patients, *Am. J. Med.* **50**:218.

Eichner, E. R., and Hillman, R. S., 1973, Effect of alcohol on serum folate level, *J. Clin. Invest.* **52**:584.

England, J. M., Walford, D. M., and Waters, D. A. W., 1972, Reassessment of the reliability of the hematocrit, *Br. J. Haematol.* **23**:247.

Erbe, R. W., 1975, Inborn errors of folate metabolism, *New Engl. J. Med.* **293**:753, 807.

Faber, H. K., 1928, Value of liver extract (343) in identifying and treating certain anemias of infancy and of childhood; report of case of probable primary anemia in infant nine and one-half months old, *Am. J. Dis. Child,* **36**:1121.

FAO/WHO, 1971, Food fortification and protein-calorie malnutrition, World Health Organisation Technical Report Series No. 477.

Fehling, C., Jagerstad, M., Lindstrand, K., and Elmquist, D., 1975, Reduction of folate levels in the rat: difference in depletion between the central and the peripheral nervous system, *Z. Ernahrungswiss.* **15**:1.

Felix, J. S., and DeMars, R., 1969, Purine requirement of cells cultured from humans affected with Lesch-Nyhan syndrome (hypoxanthine-guanine phosphoribosyltransferase deficiency), *Proc. Natl. Acad. Sci. USA* **62**:536.

Fleming, A. F., 1972, Urinary excretion of folate in pregnancy, *J. Obstet. Gynaecol. Br. Commonw.* **79**:916.

Fleming, A., Knowles, J. P., and Prankerd, T. A., 1963, Pregnancy anemia, *Lancet* **1**:606.

Food and Nutrition Board/National Research Council, 1974, *Recommended Dietary Allowances,* 8th ed., National Academy of Sciences, Washington, D. C.

Food and Nutrition Board, 1977, *Folic Acid: Biochemistry and Physiology in Relation to the Human Folate Requirement; Proceedings of a Workshop on Human Folate Requirement,* National Academy of Sciences, Washington, D.C.

Fox, I. H., Dotten, D. A., Marchant, P. J., and Lacroix, S., 1976, Acquired increases of human erythrocyte purine enzymes, *Metabolism* **25**:571.

Freeman, J. M., Finkelstein, J. D. and Mudd, S. H., 1975, Folate responsive homocystinuria and "schizophrenia": a defect in methylation due to deficient 5,10-methylene-tetrahydrofolate reductase activity, *New Engl. J. Med.* **292**:491.

Frenkel, E. P., McCall, M. S., and Sheehan, R. G., 1973, Cerebrospinal fluid folate and vitamin B_{12} in anticonvulsant-induced megaloblastosis, *J. Lab. Clin. Med.* **81**:105.

Fujioka, S., and Silver, R., 1971, Leucocyte ribonucleotide reductase: Studies in normal subjects and in subjects with pernicious anemia. *J. Lab. Clin. Med.* **77**:59.

Gerson, C. D., Cohen, N., Hepner, G. W., Brown, N., Herbert, V., and Janowitz, H. D., 1971, Folic acid absorption in man: Enhancing effect of glucose, *Gastroenterology* **61**:224.

Gerson, C. D., Hepner, G. W., Brown, N., Cohen, H., Herbert, V., and Janowitz, H. D., 1972, Inhibition by diphenylhydantoin of folic acid absorption in man, *Gastroenterology* **63**:246.

Gerson, C. D., Cohen, N., Brown, N., Lindenbaum, J., Hepner, G. W., and Janowitz, H. D., 1974, Folic acid and hexose absorption in sprue, *Am. J. Dig. Dis.* **19**:911.

Girdwood, R. H., 1960, Folic acid, its analogs and antagonists, *Advan. Clin. Chem.* **3**:235.

Godwin, H. A., and Rosenberg, I. H., 1975, Comparative studies of the intestinal absorption of ^{3}H-pteroylmonoglutamate and ^{3}H-pteroylheptaglutamate in man, *Gastroenterology* **69**:364.

Goldman, D., 1971, The characteristics of the membrane transport of amethopterin and the naturally occurring folates. *Ann. NY Acad. Sci.* **186**:400.

Grant, H. C., Hoffbrand, A. V., and Wells, D. G., 1965, Folate deficiency and neurological disease, *Lancet* **2**:763.

Gross, R. L., Newberne, P. M., and Reid, J. V. O., 1974, Adverse effects on infant development associated with maternal folic acid deficiency, *Nutr. Rep. Int.* **10**:241.

Gross, R. L., Reid, J. V. O., Newberne, P. M., Burgess, B., Marston, R., and Hift, W., 1975, Depressed cell-mediated immunity in megaloblastic anemia due to folic acid deficiency, *Am. J. Clin. Nutr.* **28**:225.

György, P., 1934, Beitrag zur Pathogenese der Ziegenmilchanamie, *Z. Kinderheilk.* **56**:1.

Halsted, C. H., Robles, E. A., and Mezey, E., 1971, Decreased jejunal uptake of labelled folic acid (^{3}H-PGA) in alcoholic patients: Roles of alcohol and nutrition, *New Engl. J. Med.* **285**:701.

Halsted, C. H., Robles, E. A., and Mezey, E., 1973, Intestinal malabsorption in folate-deficient alcoholics, *Gastroenterology* **64**:526.

Halsted, C. H., Baugh, C. M., and Butterworth, C. E., 1975, Jejunal perfusion of simple and conjugated folates in man, *Gastroenterology* **68**:261.

Halsted, C. H., Cantor, D., Reisenauer, A., and Romero, J., 1976a, Effect of celiac sprue on folate digestion and absorption, *Clin. Res.* **24**:104A.

Halsted, C. H., Reisenauer, A., Back, C., and Gotterer, G. S., 1976b, *In vitro* uptake and metabolism of pteroylpolyglutamate by rat small intestine, *J. Nutr.* **106**:485.

Halsted, C. H., Reisenauer, A. M., Romero, J. J., Cantor, D. S., and Ruebner, B., 1977, Jejunal perfusion of simple and conjugated folates in celiac sprue, *J. Clin. Invest.* **59**:933.

Haltia, M., 1970, The effect of folate deficiency on neuronal RNA content. A quantitative cytochemical study, *Br. J. Exp. Pathol.* **51**:191.

Harrison, D. K., and Ibeziako, P. A., 1973, Maternal anaemia and fetal birth weight, *J. Obstet. Gynaecol. Br. Commonw.* **80**:798.

Hatwaine, B. V., 1975, Effect of folic acid on serum and liver cholesterol in rats fed casein at two levels, *Ind. J. Med. Res.* **63**:1446.

Haurani, F. I., 1973, Vitamin B_{12} and the megaloblastic development, *Science* **182**:78.

Herbert, V., 1959, *The Megaloblastic Anemias,* Grune & Stratton, New York.

Herbert, V., 1961, The assay and nature of folic acid activity in human serum, *J. Clin. Invest.* **40**:81.

Herbert, V., 1962a, Experimental nutritional folate deficiency in man, *Trans. Assoc. Am. Phys.* **75**:307.

Herbert, V., 1962b, Minimal daily adult folate requirement, *Arch. Int. Med.* **110**:649.

Herbert, V., 1963a, A palatable diet for producing experimental folate deficiency in man, *Am. J. Clin. Nutr.* **12**:17.

Herbert, V., 1963b, Current concepts in therapy: Megaloblastic anemia, *New Engl. J. Med.* **268**:201, 368.

Herbert, V., 1964, Studies of folate deficiency in man, *Proc. R. Soc. Med.* **57**:377.

Herbert, V., 1965, Excretion of folic acid in bile, *Lancet* **1**:913.

Herbert, V., 1966, The aseptic addition method of *L. casei* assay of folate activity in human serum, *J. Clin. Pathol.* **19**:12.

Herbert, V., 1967, Biochemical and hematologic lesions in folic acid deficiency, *Am. J. Clin. Nutr.* **20**:562.

Herbert, V., 1968a, Nutritional requirements for vitamin B_{12} and folic acid. *Am. J. Clin. Nutr.* **21**:743.

Herbert, V., 1968b, Megaloblastic anemia as a problem in world health, *Am. J. Clin. Nutr.* **21**:1115.

Herbert, V., 1968c, Folic acid deficiency in man, in *Vitamins and Hormones* (R. S. Harris, P. L. Munson, and E. Diczfalusy, eds.), Vol. 26, pp. 525–535, Academic Press, New York.

Herbert, V., 1970, Polymorphonuclear hypersegmentation, *New Engl. J. Med.,* **282**:1213.

Herbert, V., 1972a, Folate deficiency, *J. Am. Med. Assoc.* **222**:834.

Herbert, V., 1972b, Folate metabolism; folate deficiency in developing populations, Lectures, XIV International Congress of Hematology, p. 6, São Paulo, Brazil.

Herbert, V., 1975a, Megaloblastic anemias, in *Textbook of Medicine,* 14th ed., (P. B. Beeson and W. McDermott, eds.), pp. 1404–1413, W. B. Saunders, Philadelphia.

Herbert, V., 1975b, Drugs effective in megaloblastic anemias, in *The Pharmacological Basis of Therapeutics,* 5th ed. (L. S. Goodman and A. Gilman, eds.), 1324–49, Macmillan, New York.

Herbert, V., and Shaprio, S., 1962, The site of absorption of folic acid in the rat *in vitro, Fed. Proc.* **21**:260.

Herbert, V., and Tisman, G., 1975, Hematologic effects of alcohol, *Ann. NY Acad. Sci.* **252**:307.

Herbert, V., and Zalusky, R., 1962, Interrelations of vitamin B_{12} and folic acid metabolism: Folic acid clearance studies, *J. Clin. Invest.,* **41**:1263.

Herbert, V., Larrabee, A. R., and Buchanan, J. M., 1962, Studies on the identification of a folate compound of human serum, *J. Clin. Invest.* **41**:1134.

Herbert, V., Zalusky, R., and Davidson, C. S., 1963, Correlation of folate deficiency with alcoholism and associated macrocytosis, anemia and liver disease, *Ann. Int. Med.* **58**:977.

Herbert, V., Tisman, G., Go, L. T., and Brenner, L., 1973, The dU suppression test using ^{125}I-UdR to define biochemical megaloblastosis, *Br. J. Haematol.* **24**:713.

Herbert, V., Colman, N., Spivack, M., Ocasio, E., Ghanta, V., Kimmel, K., Brenner, L., Freundlich, J., and Scott, J., 1975, Folic acid deficiency in the United States: Folate assays in a prenatal clinic, *Am. J. Obstet. Gynecol.* **123**:175.

Hesp, R., Chanarin, I., and Tait, C., 1975, Potassium changes in megaloblastic anaemia, *Clin. Sci. Mol. Med.* **49**:77.

Hibbard, E. D., and Kenna, A. P., 1974, Plasma and erythrocyte folate levels in low-birthweight infants, *J. Pediat.* **84**:750.

Hibbard, E. D., and Smithells, R. W., 1965, Folic acid metabolism and human embryopathy, *Lancet* **1**:1254.

Hillman, R. S., McGuffin, R., and Campbell, C., 1977, Alcohol interference with the folate enterohepatic cycle, *Clin. Res.* **25**:518A.

Hines, J. D., 1975, Hematologic abnormalities involving vitamin B₆ and folate metabolism in alcoholic subjects, *Ann. NY Acad. Sci.* **252**:316.

Hines, J. D., Kamen, B., and Caston, D., 1973, Abnormal folate binding proteins in azotemic patients, *Proceedings of the 16th Annual Meeting, American Society of Hematology* p. 57.

Hoffbrand, A. V., 1970, Folate deficiency in premature infants, *Arch. Dis. Child.* **45**:441.

Hoffbrand, A. V., 1974, Anaemia in adult coeliac disease, *Clins Gastroenterol.* **3**:71.

Hoffbrand, A. V. (ed), 1976, Megaloblastic anaemia, *Clin. Haematol.* **5**:471.

Hoffbrand, A. V., and Peters, T. J., 1969, The subcellular localisation of pteroylpolyglutamate hydrolase and folate in guinea pig intestinal mucosa, *Biochim. Biophys. Acta* **192**:479.

Hoffbrand, A. V., and Peters, T. J., 1970, Recent advances in knowledge of clinical and biochemical aspects of folate, *Schweiz. Med. Woch.* **100**:1954.

Hoffbrand, A. V., Hobbs, J. R., Kremenchuzky, S., and Mollin, D. L., 1967, Incidence and pathogenesis of megaloblastic erythropoiesis in multiple myeloma, *J. Clin. Pathol.* **20**:699.

Hoffbrand, A. V., Chanarin, I., Kremenchuzky, S., Szur, L., Waters, A. H., and Mollin, D. L., 1968, Megaloblastic anaemia in myelosclerosis. *QJ Med.* **37**:493.

Hoffbrand, A. V., Necheles, R. F., Maldonado, N., Horta, E., and Santini, R., 1969, Malabsorption of folate polyglutamates in tropical sprue, *Br. Med. J.* **ii**:543.

Hoffbrand, A. V., Douglas, A. P., Fry, L., and Steward, J. S., 1970, Malabsorption of dietary folate (pteroylpolyglutamates) in adult coeliac disease and dermatitis herpetiformis, *Br. Med. J.* **iv**:85.

Hoffbrand, A. V., Ganeshaguru, K., Hooton, J. W. L., and Tripp, E., 1976, Megaloblastic anemia: Initiation of DNA synthesis in excess of DNA chain elongation as the underlying mechanism, *Clin. Haematol.* **5**:727.

Hommes, O. R., and Obbens, E. A. M. T., 1972, The epileptogenic action of Na-folate in the rate, *J. Neurol. Sci.* **16**:271.

Hommes, O. R., Obbens, E. A. M. T., and Wijffels, C. C. B., 1973, Epileptogenic activity of sodium-folate and the blood-brain barrier in the rat, *J. Neurol. Sci.* **19**:63.

Hooton, J. W. L., and Hoffbrand, A. V., 1976, Thymidine kinase in megaloblastic anaemia, *Br. J. Haematol.* **33**:527.

Hoppner, K., 1971, Free and total folate activity in strained baby foods, *Can. Inst. Food Tech. J.* **4**:51.

Hoppner, K., Lampi, B., and Perrin, D. E., 1972, The free and total folate activity of foods available on the Canadian market, *J. Inst. Can. Sci. Technol. Alim.* **5**:60.

Hoppner, K., Lampi, B., and Perrin, D. E., 1973, Folacin activity of frozen convenience foods, *J. Am. Diet. Assoc.* **63**:536.

Hryniuk, W. M., 1972, Purineless death as a link between growth rate and cytotoxicity by methotrexate, *Cancer Res.* **32**:1506.

Hurdle, A. D. F., Barton, D., and Searles, I. H., 1968, A method for measuring folate in food and its application to a hospital diet, *Am. J. Clin. Nutr.* **21**:1202.

Hyde, R. D., and Loehry, C. A. E. H., 1968, Folic acid malabsorption in cardiac failure, *Gut* **9**:717.

Ibbotson, R. M., Colvin, B. T., Colvin, M. P., 1975a, Folic acid deficiency during intensive therapy, *Br. Med. J.* **iv**:145.

Ibbotson, R. M., Colvin, B. T., Colvin, M. P., 1975b, Folic acid deficiency during intensive therapy, *Br. Med. J.* **iv**:522.

Iyengar, L., and Rajalakshmi, K., 1975, Effect of folic acid supplement on birth weights of infants, *Am. J. Obstet. Gynecol.* **122**:332.

Jacob, E., Scott, J., Brenner, L., and Herbert, V., 1973, Apparent low serum vitamin B₁₂ level in paraplegic veterans taking ascorbic acid, in *Proceedings of the 16th Annual Meeting, American Society of Hematology*, p. 125.

Jacob, E., Colman, N., and Herbert, V., 1977, Evaluation of simultaneous radioassay for two vitamins: Folate and vitamin B₁₂, *Clin. Res.* **25**:537A.

Jacob, M., Hunt, I. F., Dirige, O., and Swendseid, M. E., 1976, Biochemical assessment of the nutritional status of low-income pregnant women of Mexican descent, *Am. J. Clin. Nutr.* **29**:650.

Jagerstad, M., Lindstrand, K., and Westesson, A-K., 1975, Folates, *Scand. J. Social Med.* [*Suppl.*] **10**:86.

Jandl, J. H., and Greenberg, M. S., 1959, Bone marrow failure due to relative nutritional deficiency in Cooley's hemolytic anemia, *New Engl. J. Med.* **260**:461.

Jennette, J. C., and Goldman, I. D., 1975, Inhibition of the membrane transport of folates by anions retained in uremia, *J. Lab. Clin. Med.* **86**:834.

Kamel, K, Waslien, C. I., El-Ramly, Z., Guindy, S., Mourad, K. A., Khattab, A-K, Hashem, N., Patwardhan, V. N., and Darby, W. J., 1972, Folate requirements of children: II. Response of children recovering from protein-calorie malnutrition to graded doses of parenterally adminis-tered folic acid, *Am. J. Clin. Nutr.* **72**:152.

Kitay, D. Z., Hogan, W. J., Eberle, B., and Mynt, T., 1969, Neutrophil hypersegmentation and folic acid deficiency in pregnancy, *Am. J. Obstet. Gynecol.* **104**:1163.

Klaus, H., 1971, Quantitative criteria of folate deficiency in cervico-vaginal cytograms with a report of a new parameter, *Acta Cytol.* **15**:50.

Klipstein, F. A., 1968, Tropical sprue, *Gastroenterology* **54**:275.

Krebs, H. A., Hems, R., and Tyler, B., 1976, The regulation of folate and methionine metabolism, *Biochem. J.* **158**:341.

Krumdieck, C. L., and Baugh, C. M., 1969, The solid phase synthesis of polyglutamates of folic acid, *Biochemistry* **8**:1568.

Krumdieck, C. L., Newman, A. J., and Butterworth, C. E., 1973, A naturally occurring inhibitor of folic acid conjugase (pteroylpolyglutamate hydrolase) in beans and other pulses, *Am. J. Clin. Nutr.* **26**:460.

Krumdieck, C. L., Boots, L. R., Cornwell, P. E., and Butterworth, C. E., 1975, Estrogen stimula-tion of conjugase activity in the uterus of ovariectomized rats, *Am. J. Clin. Nutr.* **28**:530.

Krumdieck, C. L., Boots, L. R., Cornwell, P. E., and Butterworth, C. E., 1976, Cyclic variations in folate composition and pteroylpolyglutamyl hydrolase (conjugase) activity of the rat uterus, *Am. J. Clin. Nutr.* **29**:288.

Laduron, P., 1974, A new hypothesis on the origin of schizophrenia, *J. Psychiat. Res.* **11**:257.

Laffi, R., Tolomelli, B., Bovina, C., and Marchetti, M., 1972, Influence of short-term treatment with estradiol-17β on folate metabolism in the rat, *Int. J. Vit. Nutr. Res.* **42**:196.

Lane, F., Goff, P., McGuffin, R., Eichner, E. R., and Hillman, R. S., 1976, Folic acid metabolism in normal, folate deficient and alcoholic man, *Br. J. Haematol.* **34**:489.

Lanzkowsky, P., Erlandson, M. E., and Bezan, A. I., 1969, Isolated defect of folic acid absorption associated with mental retardation and cerebral calcification, *Blood* **34**:452.

Larrabee, A. R., Rosenthal, S., Cathou, R. E., and Buchanan, J. M., 1961, A methylated derivative of tetrahydrofolate as an intermediate of methionine biosynthesis, *J. Am. Chem. Soc.* **83**:4094.

Lavoie, A., and Cooper, B. A., 1974, Rapid transfer of folic acid from blood to bile in man, and its conversion into folate coenzymes and into a pteroylglutamate with little biological activity, *Clin. Sci. Molec. Med.* **46**:729.

Lavoie, A., Tripp, E., and Hoffbrand, A. V., 1974, The effect of vitamin B_{12} deficiency on methylfolate metabolism and pteroylpolyglutamate synthesis in human cells, *Clin. Sci. Molec. Med.* **47**:617.

Lawson, D. H., Murray, R. M., and Parker, J. L. W., 1972, Early mortality in the megaloblastic anemias, *QJ Med.* **41**:1.

Leslie, G. I., and Rowe, P. B., 1972, Folate binding by the brush border membrane proteins of small intestinal epithelial cells, *Biochemistry* **11**:1696.

Lindenbaum, J., Whitehead, N., and Reyner, F., 1975, Oral contraceptive hormones, folate metabolism, and the cervical epithelium, *Am. J. Clin. Nutr.* **28**:346.

Liu, Y. K., 1974, Folate deficiency in children with sickle cell anemia. *Am. J. Dis. Child* **127**:389.

Longo, D. L., and Herbert, V., 1976, Radioassay for serum and red cell folate, *J. Lab. Clin. Med.* **87**:138.

Lowenstein, L., Cantlie, G., Ramos, O., and Brunton, L., 1966, The incidence and prevention of folate deficiency in a pregnant clinic population, *Can. Med. Assoc. J.* **95**:797.

Luhby, A. L., and Cooperman, J. M., 1967, Congenital megaloblastic anemia and progressive central nervous system degeneration, *Proceedings of the American Pediatric Society, Atlantic City, April 26–29, 1967.*

Manzoor, M., and Runcie, J., 1976, Folate-responsive neuropathy: Report of ten cases, *Br. Med. J.* **i**:1176.

Margo, G., Barker, M., Fernandes-Costa, F., Colman, N., Green, R., and Metz, J., 1975, Prevention of folate deficiency by food fortification. VII. The use of bread as a vehicle for folate supplementation, *Am. J. Clin. Nutr.* **28**:761.

Margo, G., Baroni, Y., Green, R., and Metz, J., 1977, Anemia in urban underprivileged children. Iron, folate, and vitamin B_{12} nutrition, *Am. J. Clin. Nutr.* **30**:947.

McGuffin, R., Goff, P., and Hillman, R. S., 1975, The effect of diet and alcohol on the development of folate deficiency in the rat, *Br. J. Haematol.* **31**:185.

McPhedran, P., Barnes, M. G., Weistein, J. S., and Robertson, J. S., 1973, Interpretation of electronically determined macrocytosis. *Ann. Int. Med.* **78**:677.

Melamed, E., Reches, A., and Hershko, C., 1975, Reversible central nervous system dysfunction in folate deficiency, *J. Neurol. Sci.* **25**:93.

Metz, J., 1970, Folate deficiency conditioned by lactation, *Am. J. Clin. Nutr.* **23**:843.

Metz, J., Kelly, A., Swett, V. C., Waxman, S., and Herbert, V., 1968a, Deranged DNA synthesis by bone marrow from vitamin B_{12}-deficient humans, *Br. J. Haematol.* **14**:575.

Metz, J., Zalusky, R., and Herbert, V., 1968b, Folic acid binding by serum and milk, *Am. J. Clin. Nutr.* **22**:289.

Metz, J., Lurie, A., and Konidaris, M., 1970, A note on the folate content of uncooked maize, *S. Afr. Med. J.* **44**:539.

Mitchell, H. K., Snell, E. E., and Williams, R. J., 1941, The concentration of folic acid, *J. Am. Chem. Soc.* **63**:2284.

Momtazi, S., and Herbert, V., 1973, Intestinal absorption using vibration-obtained individual small bowel epithelial cells of the rat: Folate absorption, *Am. J. Clin. Nutr.* **26**:23.

Moore, C. V., Bierbaum, O. S., Welch, A. D., and Wright, L. D., 1945, Activity of synthetic *Lactobacillus casei* factor ("folic acid") as antipernicious anemia substance; observations on four patients: two with Addisonian pernicious anemia, one with nontropical sprue, and one with pernicious anemia of pregnancy, *J. Lab. Clin. Med.* **30**:1056.

Moscovitch, L. F., and Cooper, B. A., 1973, Folate content of diets in pregnancy: Comparison of diets collected at home and diets prepared from dietary records, *Am. J. Clin. Nutr.* **26**:707.

Mudd, S. H., and Freeman, J. M., 1974, $N^{5,10}$-methylenetetrahydrofolate reductase deficiency and schizophrenia: a working hypothesis, *J. Psychiat. Res.* **11**:259.

Murphy, M., Keating, M., Boyle, P., Weir, D. G., and Scott, J. M., 1976, The elucidation of the mechanism of folate catabolism in the rat, *Biochem. Biophys. Res. Commun.* **71**:1017.

Nakao, K., and Fujioka, S., 1968, Thymidine kinase activity in the human bone marrow from various blood diseases, *Life Sci.* **7**:395.

Nelson, E. W., Streiff, R. R., and Cerda, J. J., 1975, Comparative bioavailability of folate and vitamin C from a synthetic and a natural source, *Am. J. Clin. Nutr.* **28**:1014.

Nelson, M. M., Asling, C. W., Evans, H. M., 1952, Production of multiple congenital abnormalities in young by maternal pteroylglutamic acid deficiency during gestation, *J. Nutr.* **48**:61.

Noronha, J. M., and Silverman, M., 1962, On folic acid, vitamin B_{12}, methionine and formiminoglutamic acid metabolism, in *Vitamin B_{12} und Intrinsic Factor:* 2nd European Sympoisum, 728 pp., (H. C. Heinrich, ed.), Enke, Stuttgart.

Obbens, E. A. M.-T., 1973, Experimental Epilepsy Induced by Folate Derivatives, Doctoral thesis, Catholic Univ. of Nijmegen, Jannsen, Nijmegen, Netherlands.

Obbens, E. A. M.-T. and Hommes, O. R., 1973, The epileptogenic effects of folate derivatives in the rat, *J. Neurol. Sci.* **20**:223.

O'Broin, J. D., Temperley, I. J., Brown, J. P., and Scott, J. M., 1975, Nutritional stability of various naturally occurring monoglutamate derivatives of folic acid, *Am. J. Clin. Nutr.* **28**:438.

Ordonez, L. A., and Wurtman, R. J., 1974, Folic acid deficiency and methyl group metabolism in rat brain: Effects of L-dopa, *Arch. Biochem. Biophys.* **160**:372.

Osler, W., 1919, The severe anaemias of pregnancy and the post-partum state, *Br. Med. J.* **i**:1.

Parry, T. E., and Blackmore, J. A., 1976a, Serum 'uracil + uridine' levels in pernicious anaemia, *Br. J. Haematol.* **34**:567.

Parry, T. E., and Blackmore, J. A., 1976b, Serum 'uracil + uridine' levels before and after vitamin B_{12} therapy in pernicious anaemia, *Br. J. Haematol.* **34**:575.

Paukert, J. L., Straus, L. D., and Rabinowitz, J. C., 1976, Formyl-methenyl-methylenetetra-hydrofolate synthetase—(combined), *J. Biol. Chem.* **251**:5104.

Pearson, A. G. M., and Turner, A. J., 1975, Folate-dependent 1-carbon transfer to biogenic amines mediated by methylenetetrahydrofolate reductase, *Nature (London)* **258**:173.

Perloff, B. P., and Butrum, R. R., 1977, Folacin in selected foods, *J. Amer. Dietet. Ass.* **70**:161.

Perry, J., and Chanarin, I., 1968, Absorption and utilization of polyglutamyl forms of folate in man, *Br. Med. J.* **iv**:546.

Perry, J., and Chanarin, I., 1973, Formylation of folate as step in physiological folate absorption, *Br. Med. J.* **ii**:588.

Pfiffner, J. J., Binkley, S. B., Bloom, E. S., Brown, R. A., Bird, O. D., Emmett, A. D., Hogan, A. G., and O'Dell, B. L., 1943, Isolation of antianemia factor (vitamin B_c) in crystalline form from liver, *Science* **97**:404.

Pritchard, J. A., Scott, D. E., Whalley, P. J., and Haling, R. F., 1970, Infants of mothers with megaloblastic anemia due to folate deficiency, *J. Am. Med. Assoc.* **211**:1982.

Reed, B., Weir, D., and Scott, J., 1976, The fate of folate polyglutamates in meat during storage and processing, *Am. J. Clin. Nutr.* **29**:1393.

Reisenauer, A. M., Krumdieck, C. L., and Halsted, C. H., 1977, Evidence for brush border folate conjugase in human intestine, *Fed. Proc.* **36**:1120.

Reynolds, E. H., 1976, Neurological aspects of folate and vitamin B_{12} metabolism, *Clin. Haematol.* **5**:661.

Roberts, P. D., St. John, D. J. B., Sinha, R., Stewart, J. S., Baird, I. M., Coghill, N. F., and Morgan, J. O., 1971, Apparent folate deficiency in iron-deficiency anaemia, *Br. J. Haematol.* **20**:165.

Rominger, E., Meyer, H., and Bomskov, C., 1933, Anamiestudien am wachsendem Organismus; uber die Pathogenese der Ziegenmilchanamie, *Z. Ges. Exp. Med.* **89**:786.

Rose, J. A., 1971, Folic acid deficiency as a cause of angular cheilosis, *Lancet* **2**:453.

Rosenberg, I. H., 1976, Absorption and malabsorption of folates, *Clin. Haematol.* **5**:589.

Rosenberg, I. H., Streiff, R. R., Godwin, H. A., and Castle, W. B., 1969, Absorption of poly-glutamic folate: Participation of deconjugating enzymes of the intestinal mucosa, *New Engl. J. Med.* **280**:985.

Rosenberg, I. H., and Godwin, H. A., 1971, The digestion and absorption of dietary folate, *Gastroenterology* **50**:445.

Rosenblatt, D. S., and Erbe, R. W., 1972, Methylene-tetrahydrofolate reductase in human cells from normals and from a family with reductase deficiency, *Am. J. Human Genet.* **24**:65a.

Rosensweig, N. S., 1975, Diet and intestinal enzyme adaptation: Implications for gastrointestinal disorders, *Am. J. Clin. Nutr.* **28**:648.

Rothman, D., 1970, Folic acid in pregnancy, *Am. J. Obstet. Gynecol.* **108**:149.

Saary, M., and Hoffbrand, A. V., 1976, Folic acid deficiency during intensive therapy, *Br. Med. J.* **i**:461.

Saary, M., Sissons, J. G. P., Davies, W. A., and Hoffbrand, A. V., 1975, Unusual megaloblastic anaemia with multinucleate erythroblasts: Two cases with septicaemia and acute renal failure, *J. Clin. Pathol.* **28**:324.

Sakamoto, S., and Takaku, F., 1973, Thymidylate synthetase activity in bone marrow cells of folic acid deficient rats, *Acta Haematol. Jap.* **36**:556.

Sakamoto, S., Niina, M., and Takaku, F., 1975, Thymidylate synthetase activity in bone marrow cells in pernicious anemia, *Blood* **46**:699.

Samuel, P. D., Burland, W. L., and Simpson, K., 1973, Response to oral administration of pteroyl-monoglutamic acid or pteroylpolyglutamate in newborn infants of low birth weight, *Br. J. Nutr.* **30**:165.

Sandstead, H. H., and Pearson, W. N., 1973, Clinical evaluation of nutrition status, in *Modern Nutrition in Health and Disease* (R. S. Goodhart and M. E. Shils, eds.), 5th ed., pp. 572–592, Lea and Febiger, Philadelphia.

Santiago-Borrero, P. J., Santini, R., Perez-Santiago, E., and Maldonado, N., 1973, Congenital isolated defect of folic acid absorption, *J. Pediatr.* **82**:450.

Sapira, J. D., Tullis, S., and Mullaly, R., 1975, Reversible dementia due to folic acid deficiency, *Southern Med. J.* **68**:776.

Schwartz, S. O., Kaplan, S. R., and Armstrong, B. E., 1950, The long-term evaluation of folic acid in the treatment of pernicious anemia, *J. Lab. Clin. Med.* **35**:894.

Scott, D. E., Whalley, P. J., and Pritchard, J. A., 1970, Maternal folate deficiency and pregnancy wastage. II. Fetal malformation, *Obstet. Gynecol.* **36**:26.

Scott, J. M., and Weir, D. G., 1976, Folate composition, synthesis and function in natural materials, *Clin. Haematol.* **5**:547.

Scott, J. M., Ghanta, V., and Herbert, V., 1974, Trouble-free microbiologic serum and red cell folate assays, *Am. J. Med. Technol.* **40**:125.

Sevitt, L. H., and Hoffbrand, A. V., 1969, Serum folate and vitamin B_{12} levels in acute and chronic renal disease. Effect of peritoneal dialysis. *Br. Med. J.* **2**:18.

Shinton, N. K., and Wilson, C. I. D., 1975, Effects of storage and cooking on folates in green vegetables, *3rd Meeting, European and African Division, International Society of Hematology, London, August 24–28, 1975*, **22**:09, Abstr.

Silver, H., and Frankel, S., 1971, Normal values for mean corpuscular volume as determined by the model S Coulter counter, *Am. J. Clin. Pathol.* **55**:438.

Smith, L. H., and Baker, F. A., 1960, Pyrimidine metabolism in man. III. Studies on erythrocytes and leucocytes in pernicious anemia, *J. Clin. Invest.* **39**:15.

Smith, R. M. and Osborne-White, W. S., 1973, Folic acid metabolism in vitamin B_{12} deficient sheep, *Biochem. J.* **136**:279.

Sood, S. K., Ramachandran, K., Mathur, M., Gupta, K., Ramalingaswamy, V., Swarnabai, C., Ponniah, J., Mathan, V. I., and Baker, S. J., 1975, W.H.O. sponsored collaborative studies on nutritional anemia in India. I. The effects of supplemental oral iron administration to pregnant women *QJ Med.* **44**:241.

Spector, R. G., 1972, Effects of formyltetrahydrofolic acid and noradrenalin on the oxygen consumption of rat brain synaptosome-mitochondrial preparations, *Br. J. Pharmacol.* **44**:279.

Spector, R., and Lorenzo, A. V., 1975a, Folate transport by the choroid plexus *in vitro, Science* **187**:540.

Spector, R., and Lorenzo, A. V., 1975b, Folate transport in the central nervous system, *Am. J. Physiol.* **229**:777.

Stokes, P. L., Melikian, V., Leeming, R. L., Portman-Graham, H., Blair, J. A., and Cooke, W. T., 1975, Folate metabolism in scurvy, *Am. J. Clin. Nutr.* **28**:126.

Stokstad, E. L. R., 1943, Some properties of growth factor for Lactobacillus casei, *J. Biol. Chem.* **149**:573.

Stokstad, E. L. R., 1976, Vitamin B_{12} and folic acid, in *Present Knowledge in Nutrition* (D. M. Hegsted, ed.), pp. 204–216, The Nutrition Foundation, New York–Washington, D. C.

Strachan, R. W., and Henderson, J. G., 1967, Dementia and folate deficiency, *QJ Med.* **36**:189.

Streiff, R. R., 1971, Folate levels in citrus and other juices, *Am. J. Clin. Nutr.* **24**:1390.

Sullivan, L. W., and Herbert, V., 1964, Suppression of hematopoiesis by ethanol, *J. Clin. Invest.* **43**:2048.

Swendseid, M. E., Bird, O. D., Brown, R. A., and Bethell, F. H., 1947, Metabolic function of pteroylglutamic acid and its hexaglutamyl conjugate. II. Urinary excretion studies on normal persons. Effect of a conjugase inhibitor, *J. Lab. Clin. Med.* **32**:23.

Tamura, T., and Stokstad, E. L. R., 1973, The availability of food folate in man, *Br. J. Haematol.* **25**:513.

Tamura, T., Shin, Y. S., Williams, M. A., and Stokstad, E. L. R., 1972, Lactobacillus casei response to pteroylpolyglutamates, *Anal. Biochem.* **49**:517.

Tamura, T., Shin, Y. S., Buehring, K. U., and Stokstad, E. L. R., 1976, The availability of folates in man: effect of orange juice supplement on intestinal conjugase, *Br. J. Haematol.* **32**:123.

Tauro, G. P., Danks, D. M., Rowe, P. B., van der Weyden, M. B., Schwartz, M. A., Collins, V. L., and Neal, B. W., 1976, Dihydrofolate reductase deficiency causing megaloblastic anemia in two families, *New Engl. J. Med.* **294**:466.

Thenen, S. W., 1975, Food folate values, *Am. J. Clin. Nutr.* **28**:1341.

Thenen, S. W., and Stokstad, E. L. R., 1973, Effect of methionine on specific folate coenzyme pools in vitamin B_{12} deficient and supplemented rats, *J. Nutr.* **103**:363.

Thomson, A. D., Frank, O., deAngelis, B., and Baker, H., 1972, Thiamine deficiency induced by dietary folate deficiency in rats, *Nutr. Rep. Int.* **6**:107.

Tisman, G., and Herbert, V., 1971, Inhibition by human serum, Dilantin (diphenylhydantoin), and methotrexate (MTX), and enhancement by 2-deoxyglucose, of ^{3}H-pteroylglutamic acid (^{3}HPGA) and ^{3}H-5 methyl tetrahydrofolate (M-THF) uptake by human bone marrow cells *in vitro*, *Clin. Res.* **19**:433.

Tisman, G., and Herbert, V., 1973, B_{12} dependence of cell uptake of serum folate: An explanation for high serum folate and cell folate depletion in B_{12} deficiency, *Blood* **41**:465.

Toepfer, E. W., Zook, E. G., Orr, M. L., and Richardson, L. R., 1951, Folic Acid Content of Foods. Agriculture Handbook No. 29, U.S. Department of Agriculture, Washington, D. C.

Tolomelli, B., Bovina, C., Robinetti, C., and Marchetti, M., 1972, Folate coenzyme metabolism in the castrated rat and treated with 17β-estradiol, *Proc. Soc. Exp. Biol. Med.* **141**:436.

Toskes, P. P., Smith, G. W., Bensinger, T. A., Giannella, R. A., and Conrad, M. E., 1974, Folic acid abnormalities in iron deficiency: The mechanism of decreased serum folate levels in rats, *Am. J. Clin. Nutr.* **27**:355.

U.S., DHEW, 1972, Ten-State Nutrition Survey 1968–70, Center for Disease Control, Atlanta, Ga., DHEW Publications No. (HSM) 72-8130–8134.

Van Niekerk, W. A., 1962, Cervical cells in megaloblastic anaemia of the puerperium, *Lancet* **1**:1277.

Van Niekerk, W. A., 1966, Cervical cytological abnormalities caused by folic acid deficiency, *Acta Cytol.* **10**:67.

Veeger, W., Ten Thige, O. J., Hellemans, N., Mandema, E., and Nieweg, H. O., 1965, Sprue with a characteristic lesion of the small intestine associated with folic acid deficiency, *Acta Med. Scand.* **177**:493.

Vogel, R. I., Fink, R. A., Schneider, L. C., Frank, O., and Baker, H., 1976, Effect of folic acid on gingival health, *J. Periodontol.* **47**:667.

Wagonfeld, J. B., Didzinsky, D., and Rosenberg, I. H., 1975, Analysis of rate-controlling processes in polyglutamyl folate absorption, *Clin. Res.* **23**:259A.

Wardrop, C. A. J., Heatley, R. V., Tennant, G. B., and Hughes, L. E., 1975a, Acute folate deficiency in surgical patients on aminoacid/ethanol intravenous nutrition, *Lancet* **2**:641.

Wardrop, C. A. J., Heatley, R. V., Williams, R. H. P., Lewis, M. H., Tennant, G. B., and Hughes, L. E., 1975b, Folic acid deficiency during intensive therapy, *Br. Med. J.* **iv**:344.

Waslien, C. I., 1977, Folic acid requirement in children in Food and Nutrition Board, 1977.

Watson-Williams, E. J., 1962, Folic acid deficiency in sickle cell anemia, *East Afr. Med. J.* **39**:213.

Waxman, S., and Schreiber, C., 1974, The role of folic acid binding proteins in the cellular uptake of folates, *Proc. Soc. Exp. Biol. Med.* **147**:760.

Waxman, S., Corcino, J., and Herbert, V., 1970, Drugs, toxins, and dietary amino acids affecting vitamin B_{12} or folic acid absorption or utilization, *Am. J. Med.* **48**:599.

Waxman, S., Schreiber, C., and Herbert, V., 1971, Radioisotopic assay for measurement of serum folate levels, *Blood* **38**:219.

Weir, D. G., Brown, J. P., Freedman, D. S., and Scott, J. M., 1973, The absorption of the diastereoisomers of 5-methyltetrahydropteroylglutamate in man: A carrier-mediated process, *Clin. Sci. Mol. Med.* **45**:625.

Whitehead, V. M., and Cooper, B. A., 1967, Absorption of unaltered folic acid from the gastrointestinal tract in man, *Br. J. Haematol.* **13**:679.

Whitehead, V. M., Comty, C. H., Posen, G. A., and Kaye, M., 1968, Homeostasis of folic acid in patients undergoing maintenance hemodialysis, *New Engl. J. Med.* **279**:970.

Whitehead, V. M., Pratt, R., Viallet, A., and Cooper, B. A., 1972, Intestinal conversion of folinic acid to 5-methyltetrahydrofolate in man, *Br. J. Haematol.* **22**:63.

Whitehead, N., Reyner, F., and Lindenbaum, J., 1973, Megaloblastic changes in the cervical epithelium. Association with oral contraceptives therapy and reversible with folic acid, *J. Am. Med. Assoc.* **226**:1421.

WHO, 1975, Control of nutritional anaemia with special reference to iron deficiency. Report of an IAEA/USAID/WHO joint meeting, World Health Organization Technical Report Series No. 580.

Wickramasinghe, S. N., Williams, G., Saunders, J., and Durston, J., 1975, Megaloblastic erythropoiesis and macrocytosis in patients on anticonvulsants, *Br. Med. J.* **iv**:136.

Will, J. J., Mueller, J. F., Brodine, C., Kiely, C. E., Friedman, B., Hawkins, V. R., Dutra, J., and Vilter, R. W., 1959, Folic acid and vitamin B_{12} in pernicious anemia, *J. Lab. Clin. Med.* **53**:22.

Willoughby, M. L. N., and Jewell, F. J., 1966, Investigation of folic acid requirements in pregnancy, *Br. Med. J.* **ii**:1568.

Wills, L., 1931, Treatment of "pernicious anaemia of pregnancy" and "tropical anaemia", with special reference to yeast extract as curative agent, *Br. Med. J.* **i**:1059.

Wills, L., and Mehta, M. M., 1930, Studies in "pernicious anaemia of pregnancy"; preliminary report, *Ind. J. Med. Res.* **17**:777.

Winawer, S. J., Sullivan, L. W., Herbert, V., and Zamcheck, N., 1965, The jejunal mucosa in patients with nutritional folate deficiency and megaloblastic anemia, *New Engl. J. Med.* **272**:892.

Wu, A., Chanarin, I., Slavin, G., and Levi, A. J., 1975, Folate deficiency in the alcoholic—its relationship to clinical and haematological abnormalities, liver disease and folate stores, *Br. J. Haematol.* **29**:469.

Zalusky, R., and Herbert, V., 1961, Megaloblastic anemia in scurvy with response to fifty micrograms of folic acid daily, *New Engl. J. Med.* **265**:1033.

Chapter 5

Metabolic and Nutritional Consequences of Infection

William R. Beisel

1. Introduction

Generalized infectious illnesses initiate a consistent array of complex metabolic and nutritional responses in the host (Scrimshaw *et al.*, 1968; Taylor and De-Sweemer, 1973; Latham, 1975; Beisel, 1972, 1975, 1976a). The magnitude and type of nutritional losses caused by an infection reflect both the severity and duration of illness. Well-nourished individuals are generally able to withstand and recover from a brief, infection-induced loss of body constituents. In contrast, patients with chronic or severe nutritional problems, especially deficiencies of protein and calories, are likely to suffer from a series of recurrent infections. Furthermore, an infectious disease often runs an unusually virulent course in a severely malnourished patient. Because malnutrition and infection interact adversely, especially in children, the concepts of a downward spiral or vicious cycle have been used to define the unique synergistic relationship between these two widespread disease problems.

   A great deal of the early work (Scrimshaw *et al.*, 1968; Taylor and De-Sweemer, 1973; Latham, 1975) was devoted to the documentation of increases in susceptibility of experimental animals to various infectious diseases as a consequence of either generalized starvation or the deficiency of a single nutrient. More recent research studies in laboratory animals and humans (Beisel, 1972, 1975, 1976a, 1976b; Blackburn, 1977; Long, 1977; Powanda, 1977; Wannemacher, 1977; Wilmore *et al.*, 1977) have provided new information concerning the metabolic and nutritional consequences of infection. It is this latter aspect

William R. Beisel • U.S. Army Medical Research Institute of Infectious Diseases, Fort Detrick, Frederick, Maryland 21701.

of the vicious cycle that is emphasized in the present chapter. Although much of the information now available concerning the nutritional consequences of infection is descriptive in nature (Beisel, 1976a), important insights have been gained concerning the basic pathophysiologic mechanisms through which infectious diseases initiate predictable and often stereotyped metabolic responses in the human or animal host.

2. Metabolic Responses Associated with Specific Aspects of the Infectious Process

A generalized infection is accompanied by a series of interrelated metabolic responses in tissues and cells of the host. These responses can involve or include almost all metabolic, hormonal, and physiologic mechanisms used to maintain homeostasis within the body (Beisel, 1975). It is now possible to identify many of the basic mechanistic components of the complex host response. Some of this understanding has been gained by defining the sequential patterns of change at different time periods during a generalized infectious process. Prominent features of an illness, such as fever and anorexia, produce some of the observed nutritional changes. Other metabolic changes occur as secondary consequences of key host-defensive responses, such as phagocytosis. Although metabolic responses to most generalized types of acute infection are relatively stereotyped, they may be modified subsequently if an infection becomes overwhelmingly severe, if it takes on a subacute or chronic course, if its pathogenic effects are caused by the release of organism-produced toxins, or if the infectious process becomes localized within certain body cells or organ systems (Beisel, 1972, 1975, 1976a).

2.1. Incubation Period Changes

The incubation period of a generalized infection may vary from hours to weeks depending on the dose, route, and type of invading microorganism, and is defined as the initial period that is free of demonstrable clinical symptoms or signs of illness. During the symptom-free interval, however, the interaction between invading microorganisms and host cells can give rise to subtle but demonstrable alterations in host biochemical and physiologic functions. Very early declines in the concentration of free amino acids in plasma are among the earliest changes defined in volunteers intentionally exposed to bacterial or viral microorganisms (Beisel et al., 1967; Wannemacher, 1977). Hypoaminoacidemia may be accompanied by a decline in the serum concentration of iron and zinc, owing to an accelerated uptake of plasma amino acids, zinc, and iron by the liver. This accelerated flux from plasma to liver appears to be initiated by a mediating substance(s) released into the circulation by host cells engaged in phagocytic activity (Beisel, 1972, 1975; Powanda, 1977; Wannemacher, 1977).

This hormone-like substance, termed leukocytic endogenous mediator (LEM), is analogous to the endogenous pyrogens released by phagocytic cells that initiate a febrile response by acting upon the hypothalamus (Wood, 1958).

Shortly before the onset of symptoms, multihormonal responses can be detected, including an increased secretion of adrenocorticotropic hormone (ACTH), the adrenal glucocorticoids and ketosteroids, and accelerated deiodination of thyroid hormones (Beisel, 1972, 1975). The renal excretion of sodium and chloride may increase transiently, while at the same time renal losses of phosphate and zinc are reduced (Beisel, 1972, 1975). An accelerated synthesis of proteins within the liver may also begin prior to the onset of fever (Powanda, 1977; Wannemacher, 1977).

2.2. Fever-Induced Changes

The most widely recognized metabolic and nutritional consequences of a generalized infectious disease begin soon after the onset of the febrile response (Beisel et al., 1967). Many of the biochemical and physiologic changes seen in the host during an acute infection can be attributed to the presence of fever per se, for they occur to a comparable degree when artificial types of fever are induced in healthy normal subjects (Beisel et al., 1968). The febrile response is accompanied by an increased rate of basal metabolism of approximately 10–15% per degree Centigrade.

A febrile illness is generally accompanied by an accelerated respiratory rate, with a decline in plasma P_{CO_2} and an increase in pH. If fever persists, early respiratory alkalosis may give way to metabolic acidosis secondary to an accelerated cellular production of acid metabolites.

The febrile period is accompanied by further increase of adrenal glucocorticoid hormone production to rates as much as sixfold above normal, although two- to threefold increases are typical. Increases in the basal concentrations of plasma glucagon, growth hormone, and insulin occur early in fever, and glucose tolerance becomes impaired (Shambaugh and Beisel, 1967; Rocha et al., 1973; Rayfield et al., 1973). Febrile infections are initially accompanied at their onset by accelerated gluconeogenesis, which can give rise to a modest fasting hyperglycemia (Long, 1977).

After an infection-induced fever has become fully established, its wasting effects on body tissues become evident. These are manifested biochemically by increased urinary loss of nitrogenous components including creatinine, urea, ammonia, uric acid, and α-amino nitrogen. Extensive catabolism of muscle tissues leads to the clinical picture of wasting illness. Losses of body nitrogen are accompanied by an excessive urinary loss of potassium, phosphate, sulfur, magnesium, and probably zinc, i.e., those elements normally found in highest concentrations within body cells. In contrast, the principal extracellular electrolytes are retained during fever because of a marked, aldosterone-induced renal conservation of sodium and chloride. Electrolyte retention is accompanied by an accumulation of excessive amounts of body water.

Fever is also accompanied by a more rapid deiodination of thyroid hormones, especially by the liver, with the result that total concentrations of protein-bound iodine in serum may initially be reduced. A delayed secretion of thyroid stimulating hormone may then cause protein-bound iodine values to increase. This sequential combination of events gives rise to a biphasic, down–up pattern of thyroid hormone changes in serum (Beisel, 1975).

2.3. Effects of Anorexia

Acute febrile infections are generally accompanied by anorexia, nausea, and sometimes by vomiting. Although these gastrointestinal symptoms have not been explained by any clearly defined mechanisms of origin, they are sufficiently important to reduce the intake of food appreciably, even to the point of total abstinence for several days (Beisel et al., 1967). The diminished intake of nutrients and calories during infection, the heightened excretory losses, and the accelerated expenditures of body energy combine in their effects to produce the sharply negative balances of body nitrogen and other intracellular elements.

Although anorexia contributes to negative balances, the majority of measured losses must be attributed to factors other than transient food deprivation (Beisel, 1975). Dietary restriction studies performed as a control measure in noninfected healthy volunteers (Beisel et al., 1967) have shown that the infection-induced losses of body nutrients are far greater than can be accounted for by anorexia. Furthermore, during an acute infectious disease, homeostatic mechanisms apparently do not utilize the typical nitrogen-sparing devices normally employed to compensate for a reduced dietary intake.

During simple starvation, metabolic mechanisms are initiated rapidly to conserve body nitrogen (Felig, 1973). The mobilization of body fat depots is accelerated to speed up the production of ketone bodies as sources of cellular energy. This mechanism serves to reduce the wastage of amino acids for either gluconeogenesis or ketogenesis by employing fatty acids from body fat depots as the optimal substrate for ketone formation. In contrast, ketogenesis tends to be inhibited during acute severe infections (Neufeld et al., 1976; Blackburn, 1977), and heightened cellular demands for energy appear to be met through increased hepatic gluconeogenesis utilizing amino acids released from muscle as the principal substrate (Wannemacher, 1977).

2.4. Nutritional Responses during Early and Late Convalescence

The nutritional wastages of acute illness begin to subside with the disappearance of fever. Excessive adrenal glucocorticoid secretion returns abruptly to baseline while aldosterone hypersecretion tapers off somewhat more gradually (Beisel and Rapoport, 1969). As a result, early convalescence is often accompanied by diuresis of the excess water typically retained by the body during the febrile period of the most generalized infections (Beisel, 1975). Because diuresis follows immediately after the period of maximal febrile wasting of body nutrients, muscle, and tissues, body weight generally falls to its lowest values during the first days of convalescence. At the same time, cumulative balances of various body constituent elements reach their most negative values (Beisel *et al.*, 1967). Thus, early convalescence represents the time of the greatest depletion of various body nutrient stores, typified especially by nitrogen.

This low point is followed by a generalized return of anabolic activities, which initiate a gradual reconstitution of the body cell mass and nutrient stores. During this recovery period, a marked decline in the urinary losses of free amino acids serves to indicate their increased reutilization for the synthesis of new body proteins. During early recovery, hyperphagia may develop in children and young adults (Whitehead, 1977). This increases dietary intake and helps to speed up the reconstitution of prior body losses.

Complete nutritional recovery will occur if the infectious process is fully resolved, if an adequate supply of dietary nutrients is provided, and if no secondary complications arise. The period of early convalescence is, however, the time when nutritional stores of the body are most depleted and when the patient is most susceptible to the development of a secondary, superimposed infection.

2.5. Effects of Severity and Duration of Illness

Whereas the timing and pattern of metabolic response to acute generalized infections are relatively stereotyped, the severity of an illness has an important bearing on the magnitude of the metabolic responses and the nutritional losses that accompany an infection. If the host possesses effective immune mechanisms which are able to prevent the occurrence of an infection or reduce its severity, the anticipated nutritional consequences of the infection will also be prevented or reduced (Beisel *et al.*, 1967).

Severe infections have more profound impact on nutritional status than do mild ones. Although fever may have a nonspecific, beneficial role in helping the body to defend itself against certain microorganisms, i.e., *Neisseria gonorrhea or Treponema pallidum,* the majority of nutritional or metabolic evidence suggests that higher fevers are accompanied by greater caloric expenditures and nutritional wastages.

The duration of illness also influences the pattern of metabolic and nutritional consequences of an infection. If an infectious process becomes subacute or chronic, various metabolic functions of the body tend to reach new equilibrium settings. Little is known about most hormonal responses during subacute or chronic infections, although adrenal glucocorticoid and ketosteroid excretion become diminished, often declining into a subnormal range (Beisel and Rapoport, 1969).

With chronic infection the daily losses of nitrogen are minimized. A new state of relative equilibrium is established that appears to maintain the body at a chronically wasted, cachectic level (Howard *et al.*, 1946). This cachexia eventually comes to involve body fat depots as well as muscle.

2.6. Effects Caused by Selective Localization of an Infectious Process

In contrast to the relatively stereotyped patterns of metabolic response during generalized acute febrile illnesses, a localization of the infectious process within certain body organ systems will produce additional physiologic or metabolic changes that are superimposed on the generalized host response.

Thus, any infection accompanied by diarrhea will initiate an abrupt loss of water, sodium, chloride, potassium, and bicarbonate through the stool. Losses of bicarbonate through the gut can be sufficiently severe to produce acute acidosis. More chronic losses of body potassium through the stool can result in metabolic alkalosis. Severe diarrheic losses of water and electrolytes are principally derived from the extracellular fluid spaces. These losses cause hemoconcentration and hypovolemic shock, as may be seen in severe infantile or cholera-like diarrhea.

Patients with infectious hepatitis lose their ability, for a time, to secrete bile acids and bilirubin. In patients with extensive hepatocellular failure severe hypoglycemia is also likely to develop as an apparent result of the inability of the liver to maintain an adequate rate of gluconeogenic activity (Felig *et al.*, 1970).

Severe localized kidney infections may cause sufficient tubular destruction or derangement of nephron architecture to lead to uremia and renal failure. Generalized infections may also be complicated by acute renal tubular necrosis.

An exaggerated or inappropriate secretion of antidiuretic hormone from the posterior pituitary gland often develops in patients with central nervous system (CNS) infections, thus contributing to severe body water retention and overload. In patients with destruction of primary motor neurons, as in acute poliomyelitis, muscle paralysis and atrophy develop, which lead, in turn, to negative body balances of nitrogen and other intracellular components, as well as to negative body balances of calcium and phosphate caused by secondary disuse atrophy of bone.

2.7. Hormonal Responses

It had long been postulated that an increased adrenal secretion of glucocorticoid hormones constituted the major endocrine response to infection, and furthermore, that heightened adrenocortical activity was the primary factor in producing the catabolic phenomena that characterize febrile illnesses. Neither of these concepts remains valid, however, for as detailed in the preceding paragraphs, an acute infection is now known to be accompanied by a complex array of multihormonal and endocrine-like responses within the host. These responses, together with an altered availability at the cellular level of metabolizable substrates and trace nutrients, act in concert to permit host tissues to carry out the concomitant or sequential combinations of catabolic and anabolic activities that evolve during a generalized infectious illness.

A modestly increased secretion of adrenal glucocorticoid hormones and ketosteroids does begin with, or slightly before, onset of symptoms. These increases are achieved mainly by a continuing secretion of steroids throughout the day. This secretory pattern replaces the normal human circadian decline in adrenal output during afternoon and evening hours. Plasma 17-OHCS concentrations may increase appreciably above normal, but only in very severe infections, such as those accompanied by vascular collapse. Adrenoglucocorticoid secretions generally increase only two- to fivefold during the acute stages of an infection, but they then revert quickly to normal or below-normal values if the illness becomes subacute or chronic. An abrupt return to normal also occurs at the onset of recovery. Neither the magnitude nor the duration of illness-induced adrenocortical responses are sufficient to produce negative nitrogen balances; instead, heightened glucocorticoid activity may play a beneficial role by its permissive effects on the cellular synthesis of enzymes and other proteins.

Heightened adrenal secretion of aldosterone also occurs during acute infections. This begins somewhat later than the increase in glucocorticoid secretion, and persists longer. The mineralocorticoid effect is clearly evidenced by the virtual disappearance of both sodium and chloride from the urine and by the tendency for the body to retain water within the extracellular compartments in infections such as lobar pneumonia and tick-borne typhus.

The adrenal medulla may increase its output of catecholamines, especially in gram-negative sepsis or other severe infections accompanied by hypotension.

The thyroidal response to acute infection is sluggish in comparison to that of most endocrine glands. Both of the principal thyroid hormones, thyroxine and triiodothyronine, are deiodinated at an accelerated rate within body cells such as the leukocytes and hepatocytes. This causes serum protein-bound iodine values to decline during the early phases of infection. A late rebound of thyroidal activity may then cause an increase to higher-than-normal concentrations during convalescence.

Both the anterior and posterior pituitary glands are involved in these host responses. In addition to secretions of ACTH and TSH, which control the previously mentioned adrenal and thyroidal responses, growth hormone production is also increased. The posterior pituitary may secrete antidiuretic hormone in an inappropriate manner; this problem occurs most characteristically during infections of the CNS.

The pancreatic islet hormones, insulin and glucagon, are secreted in excess, and contribute to the accelerated formation, release, and utilization of glucose.

There is no present evidence to indicate any participation of gonadal steroids or the parathyroid gland hormones in the acute response to infection.

3. Nutritional Consequences of Infection

The many metabolic and biochemical changes that accompany an infection have important nutritional consequences. These include the direct (measurable) losses of body substances, an accelerated consumption or utilization of body stores, and to a lesser degree, a physiologic redistribution or functional sequestration of some nutrients. (Beisel, 1972, 1975, 1976b; Powanda, 1977).

3.1. Protein, Amino Acid, and Nitrogen Nutrition

The rates of metabolism and utilization of free amino acids will change variably in different tissues of the body during acute infection. The complex changes in amino acid metabolism result from a diminished dietary intake, altered rates of protein synthesis and degradation in different body tissues, the need to use amino acids as a substrate for gluconeogenesis, as well as from the increased formation of nitrogen-containing excretory products lost through the urine (Wannemacher, 1977).

Changes in the plasma concentration of free amino acids represent some of the earliest detectable biochemical responses to infection (Wannemacher, 1977). Any increase or decrease in the concentration of a substance in a body pool (such as the extracellular space) result from the combined effects of changing rates of entry or egress of the substance from the pool, together with any concomitant changes in pool volume. Attention to such considerations is of vital importance when attempting to understand the mechanisms that lead to altered concentrations of amino acids in plasma during the course of an infection; it is also important when evaluating the relationship of these changes to altered patterns of protein synthesis and degradation in different body tissues as well as to altered rates of synthesis of glucose, urea, ammonia, and the ketones.

The reduced plasma concentration of most free amino acids results from their egress from plasma caused by an accelerated uptake by hepatic parenchymal cells. This increased flux of free amino acids from the plasma pool into the hepatocytes is stimulated directly or indirectly by the action of LEM released from activated phagocytic cells of the host (Wannemacher, 1977). The accelerated entry of amino acids into the liver is followed sequentially by chromatin template activation within the nucleus, accelerated synthesis of RNA, RNA localization in rough endoplasmic reticulum, and ultimately accelerated synthesis of various acute-phase reactant glycoproteins which are, in turn, promptly secreted back into the plasma.

In a recent review, Wannemacher (1977) established the concept that the complex infection-induced response involving amino acid nutrition could only be understood by evaluating the individual patterns of change in the metabolism of single (or closely related) amino acids. Sufficient information has now been obtained to make important steps toward this goal.

During early infection, the liver is stimulated to accelerate its synthesis of glucose (Long, 1977), an action that requires adequate quantities of a suitable substrate. During infection, this substrate comes primarily from the deamination of gluconeogenic amino acids represented by alanine and glutamine. Their carbon skeleton is used for gluconeogenesis while nitrogen removed from these amino acids enters ureagenic pathways and contributes to the increased urinary output of urea (Wannemacher, 1977).

Concomitant with the accelerated flux of free amino acids into the liver, important metabolic changes involve the proteins of skeletal muscle, skin, and other tissues. These sites display a net loss of protein content during an infectious process. Degraded body proteins thereby contribute importantly to the free amino acids that enter the liver through the plasma. An accelerated catabolism of skeletal muscle protein during infection results first in the intracellular release of amino acid constituents. Some of these amino acids are reutilized *in situ* within muscle cells for resynthesis of cellular protein, for although cellular anabolic activity is slowed, it does not cease. Relatively large quantities of the newly released branched-chain amino acids (leucine, isoleucine, and valine) are metabolized within muscle cells and converted into alanine or other gluconeogenic amino acids which, in turn, move through plasma to liver to furnish an important source of substrate for hepatic gluconeogenesis (Wannemacher, 1977).

One amino acid released from muscle, 3-methylhistidine, is of special importance because it can serve as an index of the rate of muscle catabolism. The methyl group is attached metabolically to histidine only after histidine has first been incorporated into the amino acid structural sequence of certain contractile proteins, such as actin and myosin. Because 3-methylhistidine cannot be reutilized within the body after its release from the contractile protein, it escapes into the urine. Thus, an enhanced urinary excretion of 3-methylhistidine can be used to quantitate the magnitude and timing of skeletal muscle protein degradation during the course of an infection (Wannemacher, 1977).

Several amino acids demonstrate unique metabolic responses during the infectious process. Tryptophan is metabolized in hepatic cells at an accelerated rate via the kynurenine pathway as well as via pathways leading to the synthesis of serotonin. The accelerated metabolism through the kynurenine pathway accounts for the increased urinary output of diazo reactant metabolites observed in all infections studied, and especially in typhoid fever (Rapoport and Beisel, 1971).

Phenylalanine must also be considered unique because it generally exhibits higher-than-normal plasma values during infection in contrast to most other amino acids, which decline in concentration. Higher phenylalanine values cannot be explained by any aberrations in the normal biochemical pathways for phenylalanine metabolism (Wannemacher, 1977). Because tyrosine values generally decline concomitantly and phenylalanine can be converted to tyrosine through the action of the hepatic enzymes phenylalanine hydroxylase (E.C. 1.14.16.1) and dihydropterine reductase (1.6.99.7), it has proved useful to calculate the plasma phenylalanine/tyrosine ratio when studying an infection.

The increase in plasma phenylalanine and the decrease in tyrosine combine to produce a marked increase in their ratio during febrile infections of humans and experimental animals. This observation has been explained by Wannemacher (1977) as resulting chiefly from an increased input into plasma of free phenylalanine derived from muscle protein degradation. Inasmuch as only small amounts of phenylalanine can be used or metabolized by the liver, accelerated rates of input of phenylalanine into plasma exceed rates at which it can be removed.

Powanda (1977) suggests that the accelerated hepatic synthesis of acute-phase reactant proteins helps to strengthen a variety of nonspecific host defense mechanisms. Thus, the release of LEM from activated host phagocytic cells leads to a purposeful redistribution of amino acids from peripheral to visceral tissues, particularly for the synthesis of acute-phase globulins. Based on this concept, the catabolic degradation of muscle protein during infection, and the consumption of amino acids, carbohydrate, and cellular energy for an accelerated synthesis of hepatic glycoprotein would all appear to have teleologic value in terms of survival.

There appears to be some selectivity about the increase in protein synthesis within the liver (Powanda, 1977). Certain proteins that are normally synthesized in healthy individuals, such as haptoglobin, α_1-antitrypsin, ceruloplasmin, fibrinogen, and α_1-acid glycoprotein are increased in concentration, whereas others, such as albumin and transferrin, are decreased. Synthesis is induced of plasma proteins specifically associated with inflammation or sepsis (C-reactive protein in humans, α_2-macrofetoprotein in rats). Certain of the intracellular hepatic enzymes, such as tryptophan oxygenase (1.13.1.12) and tyrosine transaminase (2.6.1.5) (Rapoport et al., 1968), also increase in activity, perhaps in response to the accelerated influx of amino acids into the liver. Other hepatic enzymes, however, are decreased e.g., catalase (1.11.1.6), urate oxidase (1.7.3.3), glucose 6-phosphatase (3.1.3.9), and 5'-nucleotidase (3.1.3.5). It has been proposed that the accelerated synthesis of plasma acute-phase reactant proteins occurs, in part, at the expense of momentarily expendable intrahepatic enzymes (Canonico et al., 1975, 1977). Because the increases in acute-phase proteins occur during virtually every infection and even in malnourished children and in starved, protein-, or zinc-deficient rats (Powanda, 1977), this response would appear to be of fundamental importance. The acute-phase proteins could serve in host defense by contributing to the effectiveness of phagocytic activity, the development of organism-specific immunity, and the repair of tissue damage (Powanda, 1977).

Accelerated protein synthesis is required to maintain an optimal functioning efficiency level of all known host-defensive mechanisms. Amino acids are thus required during infection for the synthesis of new phagocytic cells and other rapidly reproducing body cells. Amino acids are also utilized for the synthesis of other key proteins, such as the components of complement, the kinin precursors and coagulation system proteins, and the several classes of immunoglobulins that are synthesized *de novo* to control infections by specific microorganisms.

Purine metabolism can also change during infection, especially in conditions that cause an increased turnover of body cells. For example, infectious mononucleosis is accompanied by a prominent increase in the renal excretion of uric acid (Nessan *et al.*, 1974).

3.2. Carbohydrate Nutrition

Glucose tolerance is impaired during infectious illnesses. It has long been recognized that diabetic patients must increase insulin dosages to prevent glucosurea. The relative intolerance for glucose begins very early during the course of a febrile illness, even in a nondiabetic person (Shambaugh and Beisel, 1967; Rayfield *et al.*, 1973). This change involves a complex participation of all carbohydrate-regulating hormones and their interactions with the liver and peripheral body tissues. Although disappearance rates for glucose are slowed during infection, and insulin resistance seems to occur, current evidence indicates that hepatic gluconeogenesis is markedly stimulated during the early phases of infectious illnesses (Long, 1977). Circulating concentrations of glucagon are high, and in many infections there is an enhanced secretion of the glucocorticoid hormones from the adrenal cortex, ACTH and growth hormone from the anterior pituitary gland, and an increase of catecholamine release. Accelerated gluconeogenic activity is accompanied by the depletion of glycogen stores in liver and muscle, and often by modest elevations of fasting concentrations of glucose in plasma. This accelerated hepatic output of glucose serves to meet the heightened demands for cellular energy partly brought about by elevated body temperatures.

Although an accelerated synthesis of glucose seems characteristic of the early stages of an infectious disease, hypoglycemia can develop. Severe hypoglycemia during infection generally results either from a nonavailability of substrate or from a functional breakdown in the biochemical mechanisms for allowing glucose production to continue within the liver. For example, hypoglycemia is a frequent complication in neonatal sepsis (Yeung, 1970) and may be critical for survival. The hypoglycemic mechanism in newborn infants can be ascribed principally to a lack of adequate substrate. Neonates have only minimal amounts of skeletal muscle from which to obtain substrate amino acids. In lethal experimental endotoxemia, hepatotoxic effects inactivate the key enzymes involved in gluconeogenic activity, and death is accompanied by hypothermia and a virtual lack of detectable carbohydrate in blood or tissues (Berry and Smythe, 1960). Similarly, viral infections that severely damage the hepatocytes may lead to lethal hypoglycemia. Yellow fever and severe hepatitis serve as viral examples of this mechanism (Felig *et al.*, 1970). Whereas molecular derangements in biochemical pathways for carbohydrate metabolism apparently do not occur until a disease or toxemic process causes pathologic hepatocellular defects, these may constitute the terminal events in a dying patient. The cessation of adequate gluconeogenic activity can help explain plummeting body temperatures, which can go from infection-induced hyperthermic values to severe terminal hypothermia in a matter of hours.

3.3. Lipid Nutrition

Changes in lipid metabolism during the course of infection are far more difficult to understand than are those of carbohydrate and nitrogen. Concentrations of serum cholesterol seem to change inconsistently, with increases in some infections and decreases in others, especially the viral ones. In some infections, total serum lipid values remain unchanged, whereas in others, especially gram-negative sepsis, marked hyperlipidemia may occur and give the serum a creamy appearance (Beisel, 1976a).

During experimental infections or induced bacterial endotoxemia in rhesus monkeys, hepatic cholesterogenesis is accelerated. Fatty acids are taken up by the liver at an accelerated rate and are then incorporated more rapidly than normal into triglycerides. These monkeys also show an impaired ability of peripheral tissues to remove triglycerides from circulating plasma (Kaufmann *et al.*, 1976a,b). This defect, which is greatest during gram-negative infections, is manifested by a delayed clearance of triglycerides after giving an oral or intravenous lipid load and by reduced post-heparin activity of lipolytic enzymes.

In addition, the ability of the liver to synthesize ketones is impaired during acute infectious illnesses (Neufeld *et al.*, 1976; Blackburn, 1977). In contrast to the usual ketogenic response to starvation, neither human subjects nor experimental animals can initiate or maintain a fully appropriate ketogenic response to starvation in the presence of acute infection. An infection-induced temporary derangement in ketogenic capacity, despite the increased cellular needs for metabolizable energy, may help account for the greater dependence of the body upon hepatic gluconeogenic activity and the catabolism of peripheral body proteins to supply needed substrate. The mechanism leading to defective ketogenesis within the liver has not as yet been definitely identified, but this defect, in combination with evidence for accelerated cholesterol and triglyceride production (Beisel, 1972, 1975; Canonico *et al.*, 1977) is consistent with the propensity for lipid droplets to accumulate in large quantities within liver cells during infections. These droplets produce the characteristic fatty cellular metamorphosis commonly described as a consequence of a wide variety of different kinds of generalized infectious illnesses.

3.4. Mineral and Trace Element Nutrition

The principal intracellular elements, including phosphorus, magnesium, sulfur, potassium, and zinc, all appear to be lost in a manner analogous to the absolute losses of body nitrogen, i.e., when measured by balance techniques, loss of these elements parallels those of body nitrogen, and they show maximal cumulative deficits early in convalescence.

In addition to the direct losses from the body of important elements (Beisel *et al.*, 1967), several trace elements undergo a unique redistribution within the body during acute infectious illnesses (Beisel, 1976b). Characteristically, serum concentrations of iron and zinc decline abruptly and precipitously early in the infectious process. This change, like the abrupt decline in plasma free amino acid values, has been attributed to an action of LEM, which stimulates an accelerated flux of iron and zinc into the liver.

The iron that accumulates within the liver appears to be deposited in storage granules. For some reason, tissue iron is held in a relatively nonavailable status as long as an infection persists. Stored iron is not reutilized in a normal manner for hemoglobin synthesis, and anemia can develop. The decline in serum iron may become quite marked in pyogenic bacterial infections, and it occurs without an appreciable concomitant fall in iron-binding capacity. These phenomena result in an increased concentration of unsaturated transferrin and are thought by some (Weinberg, 1974) to constitute an important defense mechanism of the body against bacterial infections in that (1) iron in storage form is rendered essentially unavailable for uptake by bacteria, and (2) the unsaturated transferrin is able to compete successfully with bacterial siderophores for circulating serum iron. This mechanism is thought to prevent bacteria from acquiring the iron they need in order to proliferate.

Serum-iron values becomes elevated during the second and third weeks of acute infectious hepatitis, which makes this illness different from most other forms of infection (Beisel, 1976b).

The basic reason for the accelerated movement of zinc from serum to liver is unknown. The zinc may be required to protect the liver by maintaining cell membrane stability, to be incorporated into metalloenzymes, or to serve as a cofactor in the synthesis of RNA and protein (Powanda, 1977).

During virtually all infections there is a marked increase in the amount of copper present in serum. This increase is attributed to an accelerated synthesis and release from the liver of ceruloplasmin. This copper-binding protein responds as an acute-phase reactant glycoprotein during inflammatory states and is also under the influence of LEM.

3.5. Electrolyte Nutrition

The extracellular electrolytes sodium and chloride, are handled differently from the principal intracellular elements during various kinds of infectious disease. Salt and water can be lost through the skin if marked diaphoresis occurs, or through the intestinal tract if vomiting or diarrhea are prominent symptoms. Because water and electrolytes are lost as an isotonic solution during severe diarrhea, the concentration of electrolytes in plasma water does not become abnormal despite a marked concomitant increase in the concentration of proteins in serum and red cells in whole blood. The resultant hemoconcentration helps slow down capillary circulation and impairs the distribution of oxygen into peripheral tissues.

Despite direct losses of body electrolytes through several routes, the onset of a generalized febrile illness is accompanied by the secretion of adrenal mineralocorticoids, which cause the kidneys to retain salt. Sodium and chloride may then virtually disappear from the urine. An inappropriate secretion of antidiuretic hormone may occur concomitantly and lead to dilutional hyponatremia. This latter problem is especially marked in various types of infectious diseases that localize within the CNS. Superimposed on these hormonal mechanisms is the additional tendency for sodium to accumulate within body cells during infections of overwhelming severity.

3.6. Vitamin Nutrition

Only limited data are available concerning changes in vitamin nutrition during infection. Although systematic or comprehensive studies are lacking, almost all of the vitamins have been influenced by one or another kind of infectious disease studied in a laboratory animal (Scrimshaw et al., 1968; Taylor and DeSweemer, 1973). Scattered reports suggest that severe infections may precipitate the classic clinical symptoms of deficiency in single vitamins, as illustrated by the development of beriberi, pellagra, xerophthalmia, and the like, during an infection (Scrimshaw et al., 1968). In such instances, however, the preexisting status of vitamin nutrition in the patient has not been known. With the exception of an increase in urinary riboflavin excretion, little change in vitamin nutrition was detected as a result of a brief viral infection in volunteers who were receiving an optimum daily vitamin intake (Beisel et al., 1972).

4. Influence of Nutritional Status on Susceptibility

Studies in experimental animals and extensive experience in humans, as reviewed elsewhere (Scrimshaw et al., 1968; Taylor and DeSweemer, 1973), give evidence that deficiencies of single nutrients or more generalized deficiencies of protein, calories, or both, reduce the ability of the host to resist invading microorganisms. The malnourished individual becomes susceptible to infections caused by opportunistic or commensal organisms that are usually nonpathogenic. This decrease in resistance also causes an infectious disease to become unusually severe in a malnourished individual. The most widely cited example of this phenomenon is the increased likelihood of death if measles occurs in a malnourished child.

The mechanisms by which nutritional deficits depress host resistance to infection are not entirely certain. Malnutrition appears to have an adverse effect on all normal host-defense mechanisms against infection. Malnutrition of protein, calories, or both in combination, is accompanied by extensive atrophy of lymphoid tissues. The ability to initiate and sustain a normal cell-mediated immune response is inhibited. The rapid *de novo* production of specific immunoglobulins appears to be markedly curtailed in patients with severe protein-calorie malnutrition. In addition, other host-defense mechanisms, such as the function of phagocytic cells, the production of various components of complement, and the function of various nonspecific host-defense mechanisms, are all impaired to some degree in patients with severe generalized malnutrition (Suskind, 1976). Less is known about the infection-related effects of deficiencies of specific single nutrients in humans. However, extensive studies in laboratory animals show that deficiencies of many individual vitamins can increase the susceptibility of the host to a variety of infectious microorganisms (Scrimshaw *et al.*, 1968).

5. Role of Nutrition in Therapy

Studies performed in the early 1900s showed that the negative nitrogen balances incurred during acute and subacute infectious illnesses could be partly corrected by dietary measures, which included the administration of diets containing an excess of carbohydrate and an adequate amount of protein (McCann, 1922). The role of optimum nutrition also received much experimental attention in attempts to determine if vitamin deficiencies might increase susceptibility to infectious microorganisms. Anorexia, nausea, and sometimes vomiting make it generally difficult to achieve an optimum dietary intake during acute illnesses.

5.1. Nutritional Therapy during Illness

Because infectious illnesses can reduce body stores of nitrogen and other intracellular elements and can accelerate the expenditure of available energy, it has long been thought that the administration of sufficient dietary nutrients during the course of illness could produce therapeutic benefit. For many years prior to the advent of the antibiotic era, diets high in protein and calories were prescribed by physicians as an important aspect in their supportive therapy for chronic infections, such as tuberculosis.

Despite the potential value of nutritional support, accurate information has not yet been developed with regard to infections to define with certainty the optimum quantities of dietary intake and the proper proportions of different nutrients (Whitehead, 1977). In the absence of formally accepted guidelines concerning nutritional needs during infection, it would seem desirable to provide sufficient protein so as to meet widely accepted minimum daily requirements plus the minimum recommended daily amounts of individual vitamins. It would also seem desirable to increase caloric intake, in the presence of fever, to perhaps double the basal requirements (Whitehead, 1977).

It is generally difficult to ensure an adequate oral intake of dietary nutrients for a patient who is sick with a febrile illness. The intravenous administration of adequate quantities of several nutrient solutions is now being utilized in severely ill surgical patients. Extensive experience with this aggressive nutritional approach has been gained in patients with severe trauma, burns, or postoperative complications. It has been possible to meet the major nutritional needs of many such patients for long periods of time by administering hypertonic dextrose, amino acids, and other appropriate micronutrients into a large central vein. This type of supportive nutritional therapy has also been used successfully in the presence of severe surgical sepsis (Long et al., 1976; Blackburn, 1977; Wilmore et al., 1977).

Depending exclusively on carbohydrate to supply nonprotein calories during total intravenous alimentation, however, can produce iatrogenic problems, such as severe hyperglycemia and hyperosmolarity (Kaminski, 1976), which can impair phagocytosis, in addition to causing dehydration and coma. Furthermore, the concept that glucose calories are necessary during parenteral nutrition has been challenged. The resulting high insulin concentrations stimulated by glucose infusions do not permit a patient to utilize fully the potentially available energy in body fat stores (Blackburn, 1977). For these reasons, some have recommended that amino acids be used in intravenous alimentation for their caloric value as well as their nitrogen content.

The optimum nutritional relationships between the amounts of protein (as amino acids) and the amounts of carbohydrate given during periods of acute illness have not as yet been entirely resolved (Wilmore et al., 1977). Recent studies suggest that the composition of amino acid mixtures must also be evaluated, especially the ratio of branched-chain amino acids to the non-branched-chain groups. The desirability of using intravenous fats to meet the increased caloric needs also remains an unsettled issue. This question is of special importance in view of the impaired ability of infected patients to synthesize ketones from fatty acids. The necessity for including "essential" fats and trace elements also requires further study.

Superimposed on these problems in the surgical patient is the recently described evidence that bacterial sepsis is metabolically unlike other stresses (Neufeld et al., 1977). Although the newer intravenous alimentation techniques are proving their value in septic surgical patients, their potential usefulness in supportive therapy of other infectious diseases has yet to be explored. This would require a careful clinical evaluation of the nutritional status of the patient and the probable course of the infectious process. One must decide if the anticipated nutritional wastage during illness constitutes a risk great enough to warrant the use of aggressive nutritional therapy, since the latter could impose some additional risk factors for the patient. Thus, the possible utilization of aggressive alimentation techniques during the febrile period of a severe acute infectious illness will require additional study before their ultimate value can be ascertained. If an infection acquired by a well-nourished patient is known to be self-limited, or if it can easily be controlled and terminated by appropriate antimicrobial therapy, the optimal period for increasing the oral intake of food to a maximum can be shifted to the time of early convalescence.

5.2. Nutritional Therapy during Convalescence

When acute infectious illnesses terminate uneventfully, negative nitrogen balances are promptly reversed. Thus, body nutrient stores generally reach their lowest levels within several days after fever has subsided. A patient then begins to replenish his or her depleted nutritional stores, although full nutritional recovery may take many weeks. Anorexia disappears along with fever in most patients, and patients regain their appetite. It seems desirable to utilize the early convalescent period following an acute illness to speed up the repletion of lost body nutrients by providing patients with large quantities of highly nutritional foods. Hyperphagia may stimulate convalescent children to consume about twice their normal daily intake of calories until a catch-up phase of weight gain has returned their weight/height ratios to a normal value (Whitehead, 1977).

Less is known about the usefulness of vigorous dietary refeeding during early convalescence for breaking through the vicious cycle of infection, malnutrition, and reinfection. On the basis of both metabolic balance studies and physiologic measurements of host defensive capabilities (such as phagocytosis), patients would seem most susceptible to a reinfection or superimposed infection during the early stages of convalescence. A secondary infection acquired at that critical time may thus inhibit mechanisms that permit catch-up nutritional recovery, and lead instead to an even greater depletion of body nutrients during the superimposed infection.

6. Summary

Acute or chronic infectious illnesses are accompanied by a complex variety of metabolic and nutritional consequences, manifested principally by the wastage of body protein and other nutrient stores. Some metabolic responses begin shortly after the invasion of the host by infectious microorganisms, but most major nutritional responses begin during fever. These are influenced importantly by the magnitude and duration of the febrile process. Concomitant anorexia causes a diminution in dietary nutrient intake, which also contributes to the wastage of body substances.

In responding to infection, the body seems willing to catabolize body proteins, such as those of muscle and skin, in order to provide amino acids as a substrate for the accelerated synthesis of other proteins or of glucose. As a major aspect of changed lipid metabolism during infection, the body appears to be unable to synthesize keto acids from its fatty acid stores for subsequent cellular utilization as a source of metabolizable energy.

Minerals may also be lost from the body during an infection or undergo a redistribution in their localization within the body. Iron and zinc are taken up by the liver, whereas copper concentrations of serum are increased because of accelerated hepatic synthesis of ceruloplasmin. Sodium and chloride may be lost through diarrhea in some infections, but are typically retained within the body during most generalized infections. The infection-related changes in salt and water metabolism are influenced importantly by the combined effects of aldosterone and antidiuretic hormone.

Far too little is known about optimum techniques for using nutrient therapy as a key supportive measure in the management of infectious diseases. Current data suggest that caloric intake should be increased to compensate for the accelerated energy requirements caused by fever. The intake of nitrogen and essential vitamins should be maintained at least at minimum recommended levels. Early convalescence seems to be the most auspicious time for providing additional dietary nutrients to replace the catabolic losses incurred by the body during and immediately after the febrile period. Nutritional deficits should be minimized or eliminated during the management of patients with a chronic infection.

References

Beisel, W. R., 1972, Interrelated changes in host metabolism during generalized infectious illness, *Am. J. Clin. Nutr.* **25**:1254.

Beisel, W. R., 1975, Metabolic response to infection, *Annu. Rev. Med.* **26**:9.

Beisel, W. R., 1976a, Effect of infection on nutritional needs, in *CRC Handbook of Nutrition and Food* (M. Rechcigl, Jr., ed.), CRC Press, Cleveland, Ohio.

Beisel, W. R., 1976b, Trace elements in infectious processes, *Med. Clin. N. Am.* **60**:831.

Beisel, W. R., and Rapoport, M. I., 1969, Inter-relations between adrenocortical functions and infectious illness, *New Engl. J. Med.* **280**:541, 596.

Beisel, W. R., Sawyer, W. D., Ryll, E. D., and Crozier, D., 1967, Metabolic effects of intracellular infections in man, *Ann. Intern. Med.* **67**:744.

Beisel, W. R., Goldman, R. F., and Joy, R. J. T., 1968, Metabolic balance studies during induced hyperthermia in man, *J. Appl. Physiol.* **24**:1.

Beisel, W. R., Herman, Y. F., Sauberlich, H. E., Herman, R. H., Bartelloni, P. J., and Canham, J. E., 1972, Experimentally induced sandfly fever and vitamin metabolism in man, *Am. J. Clin. Nutr.* **25**:1165.

Berry, L. J., and Smythe, D. S., 1960, Some metabolic aspects of host-parasite interactions in the mouse typhoid model, *Ann. NY Acad. Sci.* **88**:1278.

Blackburn, G. L., 1977, Nutritional assessment and support during infection, *Am. J. Clin. Nutr.* **30**: 1493–1497.

Canonico, P. G., White, J. D., and Powanda, M. C., 1975, Peroxisome depletion in rat liver during pneumococcal sepsis, *Lab. Invest.* **33**:147.

Canonico, P. G., Ayala, E., Rill, W. L., and Little, J. S., 1977, Effects of pneumococcal infection on rat liver microsomal enzymes and lipogenesis by isolated hepatocytes, *Am. J. Clin. Nutr.* **30**: 1359–1363.

Felig, P., 1973, The glucose-alanine cycle, *Metab. Clin. Exp.* **22**:179.

Felig, P., Brown, W. V., Levine, R. A., and Klatskin, G., 1970, Glucose homeostasis in viral hepatitis, *New Engl. J. Med.* **283**:1436.

Howard, J. E., Bigham, R. S., Jr., and Mason, R. E., 1946, Studies on convalescence. V. Observations on the altered protein metabolism during induced malarial infections, *Trans. Assoc. Am. Physicians* **59**:242.

Kaminski, M. V., Jr., 1976, Hyperosmolar hyperglycemic nonketotic dehydration. Etiology, pathophysiology and prevention during total parenteral alimentation, *Excerpta Med.* **367**:290.

Kaufmann, R. L., Matson, C. F., and Beisel, W. R., 1976a, Hypertriglyceridemia produced by endotoxin: Role of impaired triglyceride disposal mechanisms, *J. Infect. Dis.* **133**:548.

Kaufmann, R. L., Matson, C. F., Rowberg, A. H., and Beisel, W. R., 1976b, Defective lipid disposal mechanisms during bacterial infection in rhesus monkeys, *Metab. Clin. Exp.* **25**:615.

Latham, M. C., 1975, Nutrition and infection in national development, *Science* **188**:561.

Long, C. L., 1977, Energy balances in infection and sepsis, *Am. J. Clin. Nutr.* **30**:1301–1310.

Long, C. L., Kinney, J. M., and Geiger, J. W., 1976, Nonsuppressability of gluconeogenesis by glucose in septic patients, *Metab. Clin. Exp.* **25**:193.

McCann, W. S., 1922, The protein requirement in tuberculosis, *Arch. Intern. Med.* **29**:33.

Nessan, V. J., Geerken, R. C., and Ulvilla, J., 1974, Uric acid excretion in infectious mononucleosis: A function of increased purine turnover, *J. Clin. Endocrinol. Metab.* **38**:652.

Neufeld, H. A., Pace, J. A., and White, F. E., 1976, The effect of bacterial infections on ketone concentrations in rat liver and blood and on free fatty acid concentrations in rat blood, *Metab. Clin. Exp.* **25**:877.

Powanda, M. C., 1977, Changes in body balance of nitrogen and other key nutrients: description and underlying mechanisms, *Am. J. Clin. Nutr.* **30**:1254–1268.

Rapoport, M. I., and Beisel, W. R., 1971, Studies on tryptophan metabolism in experimental animals and man during infectious illness, *Am. J. Clin. Nutr.* **24**:807.

Rapoport, M. I., Lust, G., and Beisel, W. R., 1968, Host enzyme induction of bacterial infection, *Arch. Intern. Med.* **121**:11.

Rayfield, E. J., Curnow, R. T., George, D. T., and Beisel, W. R., 1973, Impaired carbohydrate metabolism during a mild viral illness, *New Engl. J. Med.* **289**:618.

Rocha, D. M., Santeusanio, F., Faloona, G. R., and Unger, R. H., 1973, Abnormal pancreatic alpha-cell function in bacterial infections, *New Engl. J. Med.* **288**:700.

Scrimshaw, N. S., Taylor, C. E., and Gordon, J. E., 1968, "Interactions of nutrition and infection," Monograph 57, World Health Organization, Geneva.

Shambaugh, G. E. III, and Beisel, W. R., 1967, Insulin response during tularemia in man, *Diabetes* **16**:369.

Suskind, R. M. (ed.), 1976, *Malnutrition and the Immune Response,* Raven Press, New York.

Taylor, C. E., and DeSweemer, C., 1973, Nutrition and infection, *World Rev. Nutr. Diet.* **16**:203.

Wannemacher, R. W., Jr., 1977, Key role of various individual amino acids in host response to infection, *Am. J. Clin. Nutr.* **30**:1269–1280.

Weinberg, E. D., 1974, Iron and susceptibility to infections, *Science* **184**:952.

Whitehead, R. G., 1977, Protein and energy requirements of young children living in the developing countries to allow for catch-up growth after infections, *Am. J. Clin. Nutr.* **30**:1545–1547.

Wilmore, D. W., McDougal, W. S., Peterson, J. P., 1977, Newer products and formulas for alimentation, *Am. J. Clin. Nutr.* **30**:1498–1505.

Wood, W. B., Jr., 1958, The role of endogenous pyrogen in the genesis of fever, *Lancet* **2**:53.

Yeung, C. Y., 1970, Hypoglycemia in neonatal sepsis, *J. Pediat.* **77**:812.

Chapter 6

Regulation of Protein Intake by Plasma Amino Acids

Gerald Harvey Anderson

1. Introduction

Since the first statement of the aminostatic theory of food-intake regulation by Mellinkoff (1957), the role of dietary and plasma amino acids in the regulation of food intake has been examined in many studies (Harper *et al.*, 1970; Harper, 1976). Generally, animals decrease their food intake when fed diets in which the protein content is very low, very high, or deficient in an indispensable amino acid, or in which the protein pattern is grossly distorted from the amino acid requirements of the animal. This food-intake depression has been attributed to changes in free amino acids in body fluids, which give rise to signals monitored in the central nervous system (CNS) (Rogers and Leung, 1973). Even though it has been shown that specific brain regions are involved in the feeding response to the amino acid content of the diet, the functional separation of a protein-intake regulatory mechanism from the energy-intake regulatory mechanism has only recently been investigated. The demonstration that protein and energy intake are regulated independently was achieved through the use of a self-selection feeding model (Booth, 1974; Musten *et al.*, 1974; Ashley and Anderson, 1975a, 1977a). This model has also permitted the identification of specific relationships between plasma amino acid patterns and food selection (Ashley and Anderson, 1975b; Anderson and Ashley, 1976). The same plasma amino acid patterns have been shown to affect brain neurotransmitter metabolism and behavior (Fernstrom, 1976). Consequently, a mechanism whereby plasma amino acids give rise to signals that are important determinants of both protein and energy intake can be proposed.

Gerald Harvey Anderson • Department of Nutrition and Food Science, FitzGerald Building, University of Toronto, Toronto, Ontario, Canada M5S 1A1.

2.   Food-Intake Regulation

2.1.   Brain Centers and the Control of Food Intake

The control of food intake and the correlated regulation of body energy balance has been studied for many years (Mayer, 1964). Although a complete and acceptable understanding of the control mechanisms is still lacking, it is generally accepted that feeding behavior is regulated by a multiplicity of inputs integrated in some way by the CNS (Baumgardt, 1974; Hamilton, 1973; Panksepp, 1974).

The central role for the hypothalamus in the control of food intake was established several years ago by the fundamental observations that ventromedial hypothalamic damage can lead to marked overeating and obesity (Hetherington and Ranson, 1940, 1942), whereas lesions of the lateral hypothalamic area at the level of the ventromedial nucleus lead to aphagia and weight loss (Anand and Brobeck, 1951). Both centers were said to be reciprocally innervated, and it has been suggested that the constant feeding signals of the lateral hypothalamus are modulated by inhibitory influences from the ventromedial nucleus to regulate feeding. However, this simplistic understanding of the hypothalamus as the feeding center has more recently been challenged (Bell, 1971). Doubt with respect to the particular anatomic focus of the feeding and satiety center arises from the observation that stimulation or inhibition of other cortical and subcortical areas can alter the feeding mechanism (Morgane, 1969). Furthermore, it is unlikely that any nervous function can be based on a self-sufficient center rather than in terms of participating neuronal systems (Panksepp, 1974; Coscina, 1977). As a result, current attention is now focused on neurochemically distinct systems that trigger feeding reflexes, rather than on the integrative sites in the CNS (Mogenson, 1976).

2.2.   The Role of Energy in the Regulation of Food Intake

Evidence for the regulation of energy intake arises indirectly from observations on the physiological constancy of body weight (Durnin, 1961). More direct evidence has been provided by studies in which experimental animals have been forced to adjust food intake in order to meet their energy requirements. For example, when adult or weanling rats are fed diets in which the energy density is decreased by dilution with non-nutritive fiber, they increase the weight of total food intake in order to maintain energy balance (Adolph, 1947; Hervey, 1969; Peterson and Baumgardt, 1971a,b). Conversely, if the energy density of the diet is increased by the addition of fat, total food intake is decreased (Cowgill, 1928; Bosshardt et al., 1946; Morrison, 1964). Food intake is also adjusted when the demand for energy is altered by a change in environmental temperature. For example, cold exposure will bring about an increase in food intake (Sellers et al.,

1954; Andik *et al.*, 1963; Vaughan *et al.*, 1966; Klain and Hannon, 1969; Musten *et al.*, 1974; Kraly and Blass, 1976). Conversely, if the environmental temperature is high enough, the animals will die of starvation in preference to eating (Hamilton, 1965). Rats subjected to periods of regular exercise increase their energy intake over nonexercising controls (Collier, 1969a).

As a result of these many tests of feeding responses to changes in energy requirements and energy content of the diet, it has been concluded that these are the primary factors that determine food intake in animals fed diets adequate in all nutrients. The concept that there are other hunger signals or regulatory mechanisms for nutrients in the diet has been virtually ignored in recent years, which is likely because of the difficulties in designing and interpreting studies of self-selected dietary components (Lat, 1967). In fact, the focus on energy regulation has resulted in the decreased food intake of rats fed equicaloric intragastric loads of the individual macronutrients (in the form of glucose, corn oil, and casein hydrolysate) being attributed to the administered calories, not to the role of the individual nutrients (Panksepp, 1974).

2.3. The Role of Protein and Amino Acids in the Regulation of Food Intake

In evaluating the role of protein and amino acids in the regulation of food intake, it is necessary to compare the data that arise from experiments in which the animal has access to only one diet (fixed protein to energy ratio) with those in which the animal has been given a choice of two or more diets that vary in protein content. In the first situation, the animal cannot alter the amount of protein it consumes in relationship to energy (or vice versa), but in the latter situation the animal has the opportunity to select its intake of both protein and energy.

2.3.1. Fixed-Ratio Feeding Studies

In the majority of studies on the effect of dietary protein on food intake, the animals have been allowed access to only a single diet that contains a defined protein and energy content.

The effect of dietary protein on energy intake has generally been related to its function in contributing to growth. The quantity, as well as the quality, of dietary protein has been shown to affect body size and therefore food intake. Increasing the quantity of protein in the diet from zero to required levels results in correspondingly graded growth responses and food consumption. In contrast, when protein concentration in the diets is 40% or greater (depending on the protein fed, and the age of the rats) food intake of rats is decreased (Harper *et al.*, 1970; Musten *et al.*, 1974).

If the concentration of protein in the diets is limiting for growth, most rats, with the possible exception of the pregnant rat (Menaker and Navia, 1973), are unable to compensate effectively by eating more of the diet (Yoshida *et al.*, 1957). Only when the rats are exposed to low temperatures or increased exercise do they consume more energy and hence more protein (Meyer and Hargus, 1959; Andik *et al.*, 1963). However, the observation that weight gained by rats fed low-protein diets is disproportionately high in fat may indicate that the rats have attempted to compensate for an inadequate supply of protein by overeating (Meyer, 1958).

The effect of the quality of dietary proteins on food intake is related to the growth responses produced by different quality proteins. Larger amounts of poorer-quality proteins are required to produce food intake and growth responses similar to that by better-quality proteins (Howard *et al.*, 1958; Allison *et al.*, 1959). Thus, optimal protein to energy ratios in diets of rats can be defined on the basis of growth and feeding response to varying quality proteins (Sibbald *et al.*, 1956, 1957). The rat does not adjust the quantity of diet consumed in order to assure intake of adequate amounts of the most limiting amino acids.

Deficiencies, excesses, and imbalances of amino acids in diets can cause marked reduction in food intake (Harper *et al.*, 1970; Harper, 1976). The deficiency of an indispensable amino acid in a diet leads to failure of appetite. Osborne and Mendel (1916) were one of the first groups to recognize that feeding supplements of the most limiting amino acid increased food intake and growth when they added lysine to 18% gliadin diets fed to rats. Generally, excessive intakes of the indispensable amino acids cause more depression of food intake than do excesses of the dispensable amino acids (Harper *et al.*, 1970). Diets that are imbalanced by the addition to a low-protein nontoxic amounts of one or more amino acids, other than the one that is growth-limiting, cause marked depression of food intake and growth (Harper *et al.*, 1970; Rogers and Leung, 1973; Harper, 1976).

All these studies indicate that dietary protein (amino acids) alter food consumption in rats but do not suggest that the quantity of protein intake is regulated in a meaningful way in relationship to the needs of the animals. In fact, the only reasonable conclusion appears to be that protein (amino acids) affect the feeding mechanisms only under certain conditions of excess, perhaps as a result of a protective mechanism against toxicity (Krauss and Mayer, 1965; Harper, 1976).

2.3.2. Self-Selection Feeding Studies

Data that offer more direct evidence that protein intake may be regulated separately from energy intake arise from studies that have used the self-selection feeding technique. This feeding method allows the animal a choice of diets so that it may choose to eat for energy or for both energy and protein.

Osborne and Mendel (1918) reported on some of the earliest experiments in which the self-selection feeding technique was used. These workers observed that rats ate little of a protein-free diet, if given a choice between it and an 18%

casein diet. They wrote, "It is difficult to imagine what the impulse is which leads the rats to make these selections. . . . It is therefore interesting to have this evidence that the desire of a young animal for food is something more than a satisfaction of its caloric needs." Despite this early observation, investigation of the concept that the rat may have a mechanism to regulate protein intake independently of the energy-intake regulatory mechanisms began almost 50 years later. Supporting evidence has recently arisen from self-selecting feeding studies in which the following three basic observations have been made. First, rats faced with a choice of two diets differing only in protein content select a constant intake of protein, irrespective of the choice offered (Musten *et al.*, 1974; Ashley and Anderson, 1975b, 1977a). Second, when energy requirements are increased by exposing rats to the cold (Leshner *et al.*, 1971; Musten *et al.*, 1974) or increased activity (Collier *et al.*, 1969a) the rats maintain a constant protein intake compared to controls, but selectively increase their energy intake. Third, self-selecting rats fed diluted protein diets will selectively adjust the total intake of the protein-containing diet in order to maintain a constant protein intake (Rozin, 1968; Booth, 1974; Musten *et al.*, 1974).

Evidence that the rat will select a constant protein and energy intake from a choice of two diets that vary only in protein concentration was most systematically demonstrated by Musten *et al.* (1974). In their experiments, weanling rats (50 g initial weight) were fed a choice of two diets that contained differing amounts of the protein source but were otherwise identical in terms of their nutrient and energy density. Table I shows an example of the results obtained when the choice was from diets that varied in casein content. All groups gained the same

Table I. Casein and Energy Intake of Self-Selecting Weanling Rats[a]

Dietary choice	Weight gain (g)	Energy (kcal)	Casein (g)	Protein energy[b] Selected (%)	Random selection (%)
5 + 45[c]	164 ± 5[a][d]	1517 ± 35[a]	135 ± 5[a]	35.6 ± 1.3[a]	23.5
15 + 55	165 ± 7[a]	1597 ± 50[ab]	127 ± 6[a]	32.4 ± 1.8[a]	32.9
25 + 65	174 ± 7[a]	1659 ± 38[b]	138 ± 5[a]	33.5 ± 1.6[a]	42.3
40 + 70	170 ± 6[a]	1540 ± 44[ab]	162 ± 6[b]	42.2 ± 1.0[b]	51.8
50 + 0	166 ± 7[a]	1534 ± 48[ab]	134 ± 5[a]	35.4 ± 0.7[a]	23.5

[a] From Musten *et al.* (1974).

[b] Protein energy $= \dfrac{\text{Protein (kcal)}}{\text{Total food consumed (kcal)}} \times 100$.

[c] Indicates percent casein content of the two diets from which the rats could choose.

[d] Mean ± SEM of 10 rats per group fed for 4 weeks. Average initial weight, 48 g. *Note:* Means with different roman-letter superscripts are significantly different from each other ($p < 0.05$).

amount of body weight and consumed similar amounts of total food energy, with the exception of the group offered 25% and 65% casein diets, which consumed more than the 5 and 45 group. Protein intake and protein energy expressed as a proportion of the total dietary energy was the same in all except the 40 and 70 group. Because of the combination of diets provided, this latter group was unable to select a diet containing the preferred 34.2% protein energy. However, they ate almost entirely from the 40% casein diet, which indicates an attempt to keep the protein intake selected from these two choices as low as possible. A further argument that the rats were regulating both protein and energy intake is shown by a comparison of the observed data with those that might be expected if the rats selected at random from the two diets. Evidently, random selection was not the determinant of the amount eaten from each diet.

Very similar evidence for regulation was reported by Peng et al. (1975), except that the amount of casein the rats preferred in the diets was much lower, approximately 15–16%. Both young (120 g) and adult (350–440 g) rats fed a choice of 3% and 10%, 25% and 10%, 50% and 10%, or 50% and 25% selected dietary protein levels of approximately 9.5%, 15%, 15%, and 25%, respectively. It is difficult to explain the differences in the preferred protein intakes between the rats studied by Peng et al. (1975) and Musten et al. (1974). The experimental diets were quite different in that Peng et al. (1975) added methionine and agar to the casein diets, and the mineral and vitamin mixes were different from those used by Musten et al. (1974).

Further evidence that protein and energy intake are under separate regulatory mechanisms has been shown from cold-exposure studies—i.e., rats exposed to the cold will alter their pattern of selection so that their energy intake increases but their protein intake remains constant (Leshner et al., 1971; Musten et al., 1974). A similar recognition of the need to increase energy intake selectively and maintain a constant intake of protein was also observed when the energy requirement of rats was increased by activity (Collier et al., 1969a).

Although the foregoing data argue that the self-selecting rat prefers to regulate protein intake either at or below a certain quantity, a regulatory mechanism of significance would be expected to also create a drive for protein in order that intake be maintained at the quantity consistent with optimal weight gains in young animals or the maintenance of body weight in adult animals. Evidence that such a drive to consume protein may exist arises from the experiments of Rozin (1968), Booth (1974), and Musten et al. (1974). Rozin (1968) demonstrated that adult rats maintained on a "cafeteria regime," consisting of liquid sources of carbohydrates, protein, and fat, compensated for a greater dilution of the protein solution by appropriately increasing their intake, even when the solution was made less palatable by the addition of quinine. Booth (1974) also came to the conclusion that the adult rat compensates rather precisely for a decrease in the protein content of the protein-containing diet. He attempted to control for possible differences in palatability between the protein-free diet and the protein-containing diets by diluting both diets with agar, water, and methylcellulose. Those rats fed a choice between a protein-free diet and a 10% casein–agar water diet consumed about 13% of the dietary energy as protein and maintained this intake when given a choice of the protein-free and 5% casein–agar–water diet. Similar results were obtained by Musten *et al.* (1974) when they fed weanling rats a choice of diets diluted with non-nutritive fiber. The rats selectively increased their intake of the diluted protein-containing diets such that they maintained a constant intake of protein, and hence ingested a constant ratio of protein to energy from all dietary choices. Finally, in these studies as well as in those of Osborne and Mendel (1918), Harper (1976), and Peng *et al.* (1975), it is evident that rats will consume the protein containing diet, and avoid the protein-free diet if the protein diet contains 20% or less of protein, which suggests that they attempt to consume protein in amounts that at least provide for their requirements.

In contrast, the quantity of protein selected does not appear to be closely related to the requirements of the rats for maximal rate of weight gain. In the studies of Musten *et al.* (1974) casein-fed weanling rats prefer to select 33% of the diet as protein energy, but require only 15% to meet protein requirements (NAS-NRC, 1972). However, it should be noted that the amount of protein energy characteristically self-selected from these casein diets was similar to the dietary level shown by Hartsook *et al.* (1973) to result in a maximum nitrogen density in the carcass. Perhaps the nitrogen concentration in the carcass serves as a signal for protein intake in much the same way as has been proposed for the relationship between adipose tissue concentration and food (energy?) intake (Lepkovsky, 1973). Challenging this concept, however, are the results obtained by Collier *et al.* (1969b), which showed that young rats selected a dietary level of only 16.5% soy protein from a choice of a protein-free diet and either isolated soy protein or soybean meal, attaining considerably less growth than in the controls fed only a single 22% soy protein diet. Similarly, young rats fed 9% casein diets and water with and without sugar drank sufficiently from the sucrose solution to result in decreased weight gains as compared to the controls, which were fed only 9% casein diets and water (Muto and Miyahara, 1972). It is doubtful that this latter study really argues against protein selection, but rather shows the consequences of diluting total nutrient intakes, and the rats preference for a sugar solution over a dry, much less palatable diet. However, both these latter studies probably illustrate the consequences of providing a choice of diets that greatly differ in palatability, and as stressed by Booth (1974), they cannot be expected to give reliable indications of regulatory mechanisms.

The quantity of protein selected is a characteristic of the particular protein fed. In the studies of Musten *et al.* (1974), the rats generally selected 43% protein energy from gluten diets, but 33% from casein diets. Similarly, Pol and den Hartog (1966) found that weanling rats allowed to select protein from a 60% protein-containing diet and a protein-free diet selected more protein from gluten than from potato protein. Henry (1968) also observed that rats given a choice of a protein-containing and a protein-free diet consumed greater amounts of gluten than of peanut or fish protein. An initial consideration of these data might suggest that if a protein regulatory mechanism operates, it is sensitive to the quality of the protein fed, i.e., it seems reasonable that rats would select more of a poor-quality (e.g., gluten) than of a high-quality (e.g., potato, fish) protein. To test this hypothesis, Ashley and Anderson (1975a) studied the effect of adding the most limiting essential amino acids of gluten, casein, and zein on the self-selection of protein and energy. The results of the addition of lysine to gluten is shown in Table II. Lysine additions to gluten caused an increased weight gain and energy consumption but decreased protein intake. This observation suggested that the rats voluntarily decreased the proportion of dietary protein-energy because of the improved nutritional quality of the gluten. In addition, the 42% protein-energy selection in the absence of added lysine approximated the concentration of dietary gluten which the rats would need to consume to satisfy their lysine requirement (NAS-NRC, 1972). Unfortunately, further studies of methionine additions to casein or attempts to balance the amino acid content of casein, gluten, or zein led to the conclusion that although amino acid additions had specific effects leading to alteration in protein and energy selection, the results were unlikely to be related simply to the effects of the amino acids on the nutritional quality of the proteins (Ashley and Anderson, 1975a, 1977a).

Still, it may be concluded that the amount of protein selected by a rat is a characteristic of the protein source, presumably related in some way to its amino acid composition. If so, then the amount of protein selected would be expected to be characteristic of a protein irrespective of the form in which it is added to the diet (e.g., purified vs. meal). Supporting this prediction is the observation that

Table II. Effect of Lysine Additions to Gluten on Food Selection[a]

Dietary choice	Lysine added (g/100 g)	Weight gain (g)	Food consumption		Protein energy[e] (%)
			Energy (kcal)	Protein (g)	
15 + 55[b]	0[c]	134 ± 5[a][d]	1221 ± 26[a]	129 ± 5[a]	42.1 ± 1.4[a]
	3	191 ± 6[b]	1514 ± 48[b]	108 ± 9[ab]	27.3 ± 2.0[b]
	6	184 ± 6[b]	1664 ± 47[b]	100 ± 4[b]	24.2 ± 1.2[b]

[a] From Ashley and Anderson (1975).
[b] Indicates percent gluten content of the two diets from which the rat was allowed to choose.
[c] Lysine was added to the gluten in the quantities indicated before mixing to the desired gluten content of the diets.
[d] Mean ± SEM of 10 rats per group fed for 4 weeks. Average initial group body weight was 47 g. *Note:* Means with roman-letter superscripts are significantly different from each other ($p < 0.05$).

[e] Protein energy $= \dfrac{\text{Protein (g)} \times 4.00 \text{ (kcal/g)}}{\text{Total diet (g)} \times 4.25 \text{ (kcal/g)}} \times 100$

when purified soy protein or soybean oil meal provided the protein in a self-selection method of feeding, protein intakes were similar (Collier *et al.,* 1969b). Furthermore, Harper *et al.* (1970) have shown that rats offered a choice of amino acid imbalanced, or amino acid-deficient diets and protein-free diets prefer the protein-free diets, which again indicates the fundamental significance of the amino acid composition of the protein (amino acid) diets.

3. Mechanisms of Energy and Protein-Intake Regulation by Proteins or Amino Acids

3.1. Plasma Amino Acids and Brain Centers

As with the regulation of energy intake, a mechanism that results in protein-intake regulation would be expected to involve higher brain centers. The direct involvement of brain centers in the feeding response to dietary proteins and amino acids has been shown by several studies (Leung and Rogers, 1969, 1971; Scharrer *et al.,* 1970; Arakawa *et al.,* 1971; Rogers and Leung, 1973). It is thought that a shift in blood plasma amino acid patterns that leads to changes in brain regions is sensed by specific brain receptors.

Mellinkoff (1957) was the first to propose what is termed the aminostatic hypothesis of food-intake regulation. After several observations of an inverse relationship between plasma amino acid levels and appetite in humans, he suggested that plasma amino acid pattern might serve as a signal to the appetite center. Of relevance to the present discussion was his comment that the pattern may be more important than the total levels of amino acid nitrogen in regulating appetite (Mellinkoff *et al.,* 1956).

In an extensive review, Harper *et al.* (1970) have documented the many observations that have associated the changes in plasma amino acids after a meal with a decreased food intake. This relationship is observed in animals fed high-protein diets and amino acid excess, or deficient diets and imbalanced diets. Although extremes in amino acid content of the diet alter food consumption, there is little evidence from any of these early studies as to the role changes in plasma amino acids after a meal containing balanced protein in amounts approximating requirements may play in feeding behavior.

Perhaps the greatest contribution of this line of research was in its demonstration that brain regions other than the hypothalamus were the primary structures involved in the feeding response to amino acid deficiencies, single amino acid excesses, or an excess of balanced protein. Lesions in the ventromedial hypothalamus have little effect on the food intake depression caused by feeding imbalanced diets (Scharrer *et al.,* 1970; Anonymous, 1971; Arakawa *et al.,* 1971; Rogers and Leung, 1973). Similarly, rats with lesions in the lateral hypothalamus respond to imbalanced diets (Krauss and Mayer, 1965; Scharrer *et al.,* 1970). If provided a choice, hyperphagic rats will eat the protein-free diet but will avoid the imbalanced diet, as do intact animals (Arakawa *et al.,* 1971).

To date, both the prepyriform cortex (Leung and Rogers, 1971) and the amygdala (Rogers and Leung, 1973) have been shown to mediate the response to imbalanced diets. That is, lesioned animals fail to show the food intake depression characteristic of the response to imbalanced diets. However, these are not the only brain regions sensitive to amino acids because high-protein diets caused food-intake depression in rats with lesions in the prepyriform cortex or the amygdala (Rogers and Leung, 1973).

A neurosensitive area responsive to excess total protein or amino acids has not been identified. Panksepp and Booth (1971) observed that microinjections of balanced amino acid solutions into the dorsolateral perifornical hypothalamic area inhibited feeding, as did injections into the ventromedial area (Panksepp, 1973). These data are in conflict with the feeding studies showing that both recovered hypothalamic aphagic and hypothalamic hyperphagic rats decrease their food intake when fed high-protein diets (Scharrer *et al.,* 1970; Rogers and Leung, 1973). These latter studies imply that the brain region sensitive to amino acid excess is outside the hypothalamus.

The feeding response to amino acid-deficient diets also appears to be mediated by the CNS, but in regions separate from the hypothalamus. Amino acid-deficient diets did not cause a depression of food intake in prepyriform cortex lesioned rats (Leung and Rogers, 1971) but did in amydgala lesioned rats (Rogers and Leung, 1973).

From these many studies of the feeding responses to imbalanced, amino acid-deficient or high-protein diets there is little understanding of which changes in plasma amino acids are instrumental in affecting the feeding response. It is now evident that plasma amino acid patterns are reflected to some degree in the amino acid pattern of the brain (Leung and Rogers, 1969; Peng *et al.,* 1973), but the association is not direct (Peng *et al.,* 1972; Harper, 1976). Although feeding adaptation to a high-protein diet involves generally increased catabolism of the amino acids and decreases in total plasma amino acids (Harper *et al.,* 1970), it is unlikely that all amino acids are of equal importance in determining the feeding response. Leung and Rogers (1969) favor the idea that the lack of the growth-limiting amino acid, rather than the excess of amino acids in the blood plasma, is responsible for the depression of food intake. The importance of the most limiting amino acids to the CNS was nicely demonstrated by Rogers and Leung (1973), who showed that the injection of a small amount of the growth limiting amino acid in a threonine imbalanced diet into the carotid artery, completely restored feeding. The same amount injected into the jugular vein was insufficient to alter the state of deficiency in the peripheral tissues and had no effect on feeding. However, it is acknowledged that the changes in plasma amino acids reflecting either the most limiting or the imbalancing amino acids in the diet ingested could be rapid enough to signal food-intake depression (Rogers and Leung, 1973; Harper, 1976).

Because many changes in plasma amino acid patterns lead to alterations in food intake, a common mechanism for the effect has been elusive. However, more recently the demonstration that diet-derived alterations in plasma amino acids affect brain neurotransmitter metabolism has opened up the possibility of a new interpretation of the significance of changes in plasma amino acid patterns to feeding behavior (Fernstrom, 1976). Furthermore, the self-selection feeding model has been used to demonstrate that specific changes in plasma amino acid patterns can be related to feeding behavior (Ashley and Anderson, 1975b; Anderson and Ashley, 1976).

3.2. Relationships among Diet, Plasma, and Brain Amino Acids and Neurotransmitters

The importance of amino acids in neurotransmission in the CNS has been realized for some time—many amino acids, or the products of amino acid catabolism, are either known or putative neurotransmitters (Agranoff, 1975). However, only recently has it been shown that physiological changes in plasma amino acids can alter neurotransmitter synthesis in brain, which indicates that peripheral metabolism may have a major influence on central metabolism. The two major neuronal systems in the CNS are those that depend on either serotonin or the catecholamines as neurotransmitters. Fernstrom and Wurtman (Fernstrom and Wurtman, 1972, 1974; Fernstrom, 1976) have shown, through a number of careful experiments during the past 8 years, that physiological changes in the precursor amino acids (tryptophan and tyrosine) in plasma and brain give rise to changes in concentration of the derived neurotransmitters, serotonin and the catecholamines, respectively.

Although the precise function of serotonin in the CNS remains unknown, it is believed to be involved in the control of a number of behavioral and neurovisceral functions (Green and Grahame-Smith, 1976). The increase in serotonin synthesis that has been observed following the consumption of food, as well as the demonstration that brain serotonin concentration is altered by dietary carbohydrate, protein, and possibly fat (Green et al., 1962; Culley et al., 1963; Dupré, 1970; Fernstrom and Wurtman, 1971a,b, 1972), has led to the postulate that serotonin plays a role in the regulation of food consumption (Fernstrom and Wurtman, 1974).

In an attempt to determine a mechanism whereby food ingestion, plasma, and brain tryptophan are related, it was shown that the administration or secretion of insulin induced by the ingestion of high-carbohydrate diets causes increases in plasma tryptophan and brain concentrations of tryptophan and serotonin (Gordon and Meldrum, 1970; Fernstrom and Wurtman, 1971a). Therefore, any food that causes insulin secretion was expected to raise plasma tryptophan and brain serotonin concentrations. Protein was expected to have a more direct effect because of its contribution of tryptophan to plasma. As expected, rats fed an 18% casein diet or an artificial mixture of amino acids showed an elevation in plasma tryptophan, but surprisingly showed no change in brain tryptophan (Fernstrom and Wurtman, 1972). The explanation for this phenomenon was recognized from studies showing that groups of amino acids are transported into the brain by different transport systems, and that within a given grouping, the member amino acids compete with each other for common transport sites (Guroff and Undenfriend, 1962; Blasberg and Lajtha, 1965). Clearly, the changes in brain tryptophan concentration and hence serotonin concentration (the tryptophan hydroxylase is never fully saturated) depend on the ratio of plasma tryptophan to its competing large neutral amino acids (NAA), namely isoleucine, leucine, valine, phenylalanine, and tyrosine (Fernstrom, 1976). This relationship may be modified to some degree by the amount of tryptophan that is free rather than protein-bound in plasma, although this issue has not been satisfactorily resolved (Green and Grahame-Smith, 1976).

In more recent investigations, the concentration of tyrosine, hence catecholamines, in the brain has been shown to depend on the relative plasma concentration of the same amino acids (Fernstrom, 1976). That is, the concentration of tyrosine relative to the other large neutrals—tryptophan, phenylalanine, isoleucine, leucine, and valine—is a major determinant of tyrosine concentration and catecholamine synthesis in the brain. Again, this may prove a significant observation in understanding the feeding response to changes in plasma amino acids. Although catecholamines have a wide range of function, Leibowitz (1976) suggested that there exists both a stimulatory and an inhibitory central catecholamine mechanism for the control of hunger. It seems reasonable,then, that changes in catecholamine function, as modified to some degree by plasma amino acid state prior to and after eating, may be of physiological significance in the determination of feeding behavior.

Undoubtedly, future research will show that other neural systems play important roles in feeding. However, the principle that diet, and hence plasma amino acids, can alter the levels and synthesis of neurotransmitters has been established.

3.3.   The Self-Selection Feeding Model and Plasma Amino Acid Patterns

Logically, if protein intake and energy intake are regulated independently of each other, the true relationship among peripheral metabolic signals, central neurotransmitter metabolism, and feeding behavior will only be appreciated if the animal is allowed to demonstrate its dietary preference. Undoubtedly, the failure to recognize this principle has contributed to the uncertainties of the role of individual metabolites and neurotransmitters in feeding. For example, if shifts in plasma and brain amino acid concentrations basically give rise to an integration of appropriate information that leads to a desire to alter protein intake, then the animal has only one choice if fed only one diet. That is, in order to alter protein intake, the animal has to alter its consumption of that one diet. Depending on the situation, either the protein intake or the energy-intake mechanism will be the predominant determinant of the response. If the diet fed is severely imbalanced or excessively high in total protein or amino acids, the animal will allow the protein-intake mechanism to regulate its energy as well as its protein intake in order to avoid further stress from an imbalance of metabolic signals that arise from the amino acids. Conversely, if the diet is low in protein, the energy-intake regulating mechanism may be fully saturated, thereby operating to control energy balance, but at the same time controlling (undesirably limiting) protein intake.

The only studies to examine plasma amino acid patterns in relationship to protein and energy intakes by the self-selecting rat have been recently reported by Ashley and Anderson (1975b) and Anderson and Ashley (1976). All their experiments weanling allowed rats to self-select from two diets that differed only in protein content. The proteins fed were zein, casein, or gluten to which various amino acid additions were made such that the quality of the protein fed was manipulated (Ashley and Anderson, 1975a, 1977a). After food selection patterns were established, the rats were allowed to eat overnight, and blood was collected between 9 and 11 A.M. the following morning. Changes in the plasma amino acid pattern were correlated with protein and energy intake by the various groups.

Those rats allowed to choose between two diets that differed only in protein content selected the same amount of protein and had similar plasma amino acid patterns, irrespective of the combination of diets offered. However, when amino acid additions were made to the various proteins the shifts in the plasma aminogram did not correlate with food selection. The plasma aminogram indicated the additions of the most limiting amino acids to the protein, but these changes failed to correlate with changes in either protein or energy intake. For example, the initial 3% increase in the level of lysine in gluten (Table II) brought about an increase of plasma lysine concentration of approximately 30 μmol/100 ml and a significant decrease in protein intake. A further increase in the lysine concentration of gluten to 6% caused a further increase in the plasma lysine concentration of approximately 30 μmol/100 ml, but protein intake was little altered. Although high-level lysine additions have been shown to depress food intake by creating a relative state of arginine deficiency in fixed-ratio-fed rats (Jones *et al.*, 1966), arginine additions to lysine-supplemented gluten diets do not affect the feeding response of the self-selecting rat (Ashley and Anderson, 1977a).

A striking relationship between plasma amino acid patterns and food selection, which was observed irrespective of the amino acid content of the proteins fed, was the inverse correlation between protein intake and the plasma TRP/NAA ratio. As a result of these observations, combined with the recognition that this ratio is instrumental in altering brain serotonin levels and perhaps function

(Fernstrom, 1976), Ashley and Anderson (1975b) proposed that the changes in the plasma TRP/NAA ratio triggers a feeding mechanism that is expressed principally as an attempt to control protein intake.

To test their hypothesis that the plasma TRP/NAA ratio results in a regulation of protein intake through its effect on altering brain serotonin concentration, Ashley and Anderson (1977b) conducted a study in which brain serotonin synthesis was depressed by drugs and food selection was followed. When weanling rats were administered p-chlorophenylalanine by mouth or injected intraventricularly with 5,7 dihydroxytryptamine, a selective decrease in protein intake, but not energy intake, was observed. This decrease in protein intake was correlated with a decrease in brain serotonin, but not catecholamine concentrations (Ashley and Anderson, 1977b). These data suggest that serotonin-containing neurons are involved in some manner in a mechanism which functions to regulate protein intake, as distinct from the energy-regulating mechanism.

Plasma amino acid changes that correlate independently with energy intake have also been observed. Anderson and Ashley (1976) found that the plasma TYR/PHE ratio, but not the TYR/NAA ratio, correlated consistently with energy intake, but not gluten intake. These correlations were observed when a number of experiments were considered in which amino acid additions to gluten had resulted in a wide range of energy intake and gluten intake. In all these experiments, the TRP/NAA ratio correlated with gluten intake, but not energy intake. These relationships between the plasma ratios and energy and protein intake are shown in Figs. 1 and 2 for four experiments in which energy and protein intakes were altered by addition of various amino acids (singly and in combination) to gluten. Generally, the correlations between the plasma TRP/NAA ratio and protein intake exceed 0.9, whereas the correlations between the TYR/PHE ratio and energy intake are somewhat less. However, it may be reasonable to suggest that changes in the plasma TRY/PHE ratio reflect or stimulate, at least in part, a mechanism operating via the CNS to control energy intake. Based on the observation that changes in plasma tyrosine can change brain tyrosine and catecholamine concentrations (Wurtman *et al.*, 1974; Fernstrom, 1976) perhaps the catecholamines have a specific role in this selective feeding response. It remains to be determined if alterations in catecholamine concentrations in the brain lead to selective alterations in energy intake, but not protein intake.

The relationships of the plasma TYR/PHE ratio to energy intake and the TRP/NAA ratio to protein intake in the self-selecting rat require more examination to detail the mechanisms involved. Of particular importance is the determination of whether these plasma ratios, and associated neurotransmitter levels, operate to control initiation or termination of a meal, or are reflecting other mechanisms or adjustment which have occurred as a result of the long-term selection patterns. Because of the demonstrated relationships between these plasma ratios and brain neurotransmitters levels (Fernstrom, 1976), it is tempting

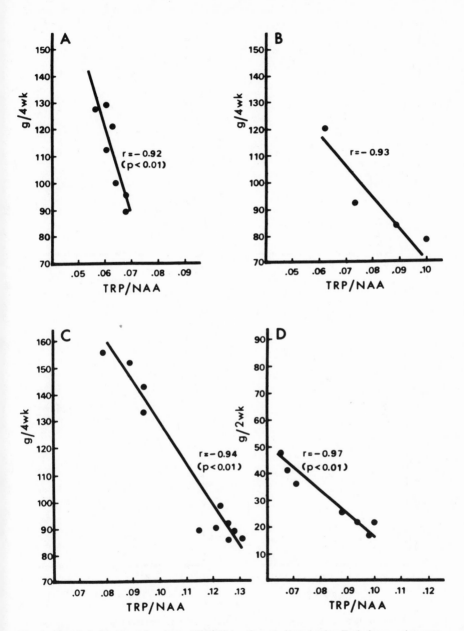

Fig. 1. Correlation between the plasma TRP/NAA ratio and gluten intake. Each datum point represents the average of values for 10 rats (initial body weight of approximately 50 g). (A, B) reported by Ashley and Anderson (1975b); (C, D) from Anderson and Ashley (1976).

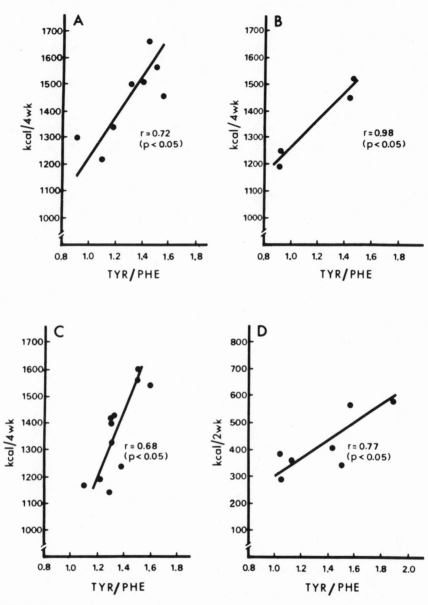

Fig. 2. Correlation between the plasma TYR/PHE ratio and energy intake. Each datum point represents the average of values for 10 rats (initial body weight of approximately 50 g). (A, B) taken from data reported by Ashley and Anderson (1975b); (C, D) from Anderson and Ashley (1976).

to concentrate on these relationships and hypothesize that herein lies a unifying concept for interpreting the relationships between plasma amino acid patterns and feeding behavior of rats fed imbalanced, amino acid-deficient or high- and low-protein diets. Unfortunately, the many studies on amino acid-deficient, imbalanced or high- and low-protein diets on food-intake depression have seldom reported plasma tryptophan concentration, and have seldom utilized self-selection as a feeding model. However, the principle of amino acid competition for entry to the brain and the synthesis of neurotransmitters from many of these amino acids, suggest that neurotransmitter function may be part of the mechanism of the feeding response.

It is unlikely that the catecholamines and serotonin are the only neuro-transmitter substances involved. For example, Lutz *et al.* (1975) showed that amino acids in surplus in the plasma compete with histidine for entry into the brain, thereby contributing to the extremely low brain histidine levels seen in rats fed histidine-imbalanced diets. *In vivo* studies on brain slices (Harper, 1976) confirmed that branched-chain and aromatic amino acids, but not lysine and arginine, have a strong inhibitory effect on histidine uptake. A mechanism whereby low brain histidine levels give rise to a signal resulting in food-intake depression is not stated, but the studies further demonstrate than an analysis of plasma amino acid ratios, rather than absolute levels of amino acids, may be more easily correlated with feeding response. It should be noted that histidine gives rise to a putative neurotransmitter in brain (Agranoff, 1975). Thus, a closer examination of those amino acids that contribute to changes in brain neuro-transmitter (proved and putative) metabolism appears to be in order. Such an examination should not preclude a consideration of the effect of protein synthesis on free amino acid concentration in blood and body fluids (Harper *et al.*, 1970). This factor, in addition to competition among amino acids for the same transport group for entry to the brain, can be expected to be an important determinant of the precursor amino acid availability for neurotransmitter synthesis.

4. Significance of the Proposed Protein-Intake Regulatory Mechanism

It should be recognized that attempts to assess the significance of a protein regulatory mechanisms in the day-to-day regulation of food intake by feeding only a single diet are unreasonable. It is only domesticated (including laboratory) animals that are subjected to such a lack of dietary choice. In contrast, on an evolutionary basis, animals have always had the opportunity for dietary selection. Indeed, human beings have this opportunity, and an indirect argument that the protein mechanism operates in normal humans might be the rather constant 11–13% of the total dietary calories selected from protein (FAO/WHO, 1973). In almost all populations in which food supplies are adequate, protein energy makes up about 11–13% of the diet, even though foods range from 4–95% of the calories as protein. This level of protein consumption is not much above the safe level of intake (Beaton and Swiss, 1974) and is much below that which can be easily tolerated by humans. Thus, the protein regulatory mechanism apparently does not operate solely to prevent excessive intake of protein.

Undoubtedly, debates on the significance of the protein regulatory mechanism to food selection patterns in normal humans will continue. More direct evidence that dietary and plasma amino acids can alter behavior in humans arises from the disease state. For example, studies of patients with hepatic encephalopathy suggest that mechanisms similar to that proposed in the regulation of protein intake in the rats may operate in humans. The realization that the symptoms of hepatic encephalopathy were related to protein metabolism arose first from the associated observations of elevated blood ammonia levels. However, this is not an acceptable explanation of the cause of the symptoms associated with protein ingestion (Schenker et al., 1974).

Recently, attention has focused on plasma amino acid patterns which have shown that the neutral amino acids phenylalanine, tryptophan, and methionine are elevated, whereas the branched-chain amino acids are depressed. The high-phenylalanine, tryptophan, and methionine levels are attributed to a decreased ability of the liver to catabolize and regulate these amino acids, whereas the low branched chains have been related to uptake in muscle as a result of high circulating insulin levels (Munro et al., 1975). Munro et al. (1975) hypothesized that these altered plasma ratios would result in major changes in neurotransmitter synthesis in the brain and hence the symptoms of the disease. The importance of these changes in the plasma amino acid ratios has been demonstrated by Fischer et al. (1976). They have succeeded in reversing hepatic coma effectively by infusion of the branched-chain amino acids. Quite likely, the neuronal systems that give rise to hepatic coma are different from those that result in feeding behavior, but these results serve to confirm that important associations exist among protein intake, plasma amino acid ratios, brain neurotransmitter levels, and behavior in humans (Kolata, 1976). Whether or not various feeding disorders of human beings can be characterized by aberrations in neurochemically distinct systems and modified by selective changes in the plasma amino acid pattern remains to be established.

5. Summary

Changes that occur in plasma amino acids in response to eating are many and are, in considerable part, related to the amino acid quantity and composition of the diet consumed. However, the shifts in plasma amino acid pattern of importance to feeding regulatory mechanisms appear to be those that affect the concentrations of neurotransmitters in the brain. As a result, it may now be proposed that consumption of protein is an important determinant of feeding behavior. That is, by altering neurochemically distinct systems, dietary amino acids influence feeding behavior with respect to both protein and energy.

References

Adolph, E. F., 1947, Urges to eat and drink in rats, *Am. J. Physiol.,* **151**:110.

Agranoff, B. W., 1975, Neurotransmitters and synaptic transmission, *Fed. Proc.* **34**:1911.

Allison, J. B., Wannemacher, R. W. Jr., Middleton, E., and Spoerlein, T., 1959, Dietary protein requirements and problems of supplementation, *Food Technol.* **13**:597.

Anand, B. K., and Brobeck, J. R., 1951, Hypothalamic control of food intake in rats and cats, *Yale J. Biol. Med.* **24**:123.

Anderson, G. H., and Ashley, D. V. M., 1976, Correlation of the plasma tryptophan and tyrosine to neutral amino acid ratios with protein and energy intakes in the self-selecting weanling rat, *Proc. Can. Fed. Biol. Soc.* **19**:24.

Andik, I., Donhoffer, Sz., Farkas, M., and Schmidt, P., 1963, Ambient temperature and survival on a protein-deficient diet, *Br. J. Nutr.* **17**:257.

Anonymous, 1971, The hypothalamus and food intake regulation in response to amino acids and protein, *Nutr. Rev.* **29**:20.

Arakawa, S. M., Standal, B. R. and Beaton, J. R., 1971, Amino acid imbalance and diet preference in the hypothalamic-hyperphagic rat, *Can. J. Physiol. Pharm.* **49**:752.

Ashley, D. V. M., and Anderson, G. H., 1975a, Food intake regulation in the weanling rat: Effects of the most limiting essential amino acids of gluten, casein, and zein on the self-selection of protein and energy, *J. Nutr.* **105**:1405.

Ashley, D. V. M., and Anderson, G. H., 1975b, Correlation between the plasma tryptophan to neutral amino acid ratio and protein intake in the self-selecting weanling rat, *J. Nutr.* **105**:1412.

Ashley, D. V. M., and Anderson, G. H., 1977a, Protein intake regulation in the weanling rat: Effects of additions of lysine, arginine, and ammonia on the selection of gluten and energy, *Life Sci.* (in press).

Ashley, D. V. M., and Anderson, G. H., 1977b, Selective decrease in protein intake following drug-induced brain serotonin depletion in the self-selecting weanling rat, *Fed. Proc.* **36**:4665.

Baumgardt, B. R., 1974, Control of feeding and the regulation of energy balance, *Fed. Proc.* **33**:1139.

Beaton, G. H. and Swiss, L. D., 1974, Evaluation of the nutritional quality of food supplies: prediction of "desirable" or "safe" protein : calorie ratios, *Am. J. Clin. Nutr.* **27**:485.

Bell, F. R., 1971, Hypothalamic control of food intake, *Proc. Nutr. Soc.* **30**:103.

Blasberg, R., and Lajtha, A., 1965, Substrate specificity of steady-state amino acid transport in mouse brain slices, *Arch. Biochem. Biophys.* **112**:361.

Booth, D. A., 1974, Food intake compensation for increase or decrease in the protein content of the diet, *Behav. Biol.* **12**:31.

Bosshardt, D. K., Paul, W., O'Doherty, K., and Barnes, R. H., 1946, The influence of caloric intake on the growth utilization of dietary protein, *J. Nutr.* **32**:641.

Collier, G., Leshner, A. I., and Squibb, R. L., 1969a, Dietary self-selection in active and non-active rats, *Physiol. Behav.* **4**:79.

Collier, G., Leshner, A. I., and Squibb, R. L., 1969b, Self-selection of natural and purified protein, *Physiol. Behav.* **4**:83.

Coscina, D. V., 1977, Brain amines in hypothalamic obesity in *Anorexia nervosa* (R. A. Vigersky, ed.), pp. 97–107, Raven Press, New York.

Cowgill, G. R., 1928, The energy factor in relation to food intake: experiments on the dog, *Am. J. Physiol.* **85**:45.

Culley, W. J., Saunders, R. N., Mertz, E. T., and Jolly, D. H., 1963, Effect of a tryptophan deficient diet on brain serotonin and plasma tryptophan level, *Proc. Soc. Exp. Biol. Med.* **113**:645.

Dupré, J., 1970, Regulation of the secretions of the pancreas, *Ann. Rev. Med.* **21**:299.

Durnin, J. G. V. A., 1961, "Appetite" and the relationships between expenditure and intake of calories in man, *J. Physiol. (London)* **156**:294.

FAO/WHO, 1973, Joint Expert Committee on Energy and Protein Requirement, World Health Organization Technical Report Series No. 522, Geneva, 1973.

Fernstrom, J. D., 1976, The effect of nutritional factors on brain amino acid levels and monoamine synthesis, *Fed. Proc.* **35**:1151.

Fernstrom, J. D., and Wurtman, R. J., 1971a, Brain serotonin content: Increase following the ingestion of carbohydrate diet, *Science* **174**:1023.

Fernstrom, J. D., and Wurtman, R. J., 1971b, Brain serotonin content: physiological dependence on plasma tryptophan levels, *Science* **173**:149.

Fernstrom, J. D., and Wurtman, R. J., 1972, Brain serotonin content: Physiological regulation by plasma neutral amino acids, *Science* **178**:414.

Fernstrom, J. D., and Wurtman, R. J., 1974, Nutrition and the brain, *Sci. Amer.* **230**:84.

Fischer, J. E., Rosen, H. M., Ebeid, A. M., James, J. H., Keane, J. M., and Soeters, P. B., 1976, The effect of normalization of plasma amino acids on hepatic encephalopathy in man, *Surgery* **80**:77.

Gordon, A. E., and Meldrum, B. S., 1970, Effects of insulin on brain 5-hydroxytryptamine and 5-hydroxy-indole-acetic acid of rats, *Biochem. Pharmacol.* **19**:3042.

Green, A. R., and Grahame-Smith, D. G., 1976, Effects of drugs on the processes regulating the functional activity of brain 5-hydroxytryptamine, *Nature (London)* **260**:487.

Green, H., Greenberg, S. M., Erickson, R. W., Sawyer, J. L., and Ellison, T., 1962, Effect of dietary phenylalanine and tryptophan upon rat brain amine levels, *J. Pharmacol. Exp. Ther.* **136**:174.

Guroff, G., and Undenfriend, S., 1962, Studies on aromatic amino acid uptake by rat brain in vivo, *J. Biol. Chem.* **237**:803.

Hamilton, C. L., 1965, Control of food intake, in *Physiological Controls and Regulations* (W. S. Yamamoto and J. R. Brobeck, eds.), pp. 274–294, W. B. Saunders, Philadelphia.

Hamilton, C. L., 1973, Physiologic control of food intake, *J. Am. Diet. Assoc.* **62**:35.

Harper, A. E., 1976, Protein and amino acids in the regulation of food intake, in *Hunger: Basic Mechanisms and Clinical Implications* (D. Novin, W. Wyrwicka, and G. Bray, eds.), pp. 103–113, Raven Press, New York.

Harper, A. E., Benevenga, N. J., and Wohlhueter, R. M., 1970, Effects of ingestion of disproportionate amounts of amino acids, *Physiol. Rev.* **50**:428.

Hartsook, E. W., Hershberger, T. V., and Nee, J. C. M., 1973, Effects of dietary protein content and ratio of fat to carbohydrate calories on energy metabolism and body composition of growing rats, *J. Nutr.* **103**:167.

Henry, Y., 1968, Libre consommation de principes énergétiques et azôté chez le rât et chez le porc selon la nature de la source azôtée, sa concentration dans le régime et le mode de présentation, *Ann. Nutr. Alim.* **22**:121.

Hervey, G. R., 1969, Regulation of energy balance, *Nature (London)* **222**:629.

Hetherington, A. W., and Ranson, S. W., 1940, Hypothalamic lesions and adiposity in the rat, *Anat. Rec.* **78**:149.

Hetherington, A. W., and Ranson, S. W., 1942, The spontaneous activity and food intake of rats with hypothalamic lesions, *Am. J. Physiol.* **136**:609.

Howard, H. W., Monson, W. J., Bauer, C. D., and Block, R. J., 1958, The nutritive value of bread flour proteins as affected by practical supplementation with lactalbumin, non-fat dry milk solids, wheat gluten and lysine, *J. Nutr.* **64**:151.

Jones, J. D., Wolters, R., and Burnett, P. C., 1966, Lysine-arginine-electrolyte relationships in the rat, *J. Nutr.* **89**:171.

Klain, G. J., and Hannon, J. P., 1969, Gluconeogenesis in cold exposed rats, *Fed. Proc.* **28**:965.

Kolata, G. B., 1976, Brain biochemistry: Effects of diet, *Science* **192**:41.

Kraly, F. S., and Blass, E. M., 1976, Increased feeding in rats in a low ambient temperature, in *Hunger: Basic Mechanisms and Clinical Implications* (D. Novin, W. Wyrwicka, and G. Bray, eds.), pp. 77–87, Raven Press, New York.

Krauss, R. M., and Mayer, J., 1965, Influence of protein and amino acids on food intake in the rat, *Am. J. Physiol.* **209**:479.

Lát, J., 1967, Self-selection of dietary components, in *Handbook of Physiology,* Sect. 6, Vol. 1, Chap. 27, pp. 367–386, American Physiological Society, Washington, D.C.

Leibowitz, S. F., 1976, Brain catecholaminergic mechanisms for the control of hunger, in *Hunger: Basic Mechanisms and Clinical Implications* (D. Novin, W. Wyrwicka, and G. Bray, eds.), pp. 1–18, Raven Press, New York.

Lepkovsky, S., 1973, Newer concepts in the regulation of food intake, *Am. J. Clin. Nutr.* **26**:271.

Leshner, A. I., Collier, G. H., and Squibb, R. L., 1971, Dietary self-selection at cold temperatures, *Physiol. Behav.* **6**:1.

Leung, P. M. B., and Rogers, Q. R., 1969, Food intake: Regulation by plasma amino acid pattern, *Life Sci.* **8**:1.

Leung, P. M. B., and Rogers, Q. R., 1971, Importance of prepyriform cortex in food-intake response of rats to amino acids, *Am. J. Physiol.* **221**:929.

Lutz, J., Tews, J. K., and Harper, A. E., 1975, Simulated amino acid imbalance and histidine transport in brain slices, *Am. J. Physiol.* **229**:229.

Mayer, J., 1964, Regulation of food intake, in *Nutrition: A Comprehensive Treatise* (G. H. Beaton and E. W. McHenry, eds.), Vol. 1, pp. 1–40, Academic Press, New York.

Mellinkoff, S., 1957, Digestive system, *Ann. Rev. Physiol.* **19**:175.

Mellinkoff, S. M., Frankland, M., Boyle, D., and Greipel, M., 1956, Relationship between serum amino acid concentration and fluctuations in appetite, *J. Appl. Physiol.* **8**:535.

Menaker, L., and Navia, J. M., 1973, Appetite regulation in the rat under various physiological conditions: The role of dietary protein and calories, *J. Nutr.* **103**:347.

Meyer, J. H., 1958, Interactions of dietary fiber and protein on food intake and body composition of growing rats, *Am. J. Physiol.* **193**:488.

Meyer, J. H., and Hargus, W. A., 1959, Factors influencing food intake of rats fed low-protein rations, *Am. J. Physiol.* **197**:1350.

Mogenson, G. J., 1976, Neural mechanisms of hunger: Current status and future prospects, in *Hunger: Basic Mechanisms and Clinical Implications* (D. Novin, W. Wyrwicka, and G. Bray, eds.), pp. 473–485, Raven Press, New York.

Morgane, P. J., 1969, The function of the limbic and rhinic forebrain-limbic midbrain systems and reticular formation in the regulation of food and water intake, *Ann. NY Acad. Sci.* **157**:806.

Morrison, A. B., 1964, Caloric intake and nitrogen utilization, *Fed. Proc.* **23**:1083.

Munro, H. N., Fernstrom, J. D., and Wurtman, R. J., 1975, Insulin, plasma amino acid imbalance and hepatic coma, *Lancet* **1**:722.

Musten, B., Peace, D., and Anderson, G. H., 1974, Food intake regulation in the weanling rat: Self-selection of protein and energy, *J. Nutr.* **104**:563.

Muto, S., and Miyahara, C., 1972, Eating behaviour of young rats: Experiment on selective feeding on diet and sugar solution, *Br. J. Nutr.* **28**:327.

NAS–NRC (USA), 1972, Nutrient Requirements of Domestic Animals, No. 10, Nutrient Requirements of Laboratory Animals, National Academy of Sciences, Washington, D.C.

Osborne, T. B., and Mendel, L. B., 1916, The amino-acid minimum for maintenance and growth, as exemplified by further experiments with lysine and tryptophane, *J. Biol. Chem.* **25**:1.

Osborne, T. B., and Mendel, L. B., 1918, The choice between adequate and inadequate diets, as made by rats, *J. Biol. Chem.* **35**:19.

Panksepp, J., 1973, Hypothalamic radioactivity after intragastric glucose-^{14}C in rats, *Am. J. Physiol.* **223**:396.

Panksepp, J., 1974, Hypothalamic regulation of energy balance and feeding behaviour, *Fed. Proc.* **33**:1150.

Panksepp, J., and Booth, D. A., 1971, Decreased feeding after injection of amino acids into the hypothalamus, *Nature (London)* **233**:341.

Peng, Y., Tews, J. K., and Harper, A. E., 1972, Amino acid imbalance, protein intake and changes in rat brain and plasma amino acids, *Am. J. Physiol.* **222**:314.

Peng, Y., Gubin, J., Harper, A. E., Vavich, M. G., and Kemmerer, A. R., 1973, Food intake regulation: Amino acid toxicity and changes in rat brain and plasma amino acids, *J. Nutr.* **103**:608.

Peng, Y., Meliza, L. L., Vavich, M. G., and Kemmerer, A. R., 1975, Effects of amino acid imbalance and protein content of diets on food intake and preference of young, adult, and diabetic rats, *J. Nutr.* **105**:1395.

Peterson, A. D., and Baumgardt, B. R., 1971a, Food and energy intake of rats fed diets varying in energy concentration and density, *J. Nutr.* **101**:1057.

Peterson, A. D., and Baumgardt, B. R., 1971b, Influence of level of energy demand on the ability of rats to compensate for diet dilution, *J. Nutr.* **101**:1069.

Pol, G., and den Hartog, C., 1966, The dependence on protein quality of the protein to calorie ratio in a freely selected diet and the usefulness of giving protein and calories separately in protein evaluation experiments, *Br. J. Nutr.* **20**:649.

Rogers, Q. R., and Leung, P. M. B., 1973, The influence of amino acids on the neuroregulation of food intake, *Fed. Proc.* **32**:1709.

Rozin, P., 1968, Are carbohydrate and protein intakes separately regulated?, *J. Comp. Physiol. Psychol.* **65**:23.

Scharrer, E., Baile, C. A., and Mayer, J., 1970, Effect of amino acids and protein on food intake of hyperphagic and recovered aphagic rats, *Am. J. Physiol.* **218**:400.

Schenker, S., Breen, K. J., and Hoyumpa, A., 1974, Hepatic encephalopathy: Current status, *Gastroenterology* **66**:121.

Sellers, E. A., You, R. W., and Moffat, N. M., 1954, Regulation of food consumption by caloric value of the ration in rats exposed to cold, *Am. J. Physiol.* **177**:367.

Sibbald, I. R., Berg, R. T., and Bowland, J. P., 1956, Digestible energy in relation to food intake and nitrogen retention in the weanling rat, *J. Nutr.* **59**:385.

Sibbald, I. R., Bowland, J. P., Robblee, A. R., and Berg, R. T., 1957, Apparent digestible energy and nitrogen in the food of the weanling rat. Influence on food consumption, nitrogen retention and carcass composition, *J. Nutr.* **61**:71.

Vaughan, D. A., Vaughan, L. N., and Stull, H. D., 1966, Dietary modification of cold induced metabolic effects, *Metabolism* **15**:781.

Wurtman, R. J., Larin, F., Mostafapour, S., and Fernstrom, J. D., 1974, Brain catechol synthesis: Control by brain tyrosine concentration, *Science* **185**:183.

Yoshida, A., Harper, A. E., and Elvehjem, C. A., 1957, Effects of protein per calorie ratio and dietary level of fat on calorie and protein utilization, *J. Nutr.* **63**:555.

Chapter 7

Metabolic Disorders of Copper Metabolism

Gary W. Evans

1. Introduction

Copper, atomic number 29, is an extremely versatile element both outside and inside living cells. The malleability, ductility, and esthetic properties of copper have promoted its use in pottery, ornaments, and currency for centuries. During the more recent decades, the electrical conductivity of copper has had an important role in the advancement of electrical technology.

Whereas elemental copper adds to the comfort and enjoyment of life, ionic copper is absolutely essential for the maintenance of life. Probably because of copper's relative abundance in the biosphere and the element's ease of undergoing oxidation reduction, organisms have developed enzyme systems that utilize copper in their catalytic action. Copper-containing enzymes have been identified in nearly every type of organism, from the simplest to the most complex, and the chemical reactions catalyzed by these enzymes cover a broad spectrum of metabolic processes (Frieden *et al.*, 1965). Because copper is an integral component of a variety of enzymes, a deficiency of this element or an abnormality in the production of a copper enzyme may result in widespread deleterious effects, and often death. In addition, because copper is a very reactive element, an excess of this ion in living cells results in marked changes in metabolism.

1.1. Copper Homeostasis

A discussion of the metabolic disorders associated with copper deficiency and copper toxicity is probably best prefaced by a short, and hopefully up-to-

Gary W. Evans • Agricultural Research Service, Human Nutrition Laboratory, United States Department of Agriculture, Grand Forks, North Dakota 58201.

date, review of our knowledge of the homeostatic mechanisms involved in the absorption, transport, storage, and excretion of copper in organisms subject to metabolic disorders of copper metabolism. Figure 1 is a schematic diagram of the molecular pathways traversed by copper ions in mammalian systems. In the intestinal lumen, copper binds to ligands, probably amino acids, before it enters the intestinal epithelial cell. The passage of copper through the intestinal cell to the blood is apparently regulated by a low-molecular-weight copper-binding protein, the synthesis of which is induced by copper (Evans and LeBlanc, 1976). After passing through the intestinal epithelial cells, copper is transported through the portal blood as a histidine–copper–albumin complex (Lau and Sarkar, 1971) to the liver.

The liver is the key organ in the metabolism of copper. In the hepatic cells, copper is stored, incorporated into ceruloplasmin, and excreted from the body through the bile (Hazelrig *et al.*, 1966). The hepatic copper-storage compartment apparently consists of a low-molecular-weight copper-binding protein (Evans *et al.*, 1975; Riordan and Gower, 1975; Winge *et al.*, 1975), and the synthesis of

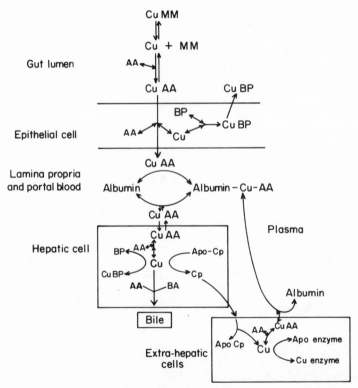

Fig. 1. Schematic illustration of copper homeostasis in mammalian systems. Abbreviations: MM, macromolecule; AA, amino acid; BP, copper-binding protein; BA, bile acid; Cp, ceruloplasmin.

this protein is regulated by the copper level of the cell (Premakumar *et al.*, 1975). Copper apparently is excreted into the bile in the form of amino acid (Evans and Cornatzer, 1971) and biliary acid complexes (Lewis, 1973) that combine with macromolecules, rendering the copper unavailable for reabsorption (Mistilis and Farrer, 1968).

The incorporation of copper into ceruloplasmin is a vital function of the liver because copper is transported to the extrahepatic tissues in the form of ceruloplasmin. Several investigations have demonstrated that the appearance of labeled copper in extrahepatic tissues coincides with the disappearance of the label from ceruloplasmin in the plasma (Owen, 1965, 1971; Marceau and Aspin, 1972). Moreover, radioactive copper from ceruloplasmin is incorporated into hepatic and brain cytochrome c oxidase (Marceau and Aspin, 1973), and ceruloplasmin has been shown to be extremely effective in restoring cytochrome oxidase activity in tissues from copper-deficient animals (Hsieh and Frieden, 1975). The copper from ceruloplasmin, which is probably taken up by pinocytosis, is apparently more readily available to certain compartments in extrahepatic tissues than copper from albumin in spite of the fact that albumin copper is easily dissociated.

The pathways outlined above indicate that a variety of ligands are involved in the transport of copper from the intestine to the organs of the body. Once within the cells of the various organs, copper is incorporated into the enzymes (Table I) that require the element for catalytic activity. Thus, metabolic disorders of copper metabolism can be classified into three categories: (1) disorders that result from a lack of dietary copper or unavailability of dietary copper caused by a high level of copper antagonists in the food; (2) disorders that result from inborn errors of metabolism that affect the ligands involved in the absorption, transport, storage, or excretion of copper; and (3) inborn errors of metabolism that affect a specific copper-dependent enzyme. The discussion in the next section is organized according to these three categories.

Table I. Copper Enzymes in Mammalian Tissues

Enzyme	Reference
Benzylamine oxidase	Buffoni and Blaschko (1964)
Ceruloplasmin (ferroxidase)	Holmberg and Laurell (1948)
	Osaki *et al.* (1966)
Cytochrome c oxidase	Peisach *et al.* (1966)
Diamine oxidase, histaminase	Mondovi *et al.* (1967)
Dopamine β-hydroxylase	Friedman and Kaufman (1965)
Lysyl oxidase	Siegel *et al.* (1970)
Spermine oxidase	Yamada and Yasunobu (1962)
Superoxide dismutase	McCord and Fridovich (1969)
Tryptophan-2,3-dioxygenase	Brady *et al.* (1972)
Tyrosinase	Pomerantz (1963)
Uricase	Mahler *et al.* (1956)

2.   Disorders Resulting from a Lack of or Unavailability of Dietary Copper

Copper deficiency occurs in animals restricted to pastures on copper-deficient soil (Underwood, 1971) and in humans restricted to low-copper milk diets (Cordano *et al.*, 1964) or total parenteral hyperalimentation (Karpel and Peden, 1972; Dunlap *et al.*, 1974). In addition, the manifestations of dietary copper deficiency occur in animals that consume diets that contain copper antagonists, such as sulfide ion, molybdenum, zinc, silver, mercury, and cadmium (Evans, 1973). The extent and types of disorders associated with copper deficiency depend on species, age, sex, and severity of deficiency, but the disorders generally include anemia; achromotrichia; lesions of the cardiovascular system, lung, skeleton, and central nervous system; changes in the growth and appearance of hair, fur, or wool; impaired growth; and reproductive failure.

2.1.   Anemia

Copper has the distinction of being the second trace element (iron was the first) discovered to be essential in mammals; this discovery emerged from experiments in rats suffering from milk anemia. In 1928, Hart *et al.* (1928) observed that an anemia developed in rats fed a diet of milk that could not be ameliorated by iron supplements alone. When both copper and iron were added to the milk diet, the hemoglobin levels of the experimental animals were restored to normal. These observations provided the first evidence that copper is an essential nutrient, and the experiments that were confirmed and extended in other species forged the first link between iron and copper metabolism.

Following the announcement of Hart *et al.* (1928), several hypotheses were proposed to explain the role of copper in iron metabolism, but very little convincing and reproducible evidence appeared until the late 1960s. Curzon and O'Reilly (1960) showed that the cupric ions of ceruloplasmin could catalyze the oxidation of ferrous ions. Later, Osaki *et al.* (1966) first proposed that ceruloplasmin promotes hematopoiesis by catalyzing the formation of ferritransferrin.

During the last decade, ample evidence, both *in vivo* and *in vitro*, has been presented to confirm the hypothesis of Osaki *et al.* (1966). Ragan *et al.* (1969) examined the effects of ceruloplasmin on iron mobilization in copper-deficient pigs supplemented with iron by intramuscular injection. The pigs were fed an iron- and copper-deficient milk diet until a severe copper deficiency developed, after which they were injected intravenously with pig ceruloplasmin, $CuSO_4$, or copper-deficient pig plasma. Ceruloplasmin injections resulted in a significant and rapid elevation of plasma iron, but $CuSO_4$ and copper-deficient pig plasma produced little or no change in plasma iron levels. In other studies with copper-deficient pigs, Roeser *et al.* (1970) demonstrated that a marked decrease in serum ceruloplasmin precedes a decrease in serum iron with an accumulation of iron in the liver. These investigators also observed that the decreased serum iron levels could be restored to normal by injecting ceruloplasmin.

Evidence for the effective mobilization of iron by ceruloplasmin was observed *in vitro* by Osaki *et al.* (1971). Livers were removed from dogs and copper-deficient pigs after which the livers were perfused with solutions that contained ceruloplasmin, apotransferrin, HCO_3, $CuSO_4$, glucose, fructose, citrate, or copper albumin. Ceruloplasmin was the only compound tested that had any appreciable effect on the mobilization of iron from the liver to the perfusate and the concentration of ceruloplasmin required to produce an effect was less than 1% of the normal ceruloplasmin concentration in plasma.

The experiments discussed above, as well as others of a similar nature (Frieden and Hsieh, 1976), indicate that ceruloplasmin is the long sought after link between copper and iron metabolism. For normal hemoglobin production to proceed, iron must be mobilized from reticuloendothelial cells to bone marrow cells. Transferrin, which binds ferric iron, is the only known protein that supplies iron to the marrow cells. Iron in cells is bound to ferritin, but the iron is released from ferritin in the ferrous state, which necessitates oxidation of the iron prior to its incorporation into apotransferrin. Because the uncatalyzed oxidation of ferrous iron is not sufficient to maintain normal hemoglobin production (Osaki *et al.*, 1966), ceruloplasmin, with copper incorporated in its structure, is essential to maintain hematopoiesis. Williams *et al.* (1974) have suggested that ceruloplasmin functions in iron mobilization as follows: (1) the ferrous iron released from ferritin binds to specific sites on reticuloendothelial cell membranes; (2) ceruloplasmin interacts with the iron-binding sites to form a ceruloplasmin–ferrous ion intermediate; and (3) iron is oxidized and transferred to apotransferrin by a specific ligand-exchange reaction (Fig. 2).

In summary, normal hemoglobin production depends on the ferroxidase activity of ceruloplasmin. Since the ferroxidase activity of ceruloplasmin is dependent on the presence of copper in the molecule, a deficiency of copper results in a marked decrease in ceruloplasmin ferroxidase activity, which leads to anemia.

2.2. Cardiovascular Defects

For several years after the announcement of Hart *et al.* (1928), the only known function for copper was its role in hematopoiesis. However, during the early 1960s several investigators began reporting experimental observations that indicated that copper had physiological functions more far-reaching that preventing anemia.

In 1961, Hill and Matrone (1961) observed a high mortality rate in copper-deficient chicks despite the fact that the anemia was not severe. Later, O'Dell *et al.* (1961) reported that the mortality observed in copper-deficient chicks is caused by rupture of the major blood vessels. O'Dell *et al.* (1961) presented

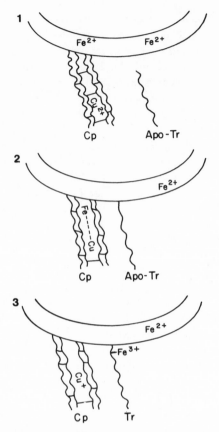

Fig. 2. Schematic illustration of ceruloplasmin (ferroxidase) function in iron mobilization. Abbreviations: Cp, ceruloplasmin; Tr, transferrin.

histological evidence that indicated that the elastic tissue in the aorta is abnormal in copper-deficient chicks.

Abnormal elastic membranes and death from aortic rupture are not limited to copper-deficient chicks. Coulson and Carnes (1963) and Shields *et al.* (1962) observed high mortality resulting from aortic rupture in pigs fed copper-deficient diets. These investigators also observed histological defects in the aortic elastic tissue from copper-deficient pigs.

The derangements observed in elastic tissue of blood vessels from copper-deficient animals prompted several investigators to examine the effect of copper deficiency on the synthesis of aortic elastin, the major component of elastic tissue. Weissman *et al.* (1963) and Kimball *et al.* (1964) isolated elastin from pig aorta and observed a marked decrease in the elastin content of aorta from copper-deficient pigs. Starcher *et al.* (1964) found a decreased quantity of elastin in the aorta from copper-deficient chicks, and these investigators extended their experiments to show that the effects of copper deficiency on aortic elastin content could be reversed by adding copper to the diet.

The experiments of Starcher *et al.* (1964) with aorta from copper-deficient chicks proved quite valuable in elucidating the function of copper in the production of elastin. First, these authors demonstrated that copper is involved in the production of elastin, rather than in the catabolism of the protein. Second, they observed a marked increase in the quantity of lysine in elastin from copper-deficient chick aorta. This observation drew attention to the work of Partridge (1966), who had earlier proposed a structure for the cross-linkage groups in elastin.

Partridge and co-workers (1964, 1966) proposed that the structural integrity of elastin results from the formation of cross-linking substances called desmosine and isodesmosine. This group also obtained experimental evidence which indicated that the cross-linking substances desmosine and isodesmosine in elastin are formed by the condensation of lysine residues in elastin precursors. Thus, the results of Starcher *et al.* (1964) could be explained by assuming that the increased lysine content of elastin from copper-deficient chick aorta resulted from a decreased conversion of lysine to desmosine. This assumption was confirmed by the experiments of Miller *et al.* (1965), who demonstrated that the desmosine content of copper-deficient tissue decreases as the lysine content increases.

The conversion of lysine to desmosine requires oxidative deamination of the ϵ-amino group of lysine residues in preelastin chains. A reaction of this complexity is not likely to occur spontaneously, and this fact sent several investigators in search of a copper-dependent enzyme capable of catalyzing the cross-linking of elastin precursors. Copper-containing amine oxidases capable of deaminating lysine had been isolated previously (Yamada and Yasunobu, 1962; Buffoni and Blaschko, 1964), but first evidence implicating an amine oxidase in aortic elastin cross-linking was presented by Hill *et al.* (1967). These investigators examined amine oxidase activity in chick aorta and found a marked decrease in enzyme activity in aorta from copper-deficient chicks. In addition, Hill *et al.* (1967) maintained aortas in tissue culture and demonstrated that the lysine to desmosine ratio was increased in copper-deficient chick aorta. When a crude preparation of amine oxidase was added to the culture medium, the ratio of lysine to desmosine was decreased in aortas from both copper-deficient and copper-supplemented chicks. These experiments were a strong indication that a copper-dependent amine oxidase is involved in the elastin cross-linking process.

An amine oxidase that catalyzes the oxidation of peptidyllysine to form α-aminoadipic-δ-semialdehyde (allysine) has been isolated from bovine aorta (Rucker *et al.*, 1970) and chick aorta (Harris *et al.*, 1974). This enzyme has been given the name lysyl oxidase to differentiate it from other amine oxidases. Lysyl oxidase contains copper and apparently requires pyridoxal phosphate as a cofactor (Chou *et al.*, 1970).

2.3. Bone Defects

The enzymatic activity of lysyl oxidase is not limited to the aorta. Bone abnormalities have been observed in several species that have been fed copper-deficient diets (Underwood, 1971). Skeletal tissue from copper-deficient animals is characterized by a lack of bone deposition in the cartilage matrix, and the bones from these animals contain more soluble collagen than bones from control animals. The structural integrity of collagen, like elastin, depends upon cross-linking between collagen precursors. The cross-linking process requires lysyl oxidase, which converts lysyl residues into allysine (Pinnell and Martin, 1968; Siegel *et al.*, 1970). Recently, O'Dell *et al.* (1976a) presented experimental results that indicate that lysyl oxidase activity is important for normal development of lung tissue.

To summarize, the defects in connective tissue development that accompany copper-deficiency result from a decrease in the activity of the copper-dependent enzyme, lysyl oxidase. Lysyl oxidase catalyzes the formation of aldehydes (allysine) from peptidyl lysine in pre-elastin and pre-collagen. These aldehyde residues condense to form the covalent cross-links in collagen and elastin (Fig. 3). In copper-deficient tissues, production of the intermediate residues, allysine, is impaired; cross-linking is prevented and the result is a fragility and loss of strength in elastin and collagen.

2.4. Disorders of the Central Nervous System

A debilitating disease has been observed in lambs grazing on copper-deficient pastures in Western Australia and other countries (Underwood, 1971). Lambs affected by this disease walk with a stiff and staggering gait, and the hind quarters sway in incoordinated movement. Many lambs are paralyzed or ataxic at birth and soon die. The condition, known as enzootic neonatal ataxia, can be prevented but not reversed by copper supplementation.

For several years neonatal ataxia was thought to be restricted to sheep but this belief is no longer accepted since the condition has been observed experimentally and under field conditions in several species (Underwood, 1971). Pathological lesions associated with neonatal ataxia vary among species, but lack of myelination of nerve tracts always accompanies the onset of the gross symptoms (O'Dell *et al.*, 1976b).

Because myelin contains a high content of phospholipid, the derangement of myelin accompanying neonatal ataxia was once thought to result from decreased activity of the enzyme α-glycerophosphate acyltransferase (Gallagher *et al.*, 1956). This enzyme catalyzes the attachment of fatty acids to α-glycerophosphate. However, subsequent investigations (Gallagher and Reeve, 1971; DiPaolo and Newberne, 1971) have demonstrated that α-glycerophosphate acyltransferase is not a copper-dependent enzyme.

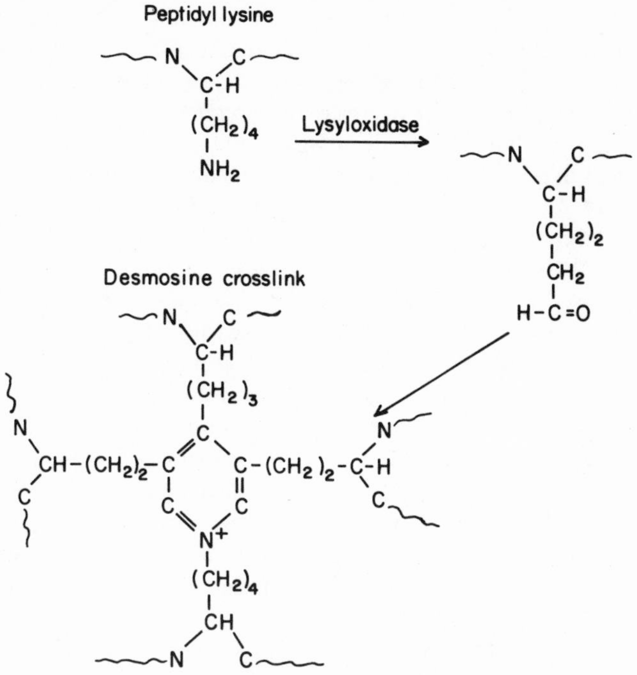

Fig. 3. Schematic illustration of the function of lysyl oxidase in the formation of desmosine crosslinks in elastin.

Several investigators (Mills and Williams, 1962; Prohaska and Wells, 1974) have observed a marked decrease in the copper content and cytochrome c oxidase activity of neural tissue from copper-deficient animals. Cytochrome c oxidase is a copper-dependent enzyme and the terminal oxidase in the respiratory chain of mitochondria. These facts have led to speculation that the primary lesion in neonatal ataxia is the depression of cytochrome c oxidase, which leads to a diminution of aerobic metabolism and a subsequent decrease in phospholipid and myelin synthesis. Although this hypothesis seems tenable, definitive evidence linking the activity of neural cytochrome c oxidase to the production of myelin is lacking. Moreover, yet to be explained are the metabolic effects of the decreased activity in superoxide dismutase and dopamine-β-hydroxylase, two copper-dependent enzymes that are decreased in brain tissue from copper-deficient rats (Prohaska and Wells, 1974).

In view of our lack of knowledge regarding the functions of copper in the central nervous system to attribute the nervous disorder of copper-deficient animals to a derangement of myelin is still premature. O'Dell et al. (1976b) have examined the level of neurotransmitters in the brains of ataxic and nonataxic lambs and found a decreased level of both dopamine and norepinephrine in the brainstem from ataxic animals. Prohaska and Wells (1974) have also detected a significant decrease in the norepinephrine concentration of brains from copper-deficient neonatal rats. The decrease in norepinephrine in copper-deficient tissues can be attributed to a decrease in the activity of the copper enzyme dopamine-β-hydroxylase. The decrease in dopamine concentration is not so easily explained because neither of the enzymes, tyrosine hydroxylase and dopadecarboxylase, involved in dopamine synthesis is known to be copper-dependent. Whether or not the decreased level of catecholamines in brain tissue from copper-deficient animals leads to nervous disorders remains to be determined. However, some nervous disorders, such as Parkinson's disease, are associated with decreased levels of catecholamines; O'Dell *et al.* (1976b) have pointed out several similarities between Parkinson's disease and neonatal ataxia.

This section may be summarized by pointing out that copper deficiency affects the central nervous system, but the primary lesion that produces the symptoms is unknown. A nervous disorder develops in copper-deficient animals that affects locomotor activity. The ataxic condition is accompanied by hypomyelination of the nerve tract, decreased activity of both neural cytochrome oxidase and superoxide dismutase, and decreased levels of dopamine and norepinephrine in the brain.

2.5. Achromotrichia

Lack of hair and wool pigmentation resulting from copper deficiency has been observed in several species (Underwood, 1971), but is particularly noticeable in black-wooled sheep and rabbits. When black-wooled sheep consume a copper-deficient diet, the wool develops an odd-looking appearance with alternating bands of pigmented and unpigmented fibers. In the rabbit, achromotrichia is a more sensitive index of copper deficiency than is anemia (Smith and Ellis, 1947).

Achromotrichia in copper-deficient animals is probably caused by decreased activity of the copper-dependent enzyme tyrosinase. Tyrosinase catalyzes the conversion of tyrosine to 3,4-dihydroxyphenylalanine (dopa) as well as the oxidation of dopa to dopaquinone. Dopaquinone is then converted through a series of reactions to melanin pigment.

2.6. Steely Wool and Hair

The wool of copper-deficient sheep grows poorly and lacks the crimp characteristic of wool from normal, healthy animals (Underwood, 1971). In addition, the hair of human infants affected with Menkes' syndrome, a genetic disease resulting in malabsorption of copper, has an abnormal, tortuous appearance (Danks *et al.*, 1972b). The wool from copper-deficient sheep, as well as the hair from Menkes' patients, contain more free sulfhydryl groups than do normal wool and hair (Burley, 1954; Danks *et al.*, 1972b). The presence of these free sulfhydryl groups indicates a reduction in the cross-linking of keratin in the wool and hair fibers, and this accounts for the stringy or steely appearance. The reduction of disulfide bond formation in wool and hair that accompanies copper-deficiency suggests that a copper-dependent enzyme may be required in the process, but no enzyme of this nature has been described to date.

2.7. Reproductive Failure and Hypercholesterolemia

Reproductive failure has been observed in both rats (Hall and Howell, 1969) and chickens (Savage, 1968) as a result of copper deficiency. Simpson *et al.* (1967) have presented evidence that suggests that reproductive failure is caused by defective formation of connective tissue and red cells.

Recently, Allen and Klevay (1976) demonstrated that copper-deficient rats have an increased concentration of plasma cholesterol. These observations suggest that a copper enzyme may be involved in the regulation of cholesterol synthesis or breakdown.

3. Metabolic Disorders Arising from Inborn Errors of Metabolism within the Copper Homeostatic System

3.1. Menkes' Steely Hair Syndrome (Trichopoliodystrophy)

The preceding sections document the widespread role of copper in mammalian metabolism. From these descriptions, one can readily see that an inborn error of metabolism that results in malabsoprtion of copper would be extremely deleterious. Such an inborn error has been discovered in human infants.

In 1962, Menkes and his colleagues (1962) wrote a detailed description of a degenerative disease of the central nervous system. The condition was inherited as a sex-linked recessive trait and the affected males were both physically and mentally retarded and had peculiar white, stubbly hair. Later, O'Brien and Sampson (1966), who first coined the name "Kinky Hair Disease," examined the fatty acid levels in brain tissue from Menkes' patients and found a decreased quantity of docosahexenoic acid, a highly unsaturated fatty acid. At that time, the significance of the peroxidation of lipids was not understood.

In the early 1970s, David Danks and his colleagues in Australia began examining infants suffering from Menkes' kinky hair disease. Acting on the suggestion of Dr. J. M. Gillespie of the Division of Protein Chemistry (CSIRO, in Melbourne), Danks and associates (1972a) examined the copper status of these infants, and a new era in copper metabolism began. In a series of publications Danks *et al.* (1972a,b, 1973b) demonstrated that infants with Menkes' disease absorb copper poorly and have a decreased concentration of copper in the plasma and liver but an elevated concentration of copper in the intestinal epithelium. These observations have been verified by Walker-Smith *et al.* (1973) and Dekaban *et al.* (1975). In addition, Goka *et al.* (1976) have demonstrated that cultured skin fibroblasts from patients with Menkes' disease contain an elevated concentration of copper.

With the exception of anemia, all the pathological abnormalities associated with copper deficiency in experimental animals have been observed in patients with Menkes' syndrome (Danks, 1975). In fact, the similarity between wool from copper-deficient sheep and hair from Menkes' patients prompted Danks *et al.* (1973a) to suggest that the term "steely hair" be used to describe the hair on these affected infants.

3.1.1. Therapy

As oral copper is poorly absorbed by infants affected with Menkes' syndrome, parenteral copper administration has been used as a therapeutic measure, but the results have been disappointing (Danks, 1975). Copper therapy restores ceruloplasmin and hepatic copper levels to normal, but the symptoms of the disease usually do not improve. This may be because most of the cases reported to date were at least three months old before they were identified and treatment was begun. At this late stage of development, the tissue damage caused by the malabsorption syndrome may be so extensive that it is irreversible. Alternatively, the mutant gene involved in Menkes' syndrome may alter the accessibility of serum copper to tissue cells (Danks, 1975).

Until research establishes exactly how copper is taken up by tissue cells, the methods now being used for copper therapy in Menkes' infants will probably be ineffective. When methods are developed to circumvent the metabolic block of copper uptake by the cells, therapeutic programs will have to be carefully analyzed and designed because Menkes' syndrome is a complex anomaly. For example, some patients exhibit symptoms at birth (Danks *et al.*, 1972a), whereas in others, clinical manifestations appear months after birth (Walker-Smith *et al.*, 1973). These observations suggest the possibility that placental copper transport may be affected, but in varying degrees, by the mutant gene that causes Menkes' disease. If the placenta is affected by the mutant gene, heterozygous carrier females will have to be identified and therapy begun to ensure adequate transport of copper to the developing fetus.

3.1.2. Genetic Heterogeneity

The apparent complexity of the steely-hair syndrome may be partly attributable to genetic heterogeneity. Whereas in most affected infants examined to date, hepatic copper uptake is normal during infusion therapy, in some patients, hepatic copper uptake is abnormal. Garnica and Fletcher (1975) have observed an abnormally high rate of urinary copper excretion in one patient both during and after copper infusion therapy. As excess copper is normally taken up by the liver and excreted through the bile, this observation suggests that hepatic copper uptake may be affected in some infants with steely-hair syndrome. In addition, Horn *et al.* (1975) examined copper distribution in a male fetus suspected of this inborn error and discovered that the liver was the only tissue that contained less copper than did specimens from control subjects. The copper concentrations of the kidneys, spleen, pancreas, and placenta of the diseased fetus were significantly higher than those from controls. Therefore, the gene mutation that occurs in the steely-hair syndrome may affect the copper transport system of organs other than the intestine.

3.1.3. Animal Model for Menkes' Syndrome

Research on the steely-hair syndrome has been greatly facilitated by the observations of Hunt (1974) who discovered that the sex-linked inherited disorder in mottled mutant (Mo^{br}) mice is nearly identical to the steely-hair syndrome in humans. Hunt found a decreased concentration of copper in livers and brains from Mo^{br} male mice, but the copper concentration of the intestines from the mutant mice was much higher than normal. Ceruloplasmin activity was significantly decreased in the Mo^{br} mice.

Experiments in our laboratory have confirmed the observations of Hunt and also demonstrate that hepatic copper uptake is impaired in Mo^{br} mice. Table II shows the results we obtained when lactating dams were injected with ^{64}Cu and the pups were analyzed 48 h later. In Mo^{br} mice suckling labeled dams, 72% of

Table II. Distribution of ^{64}Cu in Mo^{br} Male Pups, a Heterozygous Female Pup, and Normal Pups Suckling Dams That Had Been Injected with ^{64}Cu

Pups[b]	% Total ^{64}Cu[a]			
	Intestine	Liver	Kidney	Carcass
Male hemizygote (6)	72.2 ± 6.7[f]	2.6 ± 0.5[f]	1.8 ± 0.7	23.4 ± 4.3
Female heterozygote (1)	33.1[f]	13.7[f]	3.5	49.7
Normal homozygote[c] (8)	18.1 ± 1.2	45.3 ± 1.3	1.0 ± 0.05	35.6 ± 1.7
Normal homozygote[d] (5)	22.8 ± 1.1	49.2 ± 2.4	1.5 ± 0.05	26.5 ± 2.9
Normal homozygote[e] (6)	17.4 ± 1.1	50.6 ± 2.1	4.0 ± 0.1	27.9 ± 2.4

[a]Each value represents mean ± SEM. The values in parentheses refer to number of pups analyzed.
[b]The pups were 5–6 days old when analyzed.
[c]Littermates of the male hemizygotes and female heterozygote.
[d]Pups from a heterozygous dam.
[e]Pups from a homozygous dam.
[f]Differences between the value of the mutant mice and the mean value of the normal control mice were significant at the 1% probability level.

the total radioactivity in the pups was in the intestinal cells, whereas the liver contained only 3% of the radioactivity. In homozygous littermates less than 25% of the radioactivity was found in the intestine and the liver contained 50% of the radioactivity. These results demonstrate that copper accumulates in the intestine of Mo^{br} mice, and the results suggest that the copper absorbed from the intestine of these mutants is diverted to extrahepatic tissues.

During our experimentation with Mo^{br} mice, we also examined copper metabolism in the Mo^{br} heterozygous females. The kidney copper concentration of heterozygous females was significantly higher than that of homozygous females, but there was no difference in liver and brain copper concentrations. In addition, we have observed a marked decrease in copper absorption in heterozygous females (Fig. 4). The observations discussed earlier demonstrate that Mo^{br} mice will have an important role in elucidating the biochemical defects that arise from the lethal mutant gene that is present in infants affected with Menkes' steely hair disease.

3.2.   Wilson's Disease (Hepatolenticular Degeneration)

The inborn error of copper metabolism known as Wilson's disease was first described in 1912 (Wilson, 1912). Since that time, volumes of literature have appeared describing various aspects of the disorder. To cover the subject of

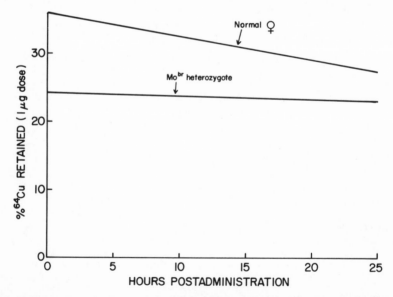

Fig. 4. Absorption and retention of oral ^{64}Cu in normal and heterozygous Mo^{br} female mice. All mice were intubated with 1 μg ^{64}Cu, after which radioactivity was measured hourly in a whole-body counter. Mice in both groups attained a constant rate of isotope excretion 3–5 hr after oral administration. The data depicted are the mean extrapolated values obtained from 10 mice in each group.

Wilson's disease adequately would require a chapter in itself. Therefore, the comments of this section will be brief and directed toward informing the uninformed regarding the pathogenesis, pathology, and treatment of Wilson's disease. For more comprehensive discussions of the disease, the reader is referred to several excellent reviews (Sass-Kortsak, 1965; Walshe, 1966; Bearn, 1972; Scheinberg and Sternlieb, 1975).

The mutant gene in individuals with Wilson's disease produces a yet-undiscovered biochemical defect that results in excess retention of hepatic copper. Most affected patients have a decreased concentration of plasma ceruloplasmin and the excretion of copper into the bile is impaired. Thus, the defect apparently involves the mechanisms that regulate the passage of copper into the ceruloplasmin-synthesizing and biliary-excretion pathways.

After several years of accumulating copper, the capacity of the liver is eventually exceeded, and copper begins to diffuse into the plasma and fluids of extrahepatic tissues. At this point in the progression of the disease, massive necrosis can be detected in the hepatic parenchymal cells, and the first clinical symptoms begin to appear in the form of liver dysfunction.

If the untreated patient survives the liver disease, pathological changes resulting from excess copper eventually appear in the central nervous system, kidneys, and cornea. At this stage, neurological and psychiatric symptoms may appear, and renal function is impaired. If untreated, the manifestations of copper toxicity result in death at an early age.

Fortunately, the manifestations of Wilson's disease can be arrested and prevented by drug therapy. The drug now being used most successfully is D-penicillamine (β,β-dimethylcysteine), which produces a marked increase in the urinary excretion of copper. Treatment with penicillamine results in a dramatic recovery of the affected patients and continuous therapy expands the lifespan of individuals who possess this mutant gene that alters copper homeostasis in the liver.

4. Metabolic Disorders Resulting from an Absence of a Specific Copper Enzyme

4.1. Albinism

The only known inborn error of metabolism that can be linked to a copper-dependent enzyme is albinism. A genetically acquired absence of the enzyme tyrosinase results in albinism, which is characterized by the complete lack of pigmentation in the eyes and integument. Tyrosinase is essential in the pigmentation process because this enzyme catalyzes the first two steps in the synthesis of melanin pigment from tyrosine. Mammals that lack pigment do not have the capability to filter the sun's ultraviolet rays, which penetrate the integument and cause molecular havoc.

4.2.   Unidentified Disorders and "Enzymes Looking for a Disease"

Although only one genetic disease has thus far been linked to a copper enzyme, this is not to say that others do not exist. In view of the number of copper enzymes present in mammalian systems, there is a good possibility that some of the diseases of unknown etiology will some day be linked to a copper enzyme.

Some of the copper enzymes that have not been associated with a specific pathological defect include cytochrome c oxidase, the terminal enzyme in the mitochondrial electron transport system; superoxide dismutase, the enzyme that catalyzes the dismutation of toxic superoxide anions; dopamine-β-hydroxylase, the enzyme that catalyzes the conversion of dopamine to norepinephrine; and uricase, the enzyme that catalyzes the oxidation of uric acid. Careful analysis of the pathological defects that accompany copper deficiency will probably lead to association of some of these enzymes with specific pathological defects. However, some of these enzymes may bind copper very tenaciously or have a slow turnover rate. Also, some of the enzymes may be present in a large excess. These characteristics would preclude use of experimental copper deficiency for associating the enzyme with a specific defect. In these cases, the effects of loss of the enzyme can only be determined when a gene mutation occurs that abolishes synthesis of the enzyme.

5.   Relevance of Research on the Role of Copper in Metabolism

After reviewing the list of pathological defects that result from copper deficiency, the reader is left with an impression of the importance of copper in growth and development. However, the foregoing discussion in this chapter would be purely academic without at least a brief discussion of the relevance of these research observations.

The recognition of the many pathological anomalies associated with copper deficiency will undoubtedly aid in the diagnosis of ailments that arise when dietary copper is low or unavailable. Support for this statement has already been provided in the section describing Menkes' steely hair syndrome. The etiology of this inborn error was discovered when the symptoms of the disease were compared with the symptoms that occur in experimental copper deficiency.

Recognizing the symptoms of copper deficiency is not limited to use as a diagnostic tool for elucidating the etiology of inborn errors of metabolism. As mentioned in the second section, copper deficiency occurs in animals grazing on soils where copper is low or unavailable. Knowledge of the symptoms of copper deficiency will enable rapid diagnosis and alleviation of the nutritional deficiency in problem areas. The economic implications of this source of knowledge are obvious.

Recognition of the symptoms of copper deficiency has been and will continue to be extremely important in maintaining the health and well-being of humans. Despite suggestions to the contrary copper deficiency does occur in human beings. As early as 1931, Josephs (1931) detected copper deficiency in infants. More recently, the symptoms of copper deficiency have been observed in infants fed low-copper milk diets (Cordano et al., 1964), in premature infants (Al-Rashid and Spangler, 1971), in infants nourished by total parenteral alimentation (Karpel and Peden, 1972) and in adults nourished by total parenteral alimentation (Dunlap et al., 1974). The latter two instances occurred in a clinical situation; recognition of the symptoms of copper deficiency enabled the clinicians to diagnose and alleviate the problem.

Our present knowledge of the nutritive value of foods suggests that most diets contain an adequate level of copper. However, to assume that all factions of society are now ingesting or will continue to ingest an adequate level of dietary copper is presumptuous. Copper pipe, once a source of dietary copper, is being replaced by plastic products; diets are continually becoming more refined; and food faddism is running rampant. If changes in our lifestyle do affect copper nutriture, our knowledge of the role of copper in mammalian metabolism will prove invaluable in diagnosing and preventing the occurrence of severe deficiency syndromes.

6. Summary

Although absolutely essential for normal growth and development but, in excess, copper is extremely toxic. Living systems have developed methods for conserving copper in the body while preventing excess accumulation.

Copper deficiency results in anemia, defects in connective tissue, abnormal functioning of the central nervous system, lack of pigmentation, poor hair or wool development, infertility, and poor growth. Although several copper enzymes have been identified, only a few have been linked to specific pathological defects that accompany copper deficiency.

Two genetic disorders that affect the balance of copper have been discovered, but the biochemical defects have not been described. Studies on the metabolism of copper are still in their infancy, and without doubt, future volumes of *Advances in Nutritional Research* will contain many significant revelations with regard to the biochemistry and physiology of this fascinating element.

References

Allen, K. G. D., and Klevay, L. M., 1976, Hypercholesterolemia in rats caused by copper deficiency, *J. Nutr.* **106**:XXII.

Al-Rashid, R. A., and Spangler, J., 1971, Neonatal copper deficiency, *New Engl. J. Med.* **285**:841.

Bearn, A. G., 1972, Wilson's disease, in *The Metabolic Basis of Inherited Disease* (J. B. Stanbury, J. C. Wyngaarden, and D. S. Fredrickson, eds.), 3rd ed., p. 1033, McGraw-Hill, New York.

Brady, F. O., Monaco, M. E., Forman, H. J., Schutz, G., and Feigelson, P., 1972, On the role of copper in activation of and catalysis by tryptophan-2,3-dioxygenase, *J. Biol. Chem.* **247**:7915.

Buffoni, F., and Blaschko, H., 1964, Benzylamine oxidase and histaminase: Purification and crystallization of an enzyme from pig plasma, *Proc. R. Soc., Ser. B* **161**:153.

Burley, R. W., 1954, Sulphydryl groups in wool, *Nature (London)* **174**:1019.

Chou, W. S., Rucker, R. B., Savage, J. E., and O'Dell, B. L., 1970, Impairment of collagen and elastin crosslinking by an amine oxidase inhibitor, *Proc. Soc. Exp. Biol. Med.* **134**:1078.

Cordano, A., Baertl, J. M., and Graham, G. G., 1964, Copper deficiency in infancy, *Pediatrics* **34**:324.

Coulson, W. F., and Carnes, W. H., 1963, Cardiovascular studies on copper-deficient swine. V. The histogenesis of the coronary artery lesions, *Am. J. Pathol.* **43**:945.

Curzon, G., and O'Reilly, S., 1960, A coupled iron ceruloplasmin oxidation system, *Biochem. Biophys. Res. Commun.* **2**:284.

Danks, D. M., 1975, Steely hair, mottled mice and copper metabolism, *New Engl. J. Med.* **293**:1147.

Danks, D. M., Campbell, P. E., Stevens, B. J., Mayne, V., and Cartwright, E., 1972a, Menkes's kinky hair syndrome. An inherited defect in copper absorption with widespread effects, *Pediatrics* **50**:188.

Danks, D. M., Campbell, P. E., Walker-Smith, J., Stevens, B. J., Gillespie, J. M., Blomfield, J., and Turner, B., 1972b, Menkes' kinky-hair syndrome, *Lancet* **1**:1100.

Danks, D. M., Cartwright, E., and Stevens, B., 1973a, Menkes' steely-hair (kinky hair) disease, *Lancet* **1**:891.

Danks, D. M., Cartwright, E., Stevens, B. J., and Townley, R. R. W., 1973b, Menkes' kinky hair disease: Further definition of the defect in copper transport, *Science* **179**:1140.

Dekaban, A. S., Aamodt, R., Rumble, W. F., Johnston, G. S., and O'Reilly, S., 1975, Kinky hair disease. Study of copper metabolism with use of ^{67}Cu, *Arch. Neurol.* **32**:672.

DePaolo, R. V., and Newberne, P. M., 1971, Copper deficiency in the newborn and postnatal rat with special reference to phosphatidic acid synthesis, in *Trace Substances in Environmental Health*. V (D. D. Hemphill, ed.), p. 177, Univ. of Missouri Press, Columbia.

Dunlap, W. M., James, G. W. III, and Hume, D. M., 1974, Anemia and neutropenia caused by copper deficiency, *Ann. Intern. Med.* **80**:470.

Evans, G. W., 1973, Copper homeostasis in the mammalian system, *Physiol. Rev.* **53**:535.

Evans, G. W., and Cornatzer, W. E., 1971, Biliary copper excretion in the rat, *Proc. Soc. Exp. Biol. Med.* **136**:719.

Evans, G. W., and LeBlanc, F. N., 1976, Copper-binding protein in rat intestine: Amino acid composition and function, *Nutr. Rep. Int.* **14**:281.

Evans, G. W., Wolentz, M. L., and Grace, C. I., 1975, Copper-binding proteins in the neonatal and adult rat liver soluble fraction, *Nutr. Rep. Int.* **12**:261.

Frieden, E., and Hsieh, H. S., 1976, Ceruloplasmin: The copper transport protein with essential oxidase activity, in *Advances in Enzymology and Related Areas of Molecular Biology* (A. Meister, ed.), Vol. 44, p. 187, Wiley, New York.

Frieden, E., Osaki, S., and Kobayaski, H., 1965, Copper proteins and oxygen. Correlations between structure and function of the copper oxidase, *J. Gen. Physiol.* **49**:213.

Friedman, S., and Kaufman, S., 1965, 3,4-Dihydroxyphenylethylamine β-hydroxylase, *J. Biol. Chem.* **240**:4763.

Gallagher, C. H., and Reeve, V. E., 1971, Copper deficiency in the rat. Effect on synthesis of phospholipids, *Aust. J. Exp. Biol. Med. Sci.* **49**:21.

Gallagher, C. H., Judah, J. D., and Rees, K. R., 1956, The biochemistry of copper deficiency. II. Synthetic process, *Proc. R. Soc. Ser. B* **145**:195.

Garnica, A. D., and Fletcher, S. R., 1975, Parenteral copper in Menkes' kinky-hair syndrome, *Lancet* **2**:659.

Goka, T. J., Stevenson, R. E., Hefferan, P. M., and Howell, R. R., 1976, Menkes disease: A biochemical abnormality in cultured human fibroblasts, *Proc. Natl. Acad. Sci. USA* **73**:604.

Hall, G. A., and Howell, J. McC., 1969, The effect of copper deficiency on reproduction in the female rat, *B. J. Nutr.* **23**:41.

Harris, E. D., Gonnerman, W. A., Savage, J. E., and O'Dell, B. L., 1974, Connective tissue amine oxidase. II. Purification and partial characterization of lysyl oxidase from chick aorta, *Biochim. Biophys. Acta* **341**:332.

Hart, E. B., Steenbock, H., Waddell, J., and Elvehjem, C. A., 1928, Iron in nutrition. VII. Copper as a supplement to iron for hemoglobin building in the rat, *J. Biol. Chem.* **77**:797.

Hazelrig, J. B., Owen, C. A., Jr., and Ackerman, E., 1966, A mathematical model for copper metabolism and its relation to Wilson's disease, *Am. J. Physiol.* **211**:1075.

Hill, C. H., and Matrone, G., 1961, Studies on copper and iron deficiencies in growing chickens, *J. Nutr.* **73**:425.

Hill, C. H., Starcher, B., and Kim, C., 1967, Role of copper in the formation of elastin, *Fed. Proc.* **26**:129.

Holmberg, C. G., and Laurell, C. B., 1948, Investigations in serum copper. II. Isolation of the copper-containing protein and description of its properties, *Acta Chem. Scand.* **2**:550.

Horn, N., Mikkelsen, M., Heydorn, K., Damsgaard, E., and Tygstrus, I., 1975, Copper and steely hair, *Lancet* **1**:1236.

Hsieh, H. S., and Frieden, E., 1975, Evidence for ceruloplasmin as a copper transport protein, *Biochem. Biophys. Res. Commun.* **67**:1326.

Hunt, D. M., 1974, Primary defect in copper transport underlies mottled mutants in the mouse, *Nature (London)* **249**:852.

Josephs, H. W., 1931, Treatment of anemia in infants with iron and copper, *Bull. Johns Hopkins Hosp.* **49**:246.

Karpel, J. T., and Peden, V. H., 1972, Copper deficiency in long-term parenteral nutrition, *J. Pediat.* **80**:32.

Kimball, D. A., Coulson, W. F., and Carnes, W. H., 1964, Cardiovascular studies on copper-deficient swine. III. Properties of isolated aortic elastin, *Exp. Mol. Pathol.* **3**:10.

Lau, S., and Sarkar, B., 1971, Ternary coordination complex between human serum albumin, copper(II), and L-histidine, *J. Biol. Chem.* **246**:5938.

Lewis, K. O., 1973, The nature of the copper complexes in bile and their relationship to the absorption and excretion of copper in normal subjects and in Wilson's disease, *Gut* **14**:221.

Mahler, H. R., Baum, H. M., and Huebscher, G., 1956, Enzymatic oxidation of urate, *Science* **124**:705.

Marceau, N., and Aspin, N., 1972, Distribution of ceruloplasmin-bound [67]Cu in the rat, *Am. J. Physiol.* **222**:106.

Marceau, N., and Aspin, N., 1973, The intracellular distribution of the radiocopper derived from ceruloplasmin and from albumin, *Biochim. Biophys. Acta* **328**:338.

McCord, J. M., and Fridovich, I., 1969, Superoxide dismutase. An enzymic function for erythrocuprien (hemocuprien), *J. Biol. Chem.* **244**:6049.

Menkes, J. H., Alter, M., Steigleder, G. K., Weakley, D. R., and Sung, J. H., 1962, A sex-linked recessive disorder with retardation of growth, peculiar hair, and focal cerebral and cerebellar degeneration, *Pediatrics* **29**:764.

Miller, E. J., Martin, E. R., Mecca, C. E., and Piez, K. A., 1965, The biosynthesis of elastin cross-links. The effect of copper deficiency and lathyrogen, *J. Biol. Chem.* **240**:3623.

Mills, C. F., and Williams, R. B., 1962, Copper concentration and cytochrome oxidase and ribonuclease activities in the brains of copper deficient lambs, *Biochem. J.* **85**:629.

Mistilis, S. P., and Farrer, P. A., 1968, The absorption of biliary and non-biliary radiocopper in the rat, *Scand. J. Gastroenterol.* **3**:586.

Mondovi, B., Rotilio, G., Costa, M. T., Finazzi-Agro, A., Chiancone, E., Hansen, R. E., and Beinert, H., 1967, Diamine oxidase from pig kidney. Improved purification and properties, *J. Biol. Chem.* **242**:1160.

O'Brien, J. S., and Sampson, E. L., 1966, Kinky hair disease. II. Biochemical studies, *J. Neuropathol. Exp. Neurol.* **25**:523.

O'Dell, B. L., Hardwick, B. C., Reynolds, G., and Savage, J. E., 1961, Connective tissue defect in the chick resulting from copper deficiency, *Proc. Soc. Exp. Biol. Med.* **108**:402.

O'Dell, B. L., Morgan, R. F., McKenzie, W. N., and Kilburn, K. H., 1976a, Copper deficient rat lung as an emphysema model, *Fed. Proc.* **35**:255.

O'Dell, B. L., Smith, R. M., and King, R. A., 1976b, Effect of copper status on brain neurotransmitter metabolism in the lamb, *J. Neurochem.* **26**:451.

Osaki, S., Johnson, D. A., and Frieden, E., 1966, The possible significance of the ferroxidase activity of ceruloplasmin in normal human serum, *J. Biol. Chem.* **241**:2746.

Osaki, S., Johnson, D. A., and Frieden, E., 1971, The mobilization of iron from perfused mammalian liver by a serum copper enzyme, ferroxidase. I., *J. Biol. Chem.* **246**:3018.

Owen, C. A., Jr., 1965, Metabolism of radiocopper (Cu^{64}) in the rat, *Am. J. Physiol.* **209**:900.

Owen, C. A., Jr., 1971, Metabolism of copper 67 by the copper-deficient rat, *Am. J. Physiol.* **221**:1722.

Partridge, S. M., 1966, Biosynthesis and nature of elastin structures, *Fed. Proc.* **25**:1023.

Partridge, S. M., Elsden, D. F., Thomas, J., Dorfman, A., Telser, A., and Ho, P. L., 1964, Biosynthesis of the desmosine and isodesmonsine cross-bridges in elastin, *Biochem. J.* **93**:30c.

Peisach, J., Aisen, P., and Blumberg, W. E. (eds.), 1966, *Biochemistry of Copper* Academic Press, New York.

Pinnell, S. R., and Martin, G. R., 1968, The crosslinking of collagen and elastin: Enzymatic conversion of lysine in peptide linkage to α-amino-adipic-δ-semialdehyde (allysine) by an extract from bone, *Proc. Natl. Acad. Sci. USA* **61**:708.

Pomerantz, S. H., 1963, Separation, purification and properties of two tyrosinases from hamster melanoma, *J. Biol. Chem.* **238**:2351.

Premakumar, R., Winge, D. R., Wiley, R. D., and Rajagopalan, K. V., 1975, Copper-induced synthesis of copper-chelatin in rat liver, *Arch. Biochem. Biophys.* **170**:267.

Prohaska, J. R., and Wells, W. W., 1974, Copper deficiency in the developing rat brain: A possible model for Menkes' Steely-Hair Disease, *J. Neurochem.* **23**:91.

Ragan, H. A., Nacht, S., Lee, G. R., Bishop, C. R., and Cartwright, G. E., 1969, Effect of ceruloplasmin on plasma iron in copper-deficient swine, *Am. J. Physiol.* **217**:1320.

Riordan, J. R., and Gower, I., 1975, Purification of low molecular weight copper proteins from copper loaded liver, *Biochem. Biophys. Res. Commun.* **66**:678.

Roeser, H. P., Lee, G. R., Nacht, S., and Cartwright, G. E., 1970, The role of ceruloplasmin in iron metabolism, *J. Clin. Invest.* **49**:2408.

Rucker, R. B., Roensch, L. F., Savage, J. E., and O'Dell, B. L., 1970, Oxidation of peptidyl lysine by an amine oxidase from bovine aorta, *Biochem. Biophys. Res. Commun.* **40**:1391.

Sass-Kortsak, A., 1965, Copper metabolism, *Advan. Clin. Chem.* **8**:1.

Savage, J. E., 1968, Trace minerals and airan reproduction, *Fed. Proc.* **27**:927.

Scheinberg, I. H., and Sternlieb, I., 1975, Wilson's disease, in *Biology of Brain Dysfunction* (G. E. Gaull, ed.), Vol. 3, p. 247, Plenum, New York.

Shields, G. S., Coulson, W. F., Kimball, D. A., Carnes, W. H., Cartwright, G. E., and Wintrobe, M. M., 1962, Studies on copper metabolism. XXXII. Cardiovascular lesions in copper deficient swine, *Am. J. Pathol.* **41**:603.

Siegel, R. C., Pinnell, S. R., and Martin, G. R., 1970, Cross-linking of collagen and elastin: Properties of lysyl oxidase, *Biochemistry* **9**:4486.

Simpson, C. F., Jones, J. E., and Harms, R. H., 1967, Ultra-structure of aortic tissue in copper-deficient and control chick embryos, *J. Nutr.* **91**:283.

Smith, S. E., and Ellis, G. H., 1947, Copper deficiency in rabbits. Achromotrichia, alopecia and dermatosis, *Arch. Biochem.* **15**:81.

Starcher, B., Hill, C. H., and Matrone, G., 1964, Importance of dietary copper in the formation of aortic elastin, *J. Nutr.* **82**:318.

Underwood, E. J., 1971, *Trace Elements in Human and Animal Nutrition,* 3rd ed., Academic Press, New York.

Walker-Smith, J. A., Turner, B., Blomfield, J., and Wise, G., 1973, Therapeutic implications of copper deficiency in Menkes' steely-hair syndrome, *Arch. Dis. Childhood* **48**:958.

Walshe, J. M., 1966, Wilson's disease, a review, in *Biochemistry of Copper* (J. Peisach, P. Aisen, and W. E. Blumberg, eds.), p. 475, Academic Press, New York.

Weissman, N., Shields, G. S., and Carnes, W. H., 1963, Cardiovascular studies on copper-deficient swine. IV. Content and solubility of the aortic elastin, collagen, and hexosamine, *J. Biol. Chem.* **238**:3115.

Williams, D. M., Lee, G. R., and Cartwright, G. E., 1974, Ferroxidase activity of rat ceruloplasmin, *Am. J. Physiol.* **227**:1094.

Wilson, S. A. K., 1912, Progressive lenticular degeneration: A familial nervous disease associated with cirrhosis of the liver, *Brain* **34**:295.

Winge, D. R., Premakumar, R., Wiley, R. D., and Rajagopalan, K. V., 1975, Copper-chelatin: purification and properties of a copper-binding protein from rat liver, *Arch. Biochem. Biophys.* **170**:253.

Yamada, H., and Yasunobu, K. T., 1962, Monoamine oxidase. II. Copper, one of the prosthetic groups of plasma monoamine oxidase, *J. Biol. Chem.* **237**:3077.

Chapter 8

The Role of Nutritional Factors in Free-Radical Reactions

Lloyd A. Witting

1. Introduction

It is not unusual in science for key discoveries to lead dramatically and rapidly to the elucidation of previously complex and awkward problems. The literature on vitamin E, polyunsaturated fatty acids, and biologically available selenium is replete with extensive, competently conducted, and factually reported studies which have led investigators to formulate divergent interpretations of the interrelationships among these factors. The missing links needed to begin reconciling the apparent contradictions into a coherent, rational pattern now appear to be available.

Hamilton (1974) has discussed the sluggish reaction of molecular oxygen with organic compounds at room temperature in terms of the triplet ground state, electron spins, and the conservation of angular momentum. The interested reader is referred to reviews of the physicochemical properties of molecular oxygen (Samuel and Steckel, 1974) and the electronic structure of coordinated molecular oxygen (Lapidot and Irving, 1974). The point made is that for practical purposes oxygen must either be activated through complexing with a transition metal ion or react through a free-radical mechanism. Unless some outside factor intervenes, such combustible materials as ethyl ether, gasoline and trinitrotoluene are stable in air at room temperature, but ethyl ether will slowly form peroxides on exposure to light (Rosenberger and Johnson, 1970).

Life in an oxygen-containing atmosphere has required each organism to develop multiple lines of defense against unwanted free-radical initiated oxidative reactions. Separate and disparate lines of defense which prevent, minimize

Lloyd A. Witting • Supelco, Inc., Supelco Park, Bellefonte, Pennsylvania 16823.

or ameliorate tissue damage may seem to produce a similar end result, although each may act at a different site and/or via totally different mechanisms. This review summarizes the evidence relating to the involvement of nutritional factors in free-radical reactions.

2.   Free Radicals *In Vivo*

There is an extensive literature on biological oxidations involving molecular oxygen which it would be neither appropriate nor practical to attempt to review here. Extensive reviews appear in the two volumes edited by Hayaishi (1974a,b). Certain information, however, must be briefly noted because of its importance to the review that follows.

In the four electron transfers between O_2 and $2H_2O$, three important states occur: HO_2^-, H_2O_2, and $\cdot OH$. Certain oxidases, such as xanthine oxidase, generate the superoxide anion radical, O_2^-, which is the conjugate base of the hydroperoxyl radical HO_2^-:

$$XH + O_2 \rightarrow X + HO_2^-$$

Other oxidases, such as D-amino acid oxidase, generate hydrogen peroxide:

$$XH_2 + O_2 \rightarrow X + H_2O_2$$

The Haber-Weiss reaction (see review by Bors *et al.*, 1974) is frequently cited as the source of the extremely reactive hydroxyl radical, $\cdot OH$:

$$O_2^- + Fe^{3+} \rightarrow O_2 + Fe^{2+}$$

$$H_2O_2 + Fe^{2+} \rightarrow \cdot OH + OH^- + Fe^{3+}$$

Singlet oxygen, $^1O_2$, also appears to be produced secondary to the formation of the superoxide anion radical, possibly by a reaction such as

$$\cdot OH + O_2^- \rightarrow {}^1O_2 + OH^-$$

The various activated states are well known, as they may be readily generated by pulse radiolysis and purified by use of selective trapping agents for chemical characterization and study as reviewed by Schafferman and Stein (1975). The redox potentials of the oxygen/superoxide system have been studied by this technique (Ilam *et al.*, 1976).

The reactions conducted by various oxygenases

$$XH_2 + O_2 \rightarrow XO + H_2O$$

$$X + O_2 + DH_2 \rightarrow XO + H_2O + D$$

$$X + O_2 + DH_2 \rightarrow HOXOH + D$$

also involve free radicals. These enzymes include flavoprotein oxygenases (Flashner and Massey, 1974), pterin-requiring hydroxylases (Kaufman and Fisher, 1974), α-ketoglutarate-coupled dioxygenases (Abbott and Udenfriend, 1974), and cytochrome P450-linked monooxygenases (Orrenius and Ernster, 1974). Oxygenase-catalyzed hydroxylations have also been reviewed by Gunsalus *et al.* (1974, 1975). Intermediates such as

have been depicted in reviews by Abbott and Udenfriend (1974) and Hamilton (1974), respectively, and in a recent paper by Entsch *et al.* (1976). Hamilton (1971) has described one of the valence structures of the flavin derivative as a vinylogous ozone:

Recently, Nordblom *et al.* (1976) have shown that highly purified cytochrome P450 catalyzes the hydroperoxide-dependent hydroxylation of a variety of substrates in the absence of NADPH, NADPH-cytochrome P450 reductase, and molecular oxygen. Flavoproteins and cytochrome b_5 were absent from this preparation, and phosphatidyl choline was needed for maximum activity. The oxygen added to the substrate was derived from the hydroperoxide.

The monooxygenases are frequently referred to as mixed-function oxidases, particularly in studies of drugs, toxic materials, and xenobiotics, because they produce both XO and H_2O. The enzyme containing a reduced prosthetic group may react slowly with oxygen in the absence of substrate for oxygenation (Boveris et al., 1972) and "leak" detectable quantities of free radicals into the incubation mixture. Such "leakage" appears to stop when a suitable substrate is provided (Gram and Fouts, 1966). In the case of some flavin enzymes, reduction of the prosthetic group occurs only after substrate binding (Flashner and Massey, 1974). Oxygenase and oxidase activities are to some extent separable in vitro. If the substrate is a poor fit, oxygen may still be bound, and the oxidase reaction occurs without oxygenation of the substrate (Yamamoto et al., 1972). Enzymes are not necessary for free-radical production. Combinations of ascorbic acid, iron salts and flavin, for instance, are known to generate free radicals. This phenomenon may be important in tissue damage or injury.

"Activated" forms of oxygen are used for various biological purposes. Tryptophan dioxygenase utilizes O_2^- (Hirata and Hiyaishi, 1971) and peroxidases utilize H_2O_2 to oxidize various substrates to free radicals (Yamazaki et al., 1960; Yamazaki, 1974):

$$2XH_2 + H_2O_2 \rightarrow 2XH\cdot + 2H_2O$$

Weves et al. (1976) reported that during phagocytosis human granulocytes generate $^1O_2$ and O_2^-, which are potent bactericidal agents. Radical generation appears to involve myeloperoxidase in this case. Myeloperoxidase plus Cl^- appears to be a source of singlet oxygen, $^1O_2$ (Krinsky, 1974; Allen, 1975), and lactoperoxidase has been suggested as a source of ongoing lipid peroxidation in vivo (Beuge and Aust, 1976). Hence free-radicals are normal products of many biochemical reactions.

Recent work on prostaglandins and thromboxanes has strongly suggested that controlled lipid peroxidation is involved in essential biochemical processes (Hamberg and Samuelsson, 1974a,b; Hamberg et al., 1974a,b; Malmsten et al., 1975, 1976; Rahintula and O'Brien, 1976; Wolfe et al., 1976).

3. Adventitious Reactions Initiated by Free Radicals

It follows that when a reactive free radical is the end product of an enzymatic reaction, the possibility must be considered that adventitious, nonenzymatic reactions will ensue thereafter. One such potential reaction is the peroxidation of lipid-bound polyunsaturated fatty acids (RH). This reaction (Uri, 1961a) is unusual in that peroxy free radical ($RO_2\cdot$) reacts with another polyunsaturated fatty acid to produce a lipid hydroperoxide (ROOH) and regenerate the original fatty acid free radical ($R\cdot$):

$$CH_3(CH_2)_a\overset{\textstyle\cdot}{C}H \quad H$$

$$C=C \quad CH_2(CH_2)_b CO_2R$$

$$H \quad C=C$$

$$H \quad H$$

$$RH + \cdot OH \rightarrow R\cdot + H_2O$$
$$R\cdot + O_2 \rightarrow RO_2\cdot$$
$$RO_2\cdot + RH \rightarrow ROOH + R\cdot$$

This self-perpetuating cyclic chain reaction is thus capable of resulting in large amounts of lipid peroxidation per single free-radical initiation. In addition, the product of the reaction may give rise to new free radicals, resulting in chain branching or free-radical multiplication:

$$2ROOH \rightarrow RO_2\cdot + RO\cdot + H_2O$$

The conjugated diene hydroperoxide is a reactive molecule that, through further oxidation, gives rise to products that may damage various cellular constituents. Formation of lipid hydroperoxides in the membrane lipids results in deleterious alterations in membrane properties (Tam and McCay, 1970; Bidlack and Tappel, 1973). The seriousness of these problems and the probability of their actual occurrence is suggested by the number of known mechanisms that exist to prevent such damage.

The first of these is the enzyme superoxide dismutase, which converts the hydroperoxyl radical to oxygen and hydrogen peroxide (Fridovich, 1975):

$$2HO_2^- \rightarrow O_2 + H_2O_2$$

The conversion of hydrogen peroxide to oxygen and water by catalase seems to be somewhat less important. This may be attributable to the very high K_m (1.10 M) of H_2O_2 for catalase (Ogura, 1955), which is not an efficient H_2O_2 scavenger (Misra, 1974). The low K_m (1 μM) of H_2O_2 for glutathione peroxidase (Flohe and Brand, 1969) may make this enzyme important at low peroxide levels. McCay et al. (1976b) mention unpublished data that have led them to conclude that glutathione peroxidase inhibits lipid peroxidation by removing H_2O_2 and thus preventing $\cdot OH$ formation via the Haber-Weiss reaction. Glutathione peroxidase has been considered to be of importance because of its reported ability to convert fatty acid hydroperoxides to innocuous hydroxy fatty acids (Christopherson, 1968, 1969):

$$ROOH + 2GSH \rightarrow ROH + GSSG + H_2O$$

On the basis of evidence in a genetic disease, Battens disease (neuronal ceroid lipofuscinosis), and its adult variation, Kuf's disease, another peroxidase assayed originally in leukocytes is also known to be of critical importance (Armstrong *et al.*, 1974a–c; 1975; Dimmitt, 1975).

Vitamin E (AH) *competes* with polyunsaturated fatty acids for reaction with the lipid peroxy free radical (Uri, 1961b):

$$RO_2\cdot + RH \rightarrow ROOH + R\cdot$$
$$RO_2\cdot + AH \rightarrow ROOH + [A\cdot]$$

Free radicals are thus withdrawn from the system via quinone formation or dimerization of vitamin E (Boguth and Niemann, 1971; Csallany, 1971).

The known lines of defense against free-radical-initiated tissue damage therefore include (1) free-radical dismutation; (2) interception of secondary free radicals; and (3) catabolism of potentially damaging hydroperoxides, which are also a potential source of new free radicals. The role of nutritional factors in these processes is reviewed in terms of free-radical initiation, propagation of the cyclic chain reactions, chain-branching, tissue damage, and chain termination.

4. Effect of Nutritional Factors

4.1. Free-Radical Initiation

4.1.1. Control

The superoxide anion radical is a relatively weak initiating agent (Hamilton, 1974), and evidence has been presented that it is not the direct initiator of lipid peroxidation (Fong *et al.*, 1973). The actual initiators, hydroxyl radical or singlet oxygen, appear to be produced only via reactions occurring secondary to O_2^- generation. Singlet oxygen $^1O_2$ reacts very rapidly with unsaturated compounds (Foote, 1968). Hydroxyl radical, $\cdot OH$, reacts with all organic compounds typically by H abstraction or addition to π systems (Norman and Lindsay-Smith, 1965). Hydroxyl radical appears to be produced only via reactions occurring secondary to O_2^- generation. This may also be the case with singlet oxygen, but several peroxidases are thought to produce $^1O_2$ (Krinsky, 1974; Allen, 1975; Beuge and Aust, 1976). McCay and Poyer (1976) have prepared a comprehensive review of enzyme-generated free radicals as initiators of lipid peroxidation. McCay's group (McCay et al., 1976a; McCay and Poyer, 1976; King et al., 1975; Fong et al., 1976) favors initiation by ·OH, whereas Pederson and Aust (1973) and Keelog and Drifovich (1975) favor $^1O_2$.

Insofar as is now known, the organism does not appear to have enzymatic mechanisms for the control of the hydroxyl radical or singlet oxygen. The damaging effects of hydroxyl radicals produced during exposure to ionizing radiation would seem relevant to this point. Superoxide dismutase (SOD) is therefore the primary defense against potential tissue damage arising from enzymatically generated free radicals.

The critical nature of SOD is shown by the observation that oxygen is lethal to organisms, obligate anaerobes, that do not possess this enzyme (McCord et al., 1971). It appears that eukaryotic cells may have a copper and zinc containing SOD in the cytosol and a manganese containing SOD in the mitochondria. Similarly prokaryotic cells may have the mangani-enzyme in the matrix and an iron-containing SOD in the periplasmic space (see review by Fridovich, 1975).

A rare genetic variant of human Cu–Zn SOD is known to occur in northern Sweden (Marklund et al., 1976). Patients with chronic granulomatous disease appear to have defective SOD production at this site (Curnette et al., 1974). This subject is so new that very few definitive nutritional experiments have been reported.

Gregory et al. (1973) found that Escherichia coli B cells grown in an iron-deficient aerobic medium were killed by exogenously generated O_2^- and protected by bovine SOD added to the medium. Cells grown in an iron-enriched anaerobic medium were resistant to exogenous O_2^- but contained low levels of the mangani-enzyme, the concentration of which responds to the partial pressure of oxygen. These observations suggested that the mangani-enzyme protects against endogenously generated superoxide anion radicals.

Streptococcus faecalis grown under 20 atm O_2 contained 16 times as much SOD as when grown under anaerobic conditions and one-half maximal induction was attained in 90 min (Gregory and Fridovich, 1973a,b). This organism does not contain detectable levels of catalase under a variety of growth conditions (Stanier et al., 1970). Indeed, streptococci, pneumococci, and lactic acid bacteria capable of aerobic growth do not contain catalase, whereas some obligate anaerobes do contain catalase. A change in oxygen exposure (anaerobic to 5 atm O_2) resulted in a 25-fold increase in SOD in E. coli B without affecting catalase levels (Gregory and Fridovich, 1973b). The production of H_2O_2 by rat liver and pigeon heart mitochondria increased 10 to 15-fold and fourfold, respectively, at 19.5 atm O_2 (Boveris and Chance, 1973). This could either represent direct enzymatic production of H_2O_2 or H_2O_2 production from O_2^- via the Haber-Weiss reaction.

A severe anemia developed in iron-supplemented, copper-deficient swine after 8–12 weeks, which was correlated with an approximately 75–85% reduction in the level of erythrocyte SOD (Williams *et al.*, 1975). This anemia had previously been correlated with low levels of ceruloplasmin (Frieden, 1971). Superoxide dismutase has been reported to protect erythrocytes against peroxidative hemolysis (Fee and Teitelbaum, 1971) and X-irradiation (Petkau *et al.*, 1976). A copper deficiency was correlated with reduced levels (30% below controls) of SOD and abnormal-appearing mitochondria, but not with detectable lipid peroxidation, in developing rat brain in a brief 28-days experiment (Prohaska and Wells, 1975).

SOD activity (U/g tissue) of various rat tissues have been reported by Peeters-Joris *et al.* (1975): liver, 4480; adrenal, 1563; kidney, 1342; blood, 785; spleen, 441; pancreas, 300; brain, 261; lung, 200; stomach, 140; intestine, 122; ovary, 112; thymus, 88; and fat 0.

4.1.2. Factors Increasing Production of Free Radicals

The discovery of HO_2^- production originated in the observation that sulfite oxidation via a free-radical mechanism was initiated by the enzyme xanthine oxidase *in vitro* (Fridovich and Handler, 1958, 1961). A number of enzyme systems are now known to initiate lipid peroxidation *in vitro*. Such systems are considered to have not only a potential, but also a very real capacity to initiate lipid peroxidation *in vivo*. In the living organism, a series of tight controls normally prevent actual occurrence of lipid peroxidation at pathological levels but there is a distinct accumulation of lipopigment throughout life (Strehler *et al.*, 1959). As this review deals specifically with these control systems, all such systems are viewed as relevant and real rather than potential. Sources of increased free-radical production include (1) tissue injury; (2) deficiencies not directly related to control systems; (3) certain metal overloads; (4) metabolism of drugs, toxic compounds, and xenobiotics; and (5) ingestion of compounds whose toxicity derives at least in part from the overriding of control system capacities during their metabolism. Of these five sources, the last two have received the most attention.

Hartroft (1963) performed a classically simple experiment a number of years ago. A crushing injury was imposed on one of the epididymal fat pads of the rat, whereas the other fat pad served as a control. Disruption of cellular integrity and release of metalloproteins into contact with fat led to lipid peroxidation as measured by the progressive accumulation of ceroid pigments. Production of lipopigments of the lipofuscin and/or ceroid type may be a rather general phenomenon in tissue injury (Clark and Davidson, 1974).

Choline deficiency with fatty liver formation, hepatic injury and renal necrosis has also been studied by Hartroft and co-workers (Sugioka *et al.*, 1969; Ghoshal et al., 1970; Monserrat *et al.*, 1969, 1972). It is not entirely clear how the deficiency leads to tissue instability and lipid peroxidation. As noted below, however, there is a close relationship between membrane phospholipids and enzymes which generate free radicals that may be influenced by a decrease in phosphatidylcholine. In the isolated and reconstituted microsomal system, phosphatidylcholine has been reported to be required for enzymatic activity (Strobel *et al.*, 1970; Kaschnitz and Coon, 1975; Lu, 1976).

Although iron is an essential dietary constituent, toxicity is evident at excessive intakes. Injection of large doses of iron, 0.125 mg/g body weight, as iron dextran into mice resulted in an increase in the nonheme iron and total iron in liver microsomes and the rate of lipid peroxidation measured *in vitro* (Wills, 1966, 1969, 1972). The role of lipid peroxides in the induction of retinal siderosis has also been investigated (Hiramitsu *et al.*, 1975). Silver salts appear to stress the normal control systems (Shaver and Mason, 1951; Diplock *et al.*, 1967). Tissue damage, liver necrosis, and nutritional muscular dystrophy in the rat and exudative diathesis in the chick were largely, but not completely in all cases, ameliorated by the usual dietary levels of selenium or vitamin E.

Free-radical production by normal enzyme systems is a rather new topic, and whereas specific areas have been intensively studied, a comprehensive, balanced picture is not available. The microsomal mixed-function oxidase system (MFO) appears to be a major source of free radicals in the liver. Because the pharmaceutical industry has had a keen interest in this drug metabolizing system, an extensive background of relevant investigations is available. Repeated mention of this system, however, should not create the impression that other enzymatic reactions may not be of greater importance in specific tissues or that MFO is restricted to the liver. Adrenals, kidney, intestine, spleen, lung, thyroid, thymus, testes, and skin also contain MFO activity (Conney, 1967). Orrenius and Ernster (1974) tabulated cytochrome P450 levels (nmol/mg microsomal protein) in various tissues: liver, 0.60; adrenal cortex, 0.50; testes, 0.10; kidney cortex, 0.10; lung, 0.035; and spleen, 0.025. It is interesting to note the similarities to the distribution of superoxide dismutase (Peeters-Joris *et al.*, 1975). The MFO system metabolizes aliphatic and aromatic hydrocarbons, amines, alcohols, phenols, and thiophenes and carries out hydroxylations, oxidative dealkylations or deaminations and nitroxide or sulfoxide formation. It is of interest that the same substrate may be hydroxylated in a variety of positions (e.g., testosterone 1β, 2β, 6α, 6β, 7α, 15β, 16α, and 18) or undergo diverse reactions (e.g., chlorpromazine sulfoxidation, N-demethylation, or hydroxylation of the phenothiazine nucleus), (Orrenius and Ernster, 1974). The substrate fit, which must be rather general, must also quite often be rather poor.

Rat liver microsomes produce superoxide anion radical at the rate of 12 nmol min^{-1} mg^{-1} protein (Mishin *et al.*, 1976). The MFO consumes oxygen at about this rate, 6–12 nmol min^{-1} mg^{-1} protein, during the metabolism of various drugs (Staudt *et al.*, 1974). This system is readily inducible by a large variety of drugs, toxic substances, and xenobiotics to 5–10 times the activity seen in the control animals. In such cases, the cytochromes, P450 and b_5, and corresponding reductases may account for 40% of the total integral membrane proteins

(Stier, 1976). It has been proposed that NADPH-cytochrome c reductase and cytochrome b_5 are on the outside of the membrane while cytochrome P450 and NADH-cytochrome c reductase are in the membrane with an opposite transverse orientation (Orrenius and Ernster, 1971). The lipid in this fluid mosaic would seem particularly vulnerable to free-radical flux. In reviewing the induction, Conney (1967) noted that the system was responsive to dietary and nutritional factors and hormonal changes in the body as well as drugs, toxic substances, and xenobiotics. Campbell and Hayes (1974) have reviewed the role of nutrition in the drug-metabolizing enzyme system.

Nutritional experiments usually have been expressed in terms of ability to support maximal induction of MFO. It must be remembered in reading such studies that induction is not always beneficial to the subject, because some metabolites, such as heptachlor epoxide (Weatherholz *et al.*, 1969), may be more toxic than the precursor. Indeed, as noted below intoxication by some compounds requires their active metabolism.

As induction requires protein synthesis, it is impaired by starvation, low-protein diets and poor-quality protein (Webb and Miranda, 1973; Campbell and Hayes, 1976). Induction results in a proliferation of the smooth endoplasmic reticulum, which is approximately one-third lipid on a dry-weight basis. This requires a dietary source of unsaturated fatty acids (Caster *et al.*, 1970; Marshall and McLean, 1971; Norred and Wade, 1972; Rowe and Wills, 1976; Wade and Norred, 1976). Ingestion of a toxic compound (DDT) by animals fed a diet containing an otherwise adequate level of linoleate (1.5–2% of calories) may result in signs of essential fatty acid deficiency (Tinsley and Lowry, 1972; Darsie *et al.*, 1976). There are a number of conflicting reports regarding the effect of drugs or toxic compounds on fatty acid desaturation (Lyman *et al.*, 1969, 1970; Witting, 1973; Montgomery and Holtzman, 1975; Darsie *et al.*, 1976). Marginal quantities of lipotropes also restrict attainment of maximal induction (Newberne *et al.*, 1971).

Many of the micronutrients also affect MFO induction (see review by Campbell and Hayes, 1974; Zannoni and Sato, 1976; or current issues of *Biochemical Pharmacology*). Becking (1973) found MFO activity depressed in vitamin A deficiency but felt that the structure of the endoplasmic reticulum was affected. Colby *et al.* (1975), however, reported a reduction in hepatic microsomal protein and a fall in the specific activity of oxidative enzymes. Thiamine deficiency increased *in vitro* lipid peroxidation (Galdhar and Pawar, 1976), but this could be inhibited by phenobarbital treatment during the deficiency. Most deficiencies tend to impair induction. In the case of riboflavin and niacin, the basis for the requirement is quite obvious. Extra zinc has been reported to protect against carbon tetrachloride hepatotoxicity and to inhibit NADPH oxidation and oxidative metabolism of drugs (Chvapil *et al.*, 1973, 1975, 1976). Becking (1976) has considered the effect of iron, magnesium, and potassium deficiencies on hepatic drug metabolism in the rat.

The MFO system requires time to develop and highest activity is seen in the adult male. This sex difference which appears relatively early has been attributed to neonatal androgen levels (Chung *et al.*, 1975). A normal base line for MFO activity is difficult to establish. Addition of natural ingredients to the usual casein-glucose diet increases activity, and the use of pesticide sprays in animal rooms is a well-established source of induction. In the clinical setting, a comparable pitfall might be the induction of demethylases by the catechins and catechin derivatives in black tea (Babish and Stoewsand, 1975).

In addition to nutrition affecting drug metabolism, the reverse is also true. The MFO system inhibits both hepatic lipogenesis from glucose by competing for cytosolic NADPH and gluconeogenesis from lactate by diverting key intermediates, such as malate, from glucose synthesis toward NADPH generation (Thurman *et al.*, 1975).

The numerous independent demonstrations of lipid peroxidation *in vitro* require that serious consideration be given not to the potential occurrence of this reaction *in vivo*, but rather to its real occurrence, albeit under strictly "controlled" conditions. It will be of interest to learn if this system is characterized by an exaggerated form of free-radical "leakage" common to other enzyme systems or possesses a unique "flaw" related to the wide variety of substrates accepted.

Ingestion of a variety of hepatotoxic compounds including carbon tetrachloride, tri-*o*-cresyl phosphate, yellow phosphorus, ethanol, orotic acid, and monomethyl hydrazine, results in tissue damage ranging from fatty infiltration through cirrhosis to necrosis, which is prevented or ameliorated by lipid antioxidants (Hove, 1948, 1953; Recknagel and Ghoshal, 1966; DiLuzio and Costales, 1965; DiLuzio, 1966; Comporti *et al.*, 1967; Hartman and DiLuzio, 1968; Ghoshal *et al.*, 1969; Pani *et al.*, 1972; DiLuzio *et al.*, 1973). In the best-characterized case, carbon tetrachloride intoxication, failure of superoxide dismutase to prevent tissue damage may be logically explained by the production of the $\cdot CCl_3$ radical (Recknagel and Glende, 1973.) Occurrence of this radical has been demonstrated by a variety of techniques: formation of branched chain fatty acids (Gordis, 1969); detection of Cl_3CCCl_3 and $CHCl_3$ (Fowler, 1969); binding of ^{14}C-CCl_4 to microsomal lipid and protein (Rao and Recknagel, 1969); and radical trapping (McCay *et al.*, 1976c). Comparable hepatotoxin radicals have not been demonstrated with other compounds, but it is not clear if definitive experiments in this direction have been conducted. Chlorpromazine radicals and radicals of other phenothiazine drugs are known to be formed, but this radical is stated to be a specific inhibitor of the Na^+,K^+-ATPase (Lee *et al.*, 1976). These drugs are known to produce jaundice or hepatic dysfunction in humans (Ishak and Irey, 1972). As noted previously, peroxidases also produce substrate radicals free in solution (Yamazaki *et al.*, 1960). In addition to "leakage" through a primary protective mechanism controlling metabolically generated free radicals, consideration must also be given, therefore, to direct circumvention by metabolic production of radicals other than $HO_2^{\cdot}$.

5. Propagation

McCay and co-workers have shown that lipid peroxidation is dependent on enzymatic generation of free radicals and stops abruptly when the supply of NADPH is exhausted (May and McCay, 1968) or inhibitors of NADPH oxidation are added (Pfeifer and McCay, 1971). This group has also questioned the availability of sufficient molecular oxygen to sustain the reaction nonenzymatically. In their system, as much as 75% of the total microsomal polyunsaturated fatty acids (PUFA) are oxidized in 1 hr (May and McCay, 1968). In nutritional muscular dystrophy, creatinuria was noted after "loss" of 2% of the muscle phospholipid PUFA (Witting and Horwitt, 1967). Hartman and DiLuzio (1968) have also considered the quantitative aspects of this problem. Loss of 50% of the PUFA from limited sections (10%) of the endoplasmic reticulum in 5% of the liver cells over a space of 24 hr corresponds to a loss of 0.25% of the total microsomal PUFA. Although negligible or undetectable in an *in vitro* system (1/7200 of the *in vitro* rate), this level of lipid peroxidation is certainly important *in vivo*.

Availability of oxygen is, of course, a limiting factor in the conventional propagation reaction (Uri, 1961a):

$$\begin{array}{l} \longrightarrow R\cdot + O_2 \rightarrow RO_2\cdot \\ RO_2\cdot + RH \rightarrow ROOH + R\cdot \end{array}$$

The rate of oxygen consumption during drug metabolism, 6–12 nmol $min^{-1} mg^{-1}$ hepatic microsomal protein (Staudt *et al.*, 1974) is at least five orders of magnitude greater than the rate of oxygen consumption estimated to occur in lipid peroxidation in nutritional muscular dystrophy, 1–2 nmol $day^{-1} g^{-1}$ tissue (Witting, 1970).

After lipid peroxidation has been initiated, the slowest, and therefore rate limiting reaction, is the reaction of a peroxy free radical ($RO_2\cdot$) with another fatty acid. This rate is dependent on fatty acid structure, $CH_3(CH_2)_a(CH=CHCH_2)_n(CH_2)_bCO_2H$, and increases according to the ratios 0.025 : 1 : 2 : 4 : 6 : 8 as the number of double bonds n increases from 1 to 6 (Witting and Horwitt, 1964; Witting, 1965). Of greater practical interest is the ratio of the rates of propagation and termination. As the rate of the competing termination reaction

$$RO_2\cdot + AH \rightarrow ROOH + [A\cdot]$$

is dependent on antioxidant (AH) structure, the yield of lipid hydroperoxide (ROOH) per free-radical initiation increases with increasing fatty acid unsaturation. For illustrative purposes the cyclic chain reaction may be depicted as a spiral taking in fatty acids and oxygen and spewing out lipid hydroperoxides. At optimum concentrations of α-tocopherol in beakers of pure esters the average length of the chain reaction in oleate, linoleate, and linolenate is shown in Fig. 1

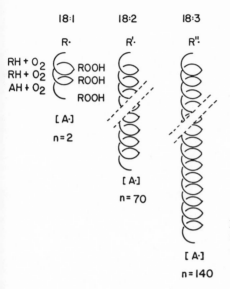

Fig. 1. Average length of the free-radical-initiated cyclic chain reaction in ethyl oleate (18 : 1), ethyl linoleate (18 : 2), and ethyl linolenate at optimal concentration of α-tocopherol *in vitro*.

(adapted from Witting, 1969). Comparable chain lengths for arachidonate, eicosa- or docasapentaenoate, and docosahexaenoate are 280, 420, and 560, respectively. Boland and Gee (1946) reported a yield of 100 molecules of hydroperoxide per initiation for a diene.

It should be quite apparent that tissue lipid PUFA content and composition are critical factors in lipid peroxidation. Because of various biochemical interactions and conversions the effect of dietary lipid on PUFA in various tissues differs in a complex manner (Witting *et al.*, 1967a). Examples include the low level (1–2%) of linoleate in brain (Witting *et al.*, 1961), which is not particularly affected by diet or by the resistance of testicular lipids to the incorporation of dietary nonessential $\omega 3$ PUFA (Witting *et al.*, 1967b). Comparisons of tissue damage in different organs or between species in terms of dietary lipid may often appear meaningless or quite confusing until recourse is had to actual tissue lipid analyses (Witting *et al.*, 1967a).

Where the experimental dietary fat differs drastically from that contained in the maternal diet, 6–8 weeks may be required to attain tissue lipid PUFA composition in equilibrium with the dietary fat even in rapidly growing weanling rats (Witting, 1972). With rats 130 days old at the time of the diet change, 17–36 weeks were required to attain a new equilibrium composition in the liver lipids (Witting, 1974). In humans the erythrocyte lipid PUFA composition changes particularly rapidly after a change in dietary fat. However, in a 52-*month* study one-third of the total observed change took place within 3 weeks, one-third between 3 weeks and 10 months, and one-third between 10 and 52 months (Witting, 1974, 1975). Other related studies in adult men at stable weight have

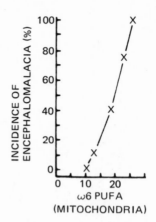

Fig. 2. Incidence of encephalomalacia in vitamin E-deficient chicks fed graded levels of linoleate to produce various levels of essential (ω6) polyunsaturated fatty acids in brain mitochondrial lipids.

shown that equilibrium composition in adipose tissue had not been reached in 6 *years* (Witting, 1970).

Unfortunately, most published work in this area, not specifically including sequential tissue lipid fatty acid analyses, is either difficult to evaluate or of no value. This situation arises because the tissue lipids were not shown to be (or frequently could not conceivably have been) in equilibrium with the composition, which would ensue after prolonged consumption of the dietary fat. Straightforward data, however, have been obtained in those few experiments that were properly designed and Green (Green *et al.*, 1967), without conceding the occurrence of lipid peroxidation *in vivo,* concedes the existence of a relationship between PUFA intake and vitamin E requirement. The incidence of encephalomalacia, for instance. in vitamin E-deficient chicks fed graded levels of linoleate has been shown to be well correlated with the levels of essential PUFA in brain mitochondria (Fig. 2).

Where the intake of ω3 PUFA is negligible or infrequent, it has been suggested that the level of adipose tissue linoleate is representative of the dietary level of linoleate with which the body is transiently in equilibrium. When the diet is changed from one high in linoleate to one normal or low in linoleate, the adipose tissue stores of linoleate may be the major factor in determining tissue PUFA levels. After several years, ingestion of "therapeutic" levels of corn oil or safflower oil may raise the level of adipose tissue linoleate from 10% to 40%. In adult men at stable weight the one-half depletion rate of adipose tissue linoleate is approximately 26 months (Witting, 1972). More than 4 *years* would therefore be required to reattain the previous level of 10% linoleate.

The availability of PUFA in foodstuffs available for consumption is known to have increased (Thompson *et al.,* 1973). In view of the well-known differences between availability and consumption, tissue lipid analyses are a better measure of actual intake. Reliable data are to be found in the literature only subsequent to the availability of commercial equipment for gas liquid chromatography. Published values for human erythrocyte (Fig. 3) and adipose tissue (Fig. 4) linoleate levels have been plotted against the year of publication of the data (Witting and Lee, 1975a). In view of commercial ventures into synthetic eggs and polyunsaturated milk and meat (Scott *et al.*, 1971) further increases are to be expected. Based on current diet analyses, a level of 20% linoleate in adipose tissue would be attained on reaching equilibrium with these diets without further additions (Witting and Lee, 1975a,b).

5.1. Desaturation and Homeostatic Mechanisms

PUFA are essential constituents of membrane lipids. Their loss during peroxidation sets off a local increase in elongation and desaturation of available dietary precursors. This repair process is a factor in propagation, as it tends to maintain the susceptibility of the tissue lipids to peroxidation, despite the ongoing destruction of those fatty acids most susceptible to peroxidation.

Lipid peroxidation may occur over a period of hours in acute intoxication, days in chronic intoxication, or months in vitamin E deficiency. In a very rapid *in vitro* system (Weddle *et al.*, 1976), phospholipids containing a peroxidized fatty acid appeared to be removed from the microsomes and appeared in cell debris. Generally, relatively rapid lipid peroxidation *in vivo* leads to observations of generalized PUFA loss (Horning *et al.*, 1962; Lyman *et al.*, 1964; Comporti *et al.*, 1971; French *et al.*, 1971). Demonstration of increased PUFA synthesis by microsomal preparations *in vitro* after feeding mild hepatotoxins is complicated by the increased susceptibility of the endoplasmic reticulum to damage during isolation (Witting, 1973). In the slow peroxidation occurring in the vitamin

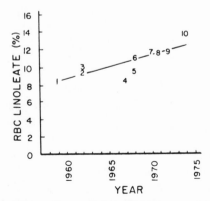

Fig. 3. Levels of linoleic acid in erythrocyte lipids of humans reported in studies published in recent years (Witting and Lee, 1976a).

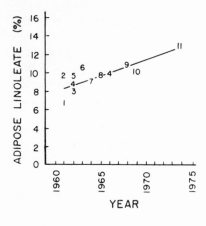

Fig. 4. Levels of linoleic acid reported in human adipose tissue lipids in studies published in recent years (Witting and Lee, 1976a).

E-deficient rat, Bernhard *et al.* (1963a,b) found increased synthesis and an actual net increase in hepatic arachidonate. These observations on the liver have been confirmed (Witting *et al.*, 1967c) and extended to muscle (Witting and Horwitt, 1967d; Witting, 1967) and testes (Witting *et al.*, 1967b).

In the lung, exposure to ozone appears to result in increased activity of the lecithin-cholesterol acyltransferase (Menzel, 1972; Shimasaki *et al.*, 1976). Arachidonate is transferred from phosphatidyl choline to cholesteryl esters. Replacement of the arachidonate in the phosphatidyl choline then proceeds via other routes. This process tends to mask the effect of ozone on fatty acid composition.

5.2. Vitamin A

As a highly unsaturated material, retinol is susceptible to lipid peroxidation and should logically therefore contribute to the propagation of this reaction. Vitamin E is known to protect tissue stores of retinol (Moore, 1940; Ames, 1968). The effect of vitamin E and selenium in diets high in retinol has most recently been investigated by Combs (1976).

6. Chain-Branching

The literature is replete with erroneous statements regarding the stability of lipid hydroperoxides and their reaction with vitamin E. Hiatt (1976) gives the thermal half-time of the reaction

$$ROOH \rightarrow RO\cdot + \cdot OH$$

as approximately 27.5 years. Linoleate hydroperoxide, for instance, is stable to purification by thin layer chromatography on silicic acid (Kokatnur *et al.*, 1965). Gruger and Tappel (1970) found the rate of tocopherol oxidation in ethanolic solutions of various pure hydroperoxides (18:2,18:3,20:4,20:5) to be negligible. The reaction

$$ROOH + AH \rightarrow RO\cdot + [A\cdot] + H_2O$$

is enhanced, however, at higher (nonphysiological) level of tocopherol (Labuza, 1969; Labuza *et al.*, 1969). This would explain the well-known "prooxidant" action of high tocopherol levels *in vitro:*

$$ROOH + Fe^{3+} \rightarrow RO_2\cdot + Fe^{2+}$$
$$ROOH + Fe^{2+} \rightarrow RO\cdot + \cdot OH + Fe^{3+}$$

When the hydroperoxides were dissociated with iron salts, it was observed that increased stability was associated with increased unsaturation (Gruger and Tappel, 1970). Metals and metalloproteins *in vitro* would be expected to contribute significantly to hydroperoxide breakdown with chain-branching and free-radical multiplication as the result.

Complete destruction of all the α-tocopherol is not necessary prior to occurrence of uncontrolled lipid peroxidation, because $ROOH + [A\cdot] \rightarrow RO_2\cdot + AH$ becomes important at relatively low peroxide levels (Mahoney, 1969). Indeed, addition of tocopherol to lipid at a high peroxide level has no effect on the reaction rate (Marbrouck and Dugan, 1961; Scott, 1965).

6.1. Fatty Acid Unsaturation

It follows logically that if the yield of lipid hydroperoxide per free-radical initiation increases with increasing fatty acid unsaturation, so too does the potential for free-radical multiplication. This is readily demonstrated *in vitro,* as the rate of tocopherol oxidation is similar to the rate of free-radical initiation. At physiological concentrations, tocopherol is oxidized approximately five times as rapidly in autoxidizing linolenate as in autoxidizing linoleate (Witting, 1969). The extra initiations arise from chain-branching, and production is related to hydroperoxide concentration. As tissue lipid fatty acid composition may be varied by diet, it follows that the potential for free-radical multiplication is influenced by diet.

6.2. Peroxidases

Lipid hydroperoxides pose a severe danger to the cell as a potential source of free radicals and, as noted in the next section, they can undergo further oxidation to even more deleterious substances. Various enzymes, of which the selenium-containing glutathione peroxidase is one of the most important, may function to prevent or minimize this reaction by converting the hydroperoxide to a hydroxy fatty acid:

$$ROOH + 2GSH \rightarrow ROH + GSSG + H_2O$$

Because selenium is an essential trace element, the level of this enzyme is sensitive to dietary composition (Chow and Tappel, 1974; Hafeman *et al.*, 1974; Lawrence *et al.*, 1974; Omaye and Tappel, 1974; Smith *et al.*, 1974; Pederson *et al.*, 1975). A comprehensive general review on selenium action and toxicity by Diplock (1976) has recently appeared. Diplock notes that the presence of an enzymatic system for hydroperoxide removal overcomes most, but not all, objections raised by Green and Bunyan (1969) to the peroxidation-antioxidant theory.

Recognition of the importance of selenium in nutrition dates to the work of Schwarz and Foltz (1957) on the prevention of liver necrosis in rats fed Torula yeast by Factor 3. The production of liver necrosis is also prevented by L-cystine (Weichselbaum, 1935) and vitamin E (Schwarz, 1948). Sixteen years of chaos and confusion followed the discovery of selenium as an essential trace metal. Numerous experiments conducted thereafter demonstrated complex interactions between dietary supplements of selenium, sulfur amino acids, and vitamin E or synthetic lipid antioxidants. Thompson and Scott (1970) made a major contribution to the resolution of these interactions which are particularly evident in the chick. Reducing the level of selenium in the diet to 0.005 ppm resulted in pancreatic fibrosis and atrophy with impaired lipid absorption. Raising this level to 0.050 ppm increased the uptake of ^{14}C-α-tocopherol from the gut to the bloodstream 100-fold. Witting *et al.* (1967f) reported that based on whole body analyses, rats fed otherwise identical diets contained significantly more tocopherol when supplemented with selenium. On the basis of similarities in distribution of radioactive selenium and α-tocopherol in the blood stream, Desai and Scott (1965) suggested the presence of a seleno carrier protein for vitamin E in the chick. In the rat, however, supplemental selenium appeared to result in lower plasma levels of vitamin E and increased turnover of hepatic α-tocopherol (Cheeke and Oldfield, 1969; 1971).

The discovery (Flohe *et al.*, 1973; Rotruck *et al.*, 1973) that selenium was a constituent of glutathione peroxidase was truly a great light in the wilderness. Subsequently Shih *et al.* (1976) showed that chick pancreas contains relatively high levels of glutathione peroxidase. Diplock (1976) categorized tissues in terms of relative glutathione peroxidase levels—high: liver and erythrocytes; medium: heart, kidney, lung, adrenal, stomach, and adipose; low: brain, muscle, testes, and lens of the eye. Activity values, nmol/mg protein, from one report (Tappel, 1974) are liver, 250–350; erythrocyte, 150–250; stomach, 100; kidney, 60–150; testes, 40–120; lung, 50–80; small intestine, 15; muscle, 10–20. It is interesting to note that erythrocytes contain large amounts of both glutathione peroxidase and catalase. Levander *et al.* (1974) noted that 60% of the total selenium in mitochondria was present in glutathione peroxidase. Cytosol, however, contains twice as much glutathione peroxidase as the mitochondria (Green and O'Brien, 1970). This distribution would seem to provide protection to both sides of the membrane.

Phagocytic cells are known to contain glutathione peroxidase (Flohe, 1971). Platelets contain relatively high levels of glutathione peroxidase, and a genetic defect in this enzyme has been stated to be associated with Glanzmann's thrombasthenia (Karpatkin and Weiss, 1972). Hemolytic anemia is associated with lack of the enzyme (Boivin *et al.*, 1964; Necheles *et al.*, 1968, 1969, 1970; Gharib *et al.*, 1969; Steinberg *et al.*, 1970; Steinberg and Necheles, 1971; Nishimura *et al.*, 1972; Hopkins *et al.*, 1974; Najman *et al.*, 1974; Bernard *et al.*, 1975) or lack of glutathione (Boivin and Galand, 1965; Prins *et al.*, 1966). The lifespan of the erythrocyte is shortened in GSH deficiency in humans and in sheep (Young *et al.*, 1975) and in man a hemolytic crisis follows drug administration (Boivin and Galand, 1965). Chronic granulomatous disease related to a deficiency of leukocyte glutathione peroxidase was thought (Holmes *et al.*, 1970) to occur only in females, but a case in a male patient has recently been reported (Matsuda *et al.*, 1976). It has been suggested that in the GSH-deficient mutants of *E. coli* K 12, glutathione is not involved in essential metabolic processes, but it is needed to respond to stress (Apontoweil and Berends, 1975).

The destruction of hydroperoxides by glutathione peroxidase provides a facile explanation to a previously thorny problem. It should be noted, however, that this reaction has been demonstrated only with free acids (Christopherson, 1968). When using either NADPH-oxidase or ascorbate to initiate lipid peroxidation, glutathione peroxidase was found to prevent *initiation* (McCay *et al.*, 1976b). Hydroxy fatty acids were not detected, because lipid peroxidation apparently did not occur. If initiation depends on production of $\cdot OH$ via $O_2^- + H_2O_2 \rightarrow O_2 + \cdot OH + OH^-$, it is logical to inquire if glutathione peroxidase can directly prevent $\cdot OH$ formation via removal of H_2O_2. An article pertinent to this subject has appeared in the Russian literature (Lankin *et al.*, 1976).

7. Termination Reactions

The preceding sections have considered free-radical production by normal enzyme reactions and adventitious initiation of lipid peroxidation. Product hydroperoxides may be "detoxified" by peroxidases, react directly with proteins, undergo further oxidation resulting in tissue damage, or break down giving rise to new free radicals. "Safe" removal of the hydroperoxides makes a major contribution toward preventing tissue damage and free-radical multiplication, but does not terminate the cyclic chain reaction

$$R\cdot + O_2 \rightarrow RO_2\cdot$$
$$RO_2\cdot + RH \rightarrow ROOH + R\cdot$$

An antioxidant (AH) *competes* with lipid bound fatty acids (RH) for reaction with the peroxy free-radical ($RO_2\cdot$) and may withdraw free radicals from the system, thereby terminating the cyclic chain reaction:

$$RO_2\cdot + AH \rightarrow ROOH + [A\cdot]$$

This is reviewed as a nonenzymatic reaction. Any nontoxic, fat-soluble, lipid antioxidant that reaches and is retained at the appropriate subcellular site is capable of competitively terminating the cyclic chain reaction.

7.1. Structural Specificity for Biological Activity

Traditionally a vitamin is either a portion of the prosthetic group of an enzyme or otherwise functions in an enzymatic reaction. A long-standing problem in the area of lipid peroxidation has been the resistance to accepting the obvious conclusion that this description does not apply in the usual sense to vitamin E. The vitamin, however, clearly functions between two enzymatic reactions, free-radical production and hydroperoxide destruction. A novel control factor requirement is generated by the unusual self-perpetuating nature of the adventitiously initiated cyclic chain reaction, which would otherwise produce large quantities of an undesirable product. A separate enzymatic function of vitamin E is not precluded, but none has been detected despite exhaustive studies.

It is still practically impossible to satisfactorily demonstrate the occurrence of lipid peroxidation *in vivo,* although the association of lipofuscin or ceroid pigments, combinations of oxidized lipid and protein, have long been known to be associated with vitamin E deficiency states. Diplock (1976) has noted that the discovery of the selenium containing enzyme glutathione peroxidase overcomes most of the previously raised objections to the biological antioxidant theory.

The basic observations relevant to the nonspecific antioxidant action of vitamin E are 20–50 years old. These include the variation in vitamin E requirement related to PUFA intake (Evans and Burr, 1927). Evans *et al.* (1939) reported that many of the over 100 compounds related somewhat in structure to α-tocopherol were active in the resorption gestation assay. Dam and co-workers (Dam *et al.,* 1948, 1951) subsequently reported partial protection against vitamin E-deficiency signs by compounds structurally unrelated to the tocopherols. Rats were maintained for three generations on a tocopherol-deficient diet supplemented with N,N'-diphenyl-p-phenylenediamine (Draper *et al.,* 1958). Subsequent experiments demonstrated the cure of proved resorption gestation by this same antioxidant (Draper *et al.,* 1964).

Whereas any lipid antioxidant potentially has biological activity, it has become apparent that there are rather critical requirements for absorption, transport, subcellular distribution, and turnover. Using rat intestinal loops, α- and γ-tocopherol were found to be absorbed better than β- or δ-tocopherol (Pearson and Barnes, 1970). After direct introduction of the tocopherols into the bloodstream, the orders of disappearance were $\delta > \beta > \gamma > \alpha$ (Chow et al., 1971). Although δ-tocopherol was absent from the blood stream after 8 h, γ-tocopherol was still detectable after 192 h. At the tissue level, γ-tocopherol appears to turn over much more rapidly than does α-tocopherol (Peake and Bieri, 1971; Peake et al., 1972). The known synthetic lipid antioxidants tend to distribute with fat per se, rather than concentrating, as does vitamin E, in the subcellular organelles (Csallany and Draper, 1960; Wiss et al., 1962), and are characterized either by poor absorption or extremely rapid turnover.

7.2. Megavitamin Fads, Fancies, and Facts

The normal vitamin E requirement is determined by tissue lipid PUFA levels, which in turn are influenced by the dietary level of PUFA. The dietary level of linoleate has doubled in the last 10–15 years and seems certain to increase still further. Tissue lipid PUFA composition in the adult at stable weight is not rapidly altered by the daily intake of 10–30 g of PUFA, most of which may be used as a source of calories, and therefore lags months or years behind the dietary changes (Witting and Lee, 1975a,b). This makes it difficult to set a specific requirement, but the recommended dietary allowance is of the order of magnitude of 12–15 IU (Food and Nutrition Board, 1974). The intake of PUFA and vitamin E tends to be balanced in any reasonable normal mixed diet. Suggestions have appeared for expressing the vitamin E requirement in terms of dietary PUFA (Harris and Embree, 1963; Horwitt, 1974) or tissue linoleate (Witting, 1972; Witting and Lee, 1975a).

What might be interpreted as minimal deficiency signs, by reference to animal models, are encountered in man only in malabsorption syndromes or after excision of portions of the gut (Binder et al., 1965; Binder and Spiro, 1967). Creatinuria and muscular weakness have been observed in cystic fibrosis patients (Nitowsky et al., 1962). Serum creatine kinase and aldolase, which are better indicators of nutritional muscular dystrophy, were elevated in only two of 50 vitamin E-deficient cystic fibrosis patients studied by Farrell et al. (1975). Actual lesions at necrosy were seen in only one of 48 children who had had cystic fibrosis (Oppenheimer, 1956). The point to be made is that even under highly adverse conditions, the pathological vitamin E deficiency signs produced in experimental animals are only minimally approached in humans. Several recent reviews have covered this topic in great depth (Bieri and Farrell, 1976; Farrell, 1976).

Various stress situations involving enhanced free-radical production can be shown to be moderated by vitamin E or preferably by a synthetic lipid antioxidant such as N,N'-diphenyl-p-phenylenediamine. Included are intoxications (Hove, 1953; DiLuzio and Costales, 1965; Recknagel, 1967; DiLuzio et al., 1973; Recknagel and Glende, 1973), hyperoxia (Kann et al., 1964; Johnson et al., 1972), photochemical smog (Thomas et al., 1967; Roehm et al., 1972), cancer (Wattenberg et al., 1976), and perhaps the deteriorative processes associated with the generalized phenomenon of aging (Tappel, 1967, 1968). The quality of the scientific data varies from extremely substantial to extremely vague. Frequently the effect can be best or only demonstrated by comparing deficient and supplemented animals rather than normal and supplemented animals.

In large quantities vitamin E appears to have a mild *transitory* antiinflammatory action presumably related to an effect on prostaglandin biosynthesis. Massive doses of the vitamin, 100–1000 times the requirement, are nontoxic even when taken for multiyear periods. Extensive claims have been made for beneficial effects of pharmacological levels of vitamin E. Farrell (1976) has recently reviewed in great detail the literature relevant to possible effects of vitamin E in (1) conditions that may tend to produce a deficiency state, e.g., prematurity, retrolental fibroplasia, cystic fibrosis, other malabsorption syndromes, A-β-lipoproteinemia, protein-calorie malnutrition, and (2) other conditions wherein pharmacological dosages have been advocated, such as ischemic heart disease, intermittent claudication, other cardiovascular disorders, spontaneous abortion, cystic mastitis, thalassemia, wound healing, gravitational ulcers, peridontal disease, porphyria, and aging. As previously noted by Marks (1962), Farrell (1976) found that most of the published studies, both negative and positive, that involve megavitamin E therapy were not adequately controlled and will not withstand careful scientific scrutiny.

The data of Losowsky et al. (1972) seem particularly pertinent when evaluating the current fad for high levels of vitamin E. In human subjects, the absorption of α-tocopherol decreased form 50–80% to 20–50% and to as little as 5% when ingested doses were 0.04, 20, and 200 mg, respectively. As is the case with most vitamins, excess intake is largely transported to the sewerage system.

8. Tissue Damage

The preceding sections have considered the multilevel control system which functions to prevent the formation and/or accumulation of lipid hydroperoxides in tissues without considering why this was desirable or necessary. In their NADPH-oxidase system Tam and McCay (1970) described hydroperoxides as transient intermediates. During lipid peroxidation the lipid-bound fatty acid free radical may exist in a number of resonance forms. The product hydroperoxide, however, will contain the structure attained by movement, with inversion of configuration, of the double bond at the site of attack into conjugation with the next adjacent double bond. This conjugated diene is much more susceptible to oxidation than is the original PUFA. Secondary oxidations and chain scission reactions rapidly become important. The evolution of pentane and/or ethane in the breath of rats has been used as an indication of the occurrence of lipid peroxidation *in vivo* (Hafeman and Hoekstra, 1975, 1976; Dillard *et al.*, 1976).

Scission products, particularly aldehydes and dialdehydes are likely candidates for reaction with protein and nucleic acids. Fluorescent pigments, combinations of oxidized lipid, and protein, have been known for decades and have been isolated by Siakotos *et al.* (1970, 1973). Model system studies have suggested malonaldehyde as a likely cross-linking agent in the production of fluorescent pigments (Chio and Tappel, 1969a,b), but this structure has never been demonstrated chemically in isolated pigment. Tappel (1975) has reviewed the topic of lipid peroxidation and what he terms fluorescent molecular damage. Fluorescence measurement on crude tissue extracts may lead to erroneous results (Csallany and Ayaz, 1976), and several investigators have failed to detect the expected correlations (Witting, 1973; Grinna, 1976).

Since relatively little is known about the actual reactions leading from lipid peroxidation to tissue damage, there has been some tendency to measure fluorescent pigment production. A shotgun approach, a mixture of selenium, sulfur amino acids, and synthetic lipid antioxidant, has been used in a variety of poorly defined systems to attempt to reduce tissue damage as noted in section 8.2.

8.1. Species Variation at the Site of Tissue Damage

Considerable confusion is apparent in the literature regarding the diversity of pathological signs associated with lipid peroxidation within or between species. Viewed in terms of new knowledge regarding the multilevel control systems for preventing free-radical-initiated tissue damage, some reports are amenable to retrospective reinterpretation. It is also evident that in a system where tissue damage is occurring a limited positive protective response may be evoked by supplementation affecting higher or lower control systems.

8.2. Aging, Genetic Diseases, and Cancer

Various lesions or degenerative processes, or both, have components suggesting that lipid peroxidation may be involved. Included are the generalized phenomena of aging, progeria, and Batten's disease. Batten's disease, neuronal ceroid lipofuscinosis in humans and English setters, is characterized by normal serum tocopherol levels (Siakotos *et al.*, 1974) and normal brain levels of glutathione peroxidase (Zeman and Rider, 1975). Massive CNS accumulation of lipopigment appears to be related to a genetic lesion in a peroxidase assayed with *p*-phenylenediamine (Armstrong *et al.*, 1974a,b, 1975; Dimmit, 1975). Shotgun therapy (α-tocopherol, ascorbic acid, methionine, and butylated hydroxytoluene,) has been stated to slow the rate of progression of the degenerative processes (Siakotos *et al.*, 1974).

Harman (1968, 1969) has considered the effect of synthetic lipid antioxidants on lifespan and average survival. Average survival time appeared to increase although maximum survival was not changed. These results in mice could not be duplicated in rats.

Epoxidation of polycyclic aromatic hydrocarbons (Sims and Glover, 1974) to "ultimate carcinogens" and inactivation by diol formation by epoxide hydrase or reaction with glutathione via glutathione-*S*-epoxide transferase (Hayakawa *et al.*, 1975) is reminiscent of the reactions discussed earlier in this review. The inhibition of cancer production by synthetic lipid antioxidants has recently been reviewed by Wattenberg *et al.* (1976) and the broader topic of nutrition and cancer has been reviewed by Wynder (1976).

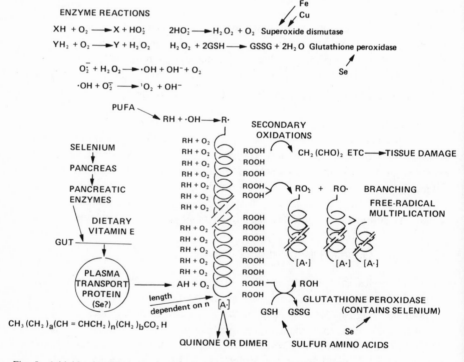

Fig. 5. A highly simplified sketch of the interrelations thought to prevail between free-radical production, initiation of adventitious reactions, and various control systems operating to prevent or minimize tissue damage. The major dietary factors, such as polyunsaturated fatty acids, vitamin E, biologically available selenium, iron, and copper, are noted near the reactions affected.

9. Summary

Figure 5 is offered as a general summary of this review. Normal enzymatic reactions generate free radicals and hydrogen peroxide, which may interact to produce hydroxy radical and then singlet oxygen. Adventitious initiation of lipid peroxidation may result forming lipid hydroperoxides which, unless converted to hydroxy fatty acids, may cause tissue damage. Control mechanisms are indicated at various levels and major nutritional factors effecting control systems are indicated. Interception of superoxide anion (Nishikimi and Machlin, 1975) or singlet oxygen (Grams, 1971) by tocopherol may need to be added at some future date.

Acknowledgments

Sincere thanks are due those friends and colleagues who made available reprints and preprints of their work, particularly Dr. J. G. Bieri, Dr. A. Diplock, Dr. P. M. Farrell, Dr. I. Fridovich, Dr. L. S. Grinna, Dr. O. Levander, Dr. L. Machlin, Dr. P. B. McCay, Dr. R. Recknagel, and Dr. A. L. Tappel.

References

Abbott, M. T., and Udenfriend, S., 1974, α-Ketoglutarate-coupled dioxygenases, in *Molecular Mechanisms of Oxygen Activation* (O. Hiyaishi, ed.), pp. 168–214, Academic Press, New York.

Allen, R. C., 1975, The role of pH in the chemiluminescent response of the myeloperoxidase-halide-HOOH antimicrobial system, *Biochem. Biophys. Res. Commun.* **63**:684–691.

Ames, S. P., 1968, Bioassay of vitamin A, in *The Biochemistry, Assay and Nutritional Value of Vitamin A* (W. J. Monson, ed.), pp. 17–31, Association of Vitamin Chemists, Chicago.

Apontoweil, P., and Berends, W., 1975, Isolation and initial characterization of glutathione-deficient mutants of *Escherichia coli* K12, *Biochim. Biophy. Acta* **399**:10–22.

Armstrong, D., Dimmitt, S., Boehme, D. H., Leonberg, S. C., and Vogel, W., 1974a, Leukocyte peroxidase deficiency in a family with a dominant form of Kuf's disease, *Science* **186**:155–156.

Armstrong, D., Dimmitt, S., and Van Wormer, D. E., 1974b, Studies on Batten's disease, I. Peroxidase deficiency in granulocytes, *Arch. Neurol.* **30**:144–152.

Armstrong, D., Van Wormer, D. E., and Dimmit, S., 1974c, Tissue peroxidase deficiency in brain and viscera in Batten's disease, *Trans. Am. Soc. Neurochem.* **5**:135.

Armstrong, D., Van Wormer, D. E. Neville, H., and Dimmit, S., 1975, Thyroid peroxidase deficiency in Batten-Spielmeyer-Vogt disease, *Arch. Pathol.* **99**(8):430–435.

Babish, J. G., and Stoewsand, G. S., 1975, Effect of tea intake on induction of hepatic microsomal enzyme activity in the rabbit, *Nutr. Rep. Int.* **12**:109–114.

Becking, G. C., 1973, Vitamin A status and hepatic drug metabolism in the rat, *Can. J. Physiol. Pharmacol.* **51**:6–11.

Becking, G. C., 1976, Hepatic drug metabolism in iron-, magnesium- and potassium-deficient rats, *Fed. Proc.* **35**:2480–2485.

Bernard, J. F., Galand, C., and Boivin, P., 1975, March hemoglobinuria. One case with erythrocyte glutathione peroxidase deficiency, *Presse Med.* **4**(15):1117–1120.

Bernhard, K., Leisinger, S., and Pedersen, W., 1963a, Vitamin E und arachidonsäure-bildung in der Leber, *Helv. Chim. Acta* **46**:1767–1772.

Bernhard, K., Lindlar, F., Schwed, P., Vuilleumier, J. P., and Wagner, H. Z., 1963b, Fettsaüre-Stoffwechsel bei vitamin E-mangel, *Ernahrungswissenschaft* **4**:42–49.

Beuge, J. A., and Aust, S. D., 1976, Lactoperoxidase catalyzed lipid peroxidation of microsomal and artificial membranes, *Biochim. Biophys. Acta* **444**:192–201.

Bidlack, W. R., and Tappel, A. L., 1973, Damage to microsomal membrane by lipid peroxidation, *Lipids* **8**:177–182.

Bieri, J. G., and Farrell, P. M., 1976, Vitamin E, *Vit. Horm.* **34**:31–75.

Binder, H. J., and Spiro, H. M., 1967, Tocopherol deficiency in man, *Am. J. Clin. Nutr.* **20**:594–601.

Binder, H. J., Herting, D. C., Hurst, V., Finch, S. C., and Spiro, H. M., 1965, Tocopherol deficiency in man, *New Engl. J. Med.* **273**:1289–1297.

Boguth, W., and Niemann, 1971, Electron spin resonance of chromanoxy free-radicals from α-, ϵ_2-, β-, γ-, δ-tocopherol and tocol, *Biochim. Biophys. Acta* **248**:121–130.

Boivin, P., and Galand, G., 1965, Synthesis of glutathione in congenital hemolytic anemia with deficient reduced glutathione—A congenital deficiency of erythrocyte glutathione synthetase?, *Nouv. Rev. Fr. Hematol.* **5**:707–720.

Bolland, J. L., and Gee, G., 1946, Thermochemistry and mechanisms of olefin oxidation, *Trans. Farraday Soc.* **42**:244–252.

Bors, W., Saran, M., Lengfelder, E., Spöttl, R., and Michel, C., 1974, The relevance of the superoxide anion radical in biological systems, *Curr. Topics Radiat. Res. Quart.* **9**:309–347.

Boveris, A., and Chance, B., 1973, The mitochondrial generation of hydrogen peroxide: General properties and effect of hyperbaric oxygen, *Biochem. J.* **134**:707–716.

Boveris, A., Oshno, N., and Chance, B., 1972, The cellular production of hydrogen peroxide, *Biochem. J.* **128**:617–630.

Campbell, T. C., and Hayes, J. R., 1974, Role of nutrition in the drug-metabolizing enzyme system, *Pharmacol. Rev.* **26**:171–197.

Campbell, T. C., and Hayes, J. R., 1976, The effect of quantity and quality of dietary protein on drug metabolism, *Fed. Proc.* **35**:2470–2474.

Caster, W. O., Wade, A. E., Norred, W. P., and Bargmann, R. E., 1970, Differential effect of dietary saturated fat on the metabolism of aniline and hexobarbital by the rat liver, *Pharmacology* **3**:117–186.

Cheeke, P. R., and Oldfield, J. E., 1969, Influence of selenium on the absorption excretion and plasma level of tritium-labelled vitamin E in the rat, *Can. J. Anim. Sci.* **49**:169–179.

Cheeke, P. R., and Oldfield, J. E., 1971, Influence of selenium on the time distribution of tritium-labelled tocopherol in selected tissues of the rat, *Can. J. Anim. Sci.* **51**:533–536.

Chio, K. S., and Tappel, A. L., 1969a, Synthesis and characterization of the fluorescent products derived from malonaldehyde and amino acids, *Biochemistry* **8**:2821–2827.

Chio, K. S., and Tappel, A. L., 1969b, Inactivation of ribonuclease and other enzymes by peroxidizing lipids and by malonaldehyde, *Biochemistry* **8**:2827–2832.

Chow, C. K., Csallany, A. S., and Draper, H. H., 1971, Relative turnover rates of the tocochromanols in rabbit plasma, *Nutr. Rep. Int.* **4**:45–48.

Chow, C. K., and Tappel, A. L., 1974, Response of glutathione peroxidase to dietary selenium in rats, *J. Nutr.* **104**:444–451.

Christopherson, B. O., 1968, Formation of monohydroxy-polyenoic fatty acids from lipid peroxides by a glutathione peroxidase, *Biochim. Biophys. Acta* **164**:35–46.

Christopherson, B. O., 1969, Reduction of linoleic acid hydroperoxide by a glutathione peroxidase, *Biochim. Biophys. Acta* **176**:463–470.

Chung, L. W. K., Raymond, G., and Fox, S., 1975, Role of neonatal androgen in the development of hepatic microsomal drug metabolizing enzymes, *J. Pharmacol. Exp. Ther.* **193**:621–630.

Chvapil, M., Ludwig, J. C., Spies, I. G., and Misiorowski, R. L., 1976, Inhibition of NADPH oxidation and related drug oxidation in liver microsome by zinc, *Biochem. Pharmacol.* **25**:1787–1791.

Chvapil, M., Ryan, J. N., Elias, S. L., and Peng, Y. M., 1973, Protective effect of zinc on carbon tetrachloride-induced liver injury in rats, *Exper. Mol. Pathol.* **19**:186–196.

Chvapil, M., Spies, I. G., Ludwig, J. C., and Halladay, S. C., 1975, Inhibition of NADPH oxidation and oxidative metabolism of drugs in liver microsomes by zinc, *Biochem. Pharmacol.* **24**:917–919.

Clark, K. G. A., and Davidson, W. M., 1974, Bone marrow lipofuscinosis, *J. Clin. Pathol.* **25**:947–950.

Colby, H. D., Kramer, R. E., Greiner, J. W., Robinson, D. A., Krause, R. F., and Canady, W. J., 1975, Hepatic drug metabolism in retinol-deficient rats, *Biochem. Pharmacol.* **24**:1644–1646.

Combs, G. F., Jr., 1976, Differential effects of high dietary levels of vitamin A on the vitamin E-selenium nutrition of young and adult chickens, *J. Nutr.* **106**:967–975.

Comporti, M., Hartman, A., and DiLuzio, N. R., 1967, Effect of in vivo and in vitro ethanol administration on liver lipid peroxidation, *Lab. Invest.* **16**:616–624.

Comporti, M., Landucci, G., and Raja, F., 1971, Changes in microsomal lipids of rat liver after chronic carbon tetrachloride intoxication, *Experientia* **27**:1155–1156.

Conney, A. H., 1967, Pharmacological implications of microsomal enzyme induction, *Pharmacol. Rev.* **19**:317–366.

Csallany, A. S., 1971, A reappraisal of the structure of a dimeric metabolite of α-tocopherol, *Int. J. Vit. Nutr. Res.* **41**:376–384.

Csallany, A. S., and Ayaz, K. L., 1976, Quantitative determination of organic solvent soluble lipofuscin pigments in tissues, *Lipids* **11**:412–417.

Csallany, A. S., and Draper, H. H., 1960, Determination of N,N'-diphenyl-*p*-phenylenediamine in animal tissues, *Proc. Soc. Exper. Biol. Med.* **104**:739–742.

Curnette, J. T., Whitten, D. M., and Babior, B. M., 1974, Defective superoxide production by granulocytes from patients with chronic granulomatous disease, *New Engl. J. Med.* **290**:593–597.

Dam, H., Kruse, I., Prange, I., and Sondergaard, E., 1951, Substances affording a partial protection against certain vitamin E deficiency symptoms, *Acta Physiol. Scad.* **22**:299–305.

Dam, H., Kruse, I., Prange, I., and Sondergaard, E., 1948, Influence of dietary ascorbic acid nordihydroguaiaretic acid and cystine in vitamin E deficiency in chicks, *Biochim. Biophys. Acta* **2**:501–513.

Darsie, J., Gosha, S. K., and Holman, R. T., 1976, Induction of abnormal fatty acid metabolism and EFA-deficiency in rats by dietary DDT, *Arch. Biochem. Biophys.* **175**:262–269.

Desai, I. D., Calvert, C. C., and Scott, M. L., 1964, A time sequence study of the relationship of peroxidation lysosomal enzymes and nutritional muscular dystrophy, *Arch. Biochem. Biophys.* **108**:60–64.

Dillard, C. J., Dumelin, E. E., and Tappel, A. L., 1977, Effect of dietary vitamin E on expiration of pentane and ethane by the rat, *Lipids* **12**:109–114.

DiLuzio, N. R., 1966, A mechanism of the acute ethanol-induced fatty liver and the modification by antioxidants, *Lab. Invest.* **15**:50–63.

DiLuzio, N. R., and Costales, F., 1965, Inhibition of ethanol and carbon tetrachloride induced fatty liver by antioxidants, *Exp. Mol. Pathol.* **4**:141–154.

DiLuzio, N. R., Stiege, T. E., and Hoffman, E. O., 1973, Protective influence of diphenyl-*p*-phenylenediamine on hydrazine induced lipid peroxidation and hepatic injury, *Exp. Mol. Pathol.* **19**:284–292.

Dimmit, S. K., 1975, The effect of biologic amines on peroxidase activity (*p*-phenylenediamine as cosubstrate in detecting peroxidase efficiency, Batten's disease), *J. Am. Med. Wom. Assoc.* **30**(12):473–483.

Diplock, A. T., 1976, Metabolic aspects of selenium action and toxicity, *CRC Crit. Rev. Toxicol.* **5**:271–329.

Diplock, A. T., Green, J., Bunyan, J., McHale, D., and Muthy, I. R., 1967, Vitamin E and stress. *3*. The metabolism of D-α-tocopherol in the rat under dietary stress with silver, *Br. J. Nutr.* **21**:115–125.

Draper, H. H., Bergan, J. G., Chiu, M., Csallany, A. S., and Boaro, A. V., 1964, A further study of the specificity of the vitamin E requirement for reproduction, *J. Nutr.* **84**:395–400.

Draper, H. H., Goodyear, S., Barbee, K. K., and Johnson, B. C., 1958, A study of the nutritional role of antioxidants in the diet of the rat, *Br. J. Nutr.* **12**:89–97.

Entsch, B., Ballou, D. P., and Massey, V., 1976, Flavin-oxygen derivatives involved in hydroxylation by *p*-hydroxybenzoate hydroxylase (for Pseudomonas fluorescense), *J. Biol. Chem.* **251**:2550–2563.

Evans, H. M. and Burr, G. O., 1927, Vitamin E: The ineffectiveness of a curative dosage when mixed with diets containing high proportions of certain fats, *J. Am. Med. Assoc.* **88**:1462–1465.

Evans, H. M., Emerson, O. H., Emerson, G. A., Smith, L. I., Ungrade, H. E., Prickard, W. W., Austin, F. L., Hoehn, H. H., Opie, J. W., and Wawzonek, S., 1939, The chemistry of vitamin E, XIII. Specificity and relationship between chemical structure and vitamin E activity, *J. Org. Chem.* **4**:376–388.

Farrel, P. M., 1977, Vitamin E in human health and disease, in *Vitamin E* (L. J. Machlin, ed.), Dekker, New York (in press).

Farrell, P. M., Fratantoni, J. C., Bieri, J. G., and di Sant'Agnese, P. A., 1975, Effects of vitamin E deficiency, *Acta Pediat. Scand.* **64**:150–159.

Fee, J. A., and Teitelbaum, H. D., 1972, Evidence that superoxide dismutase plays a role in protecting rbc against peroxidative hemolysis, *Biochem. Biophys. Res. Commun.* **49**:150–158.

Flashner, M. S., and Massey, V., 1974, Flavoprotein oxygenases, in *Molecular Mechanisms of Oxygen Activation,* Academic Press, New York.

Flohe, L., 1971, Glutathione peroxidase, enzymological and biological aspects, *Klin. Wochenschr.* **49**:669–683.

Flohe, L., and Brand, I., 1969, Kinetics of glutathione peroxidase, *Biochim. Biophys. Acta* **191**:541–549.

Flohe, L., Gunzler, W. A., and Schock, H. H., 1973, Glutathione peroxidase: a selenoenzyme, *FEBS Lett.* **32**:132–134.

Fong, K-L, McCay, P. B., Poyer, J. L., Keele, B. B., and Misra, H., 1973, Evidence that peroxidation of lysosomal membranes is initiated by hydroxyl free-radicals produced during flavin enzyme activity. *J. Biol. Chem.* **248**:7792–7797.

Fong, K-L, McCay, P. B., Poyer, J. L. Misra, H., and Keele, B. B., Evidence for superoxide-dependent reduction of Fe^{3+} and its role in enzyme-generated hydroxyl radical formation, *Chemico-biol. Interactions* **15**:77–89.

Food and Nutrition Board, 1974, Recommended Dietary Allowances, 8th rev. ed., National Academy of Sciences Publication No. 2216, 1974.

Foote, C. S., 1968, Photosensitized oxygenations and the role of singlet oxygen, *Accounts Chem. Res.* **1**:104–110.

Fowler, J. S. L., 1969, Carbon tetrachloride metabolism in the rabbit, *Br. J. Pharmacol.* **37**:733–737.

French, S. W., Ihrig, T. J., Shaw, G. P., Tanaka, T. T., and Norum, M. L., 1971, The effect of ethanol on the fatty acid composition of hepatic microsomes and inner and outer mitochondrical membranes, *Res. Commun. Chem. Pathol. Pharmacol.* **2**:567–585.

Fridovich, I., 1975, Superoxide dismutases, *Ann. Rev. Biochem.* **44**:147–159.

Fridovich, I., and Handler, P., 1958, Xanthine oxidase, III. Sulfite oxidation as an ultrasensitive assey, *Biol. Chem.* **233**:1578–1580.

Fridovich, I., and Handler, P., 1961, Detection of free radicals generated during enzymic oxidations by the initiation of sulfite oxidation, *J. Biol. Chem.* **236**:1836–1840.

Frieden, E., 1971, Ceruloplasmin, a link between copper and iron, in *Bioinorganic Chemistry* (R. F. Gould, ed.), pp. 292–321, American Chemical Society, Washington, D.C.

Galdhar, N. R., and Pawar, S. S., 1976, Hepatic drug metabolism and lipid peroxidation in thiamine deficient rats, *Int. J. Vit. Nutr. Res.* **46**:14–23.

Gharib, H., Fairbank, V. F., and Bartholomew, L. G., 1969, Hepatic failure with acanthocytosis. Association with hemolytic anemia and deficiency of erythrocyte glutathione peroxidase, *Mayo Clinic Proc.* **44**:96–105.

Ghoshal, A. K., Monserrat, A. J., Porta, E. D., and Hartroft, W. S., 1970, Role of lipoperoxidation in early choline deficiency, *Exp. Mol. Pathol.* **12**:31–35.

Ghoshal, A. K., Porta, E. A., and Hartroft, W. S., 1969, The role of lipoperoxidation in the pathogenesis fatty livers induced by phosphorus poisoning in rats, *Am. J. Pathol.* **54**:275–291.

Gordis, E., 1963, Lipid metabolites of carbon tetrachloride, *J. Clin. Invest.* **48**:203–209.

Gram, T. E., and Fouts, J. R., 1966, Effect of α-tocopherol upon lipid peroxidation and drug metabolism in hepatic microsomes, *Arch. Biochem. Biophys.* **114**:331–335.

Grams, G. W., 1971, Oxidation of α-tocopherol by singlet oxygen, *Tetrahedron Lett.* **50**:4823–4825.

Green, J., and Bunyan, J., 1969, Vitamin E and the biological antioxidant theory, *Nutr. Abstr. Rev.* **39**:321–345.

Green, J., Diplock, A. T., Bunyan, McHale, D., and Muthy, I. R., 1967, Vitamin E and stress, 1. Dietary unsaturated fatty acid stress and metabolism of α-tocopherol in the rat, *Br. J. Nutr.* **21**:69–101.

Green, R. C., and O'Brien, P. J., 1970, The cellular localization of glutathione peroxidase and its release from mitochondria during swelling, *Biochim. Biophys. Acta* **197**:31–37.

Gregory, E. M., and Fridovich, I., 1973a, Induction of superoxide dismutase by molecular oxygen, *J. Bacteriol.* **114**:543–548.

Gregory, E. M. and Fridovich, I., 1973b, Oxygen toxicity and the superoxide dismutase, *J. Bacteriol.* **114**:1193–1197.

Gregory, E. M., Yost, F. J., Jr., and Fridovich, I., 1973, Superoxide dismutases of *Escherichia coli:* Intracellular localization and functions, *J. Bacteriol.* **115**:987–991.

Grinna, L. S., 1976, Studies on the effect of dietary α-tocopherol on liver microsomes and mitochondria of aging rats, *J. Nutr.* **106**:918–929.

Gruger, E. H., and Tappel, A. L., 1970, Reactions of biological antioxidants. I. Fe(III) catalyzed reactions of lipid hydroperoxides with α-tocopherol, *Lipids* **5**:326–331.

Gunsalus, I. C., Meeks, J. R., Lipscomb, J. O., Debrunner, P., and Münck, E., 1974, Bacterial monoxygenases—The P-450 cytochrome system, in *Molecular Mechanisms of Oxygen Activation* (O. Hayaishi, ed.), pp. 559–613, Academic Press, New York.

Gunsalus, I. C., Pederson, T. C., and Sligar, S. G., 1975, Oxygenase-catalyzed biological hydroxylations, *Ann. Rev. Biochem.* **44**:377–407.

Hafeman, D. G., and Hoekstra, W. G., 1975, Protection by vitamin E and selenium against lipid peroxidation *in vivo* as measured by ethane evolution, *Fed. Proc.* **34**:939.

Hafeman, D. G., and Hoekstra, W. G., 1976, Exponentially increasing lipid peroxidation in vivo in the terminal phase of vitamin E and selenium deficiency, *Fed. Proc.* **35**:740.

Hafeman, D. G., Sunde, R. A., and Hoekstra, W. G., 1974, Effect of dietary selenium on erythrocyte and liver glutathione peroxidase in the rat, *J. Nutr.* **104**:580–587.

Hamberg, M., and Samuelsson, B., 1974a, Prostaglandin endoperoxides, VII. Novel transformations of arachidonic acid in guinea pig lung, *Biochem. Biophys. Res. Commun.* **61**(3):942–949.

Hamberg, M., and Samuelsson, B., 1974b, Prostaglandin endoperoxides, novel transformations of 20 : 4 in human platelets, *Proc. Natl. Acad. Sci. USA* **71**:3400–3404.

Hamberg, M., Svensson, J., and Samuelsson, B., 1974a, Prostaglandin endoperoxides, a new concept concerning the mode of action and release of prostaglandins, *Proc. Natl. Acad. Sci. USA* **71**:3824–3828.

Hamberg, M., Svensson, J., Wakabayashi, T., and Samuelsson, B., 1974b, Isolation and structure of two prostaglandin endoperoxides that cause platelet aggregation, *Proc. Natl. Acad. Sci. USA* **71**:345–349.

Hamilton, G. A., 1971, in *Progress in Bioinorganic Chemistry* (E. T. Kaiser and F. J. Kezdy, eds.), Vol. 1, p. 83, Wiley, New York.

Hamilton, G. A., 1974, Chemical models and mechanisms for oxygenases, in *Molecular Mechanisms of Oxygen Activation* (O. Hayaishi, ed.), pp. 405–451, Academic Press, New York.

Harman, D., 1968, Free-radical theory of aging: Effect of free-radical reaction inhibitors on the mortality rate of male LAF_1 mice, *J. Gerontol.* **23**:476–482.

Harman, D., 1969, Chemical protection against aging, *Agents Actions* **1**:3–8.

Harris, P. L., and Embree, N. D., 1963, Quantitative consideration of the effect of polyunsaturated fatty acid content of the diet upon requirement for vitamin E, *Am. J. Clin. Nutr.* **13**:385–392.

Hartman, A. D., and DiLuzio, N. R., 1968, Inhibition of chronic ethanol induced fatty liver by antioxidant administration, *Proc. Soc. Exp. Biol. Med.* **127**:270–276.

Hartroft, W. S., 1963, Pathogenesis of ceroid pigment: Preceroid a precursor, *Fed. Proc.* **22**:250.

Hayaishi, O., 1974a, *Molecular Mechanism of Oxygen Activation,* Academic Press, New York.

Hayaishi, O., 1974b, *Molecular Oxygen in Biology,* North-Holland, Amsterdam.

Hayakawa, T., Udenfriend, S., Yagi, H., and Jerina, D. M., 1975, Substrates and inhibitors of hepatic glutathione-S-epoxide transferase, *Arch. Biochem. Biophys.* **170**:438–451.

Hiatt, R. R., 1976, Hydroperoxide destroyers and how they work, CRC *Crit. Rev. Food Sci. Nutr.* **7**:1–12.

Hiramitsu, T., Hasegawa, Y., Hirata, K., Nishigaki, I., and Yagi, K., 1975, Role of lipid peroxide in the induction of the retinal siderosis, *Acta Soc. Opthalmol. Jap.* **79**(10):468–473.

Hirata, F., and Hiyaishi, O., 1971, Possible participation of superoxide anion in the intestinal tryptophan 2,3-dioxygenase reaction, *J. Biol. Chem.* **246**:7825–7826.

Holmes, B., Park, P. H., Malawista, S. E., Quie, P. G., Nelson, D. L., and Good, R. H., 1970, Chronic granulomatous disease in females, a deficiency of leukocyte glutathione peroxidase, *New Engl. J. Med.* **283**:217–221.

Hopkins, J. and Tudhope, G. R., 1974, Glutathione peroxidase deficiency with increased susceptibility to erythrocyte Heinz body formation, *Clin. Sci. Mol. Med.* **47**(6):643–647.

Horning, M. G., Earle, M. J., and Maling, H. M., 1962, Changes in fatty acid composition of liver lipids induced by carbon tetrachloride and ethionine, *Biochim. Biophys. Acta* **56**:175–177.

Horwitt, M. K., 1974, Status of human requirements for vitamin E, *Am. J. Clin. Nutr.* **27**:1182–1193.

Hove, E. L., 1948, Interrelation between α-tocopherol and protein metabolism: III. The protective effect of vitamin E and certain nitrogenous compounds against CCl_4 poisoning in rats, *Arch. Biochem.* **17**:467–473.

Hove, E. L., 1953, The toxicity of tri-*o*-cresylphosphate for rats as related to dietary casein level, vitamin E and vitamin A, *J. Nutr.* **51**:609–622.

Ilan, Y. A., Czapski, G., and Meisl, D., 1976, The one-electron transfer redox potentials of free radicals. I. The oxygen/superoxide system, *Biochim. Biophys. Acta* **430**:209–224.

Ishak, K. G., and Irey, N. S., 1972, Hepatic injury associated with the phenothiazines, *Arch. Pathol.* **93**:283–304.

Johnson, W. P., Jefferson, D., and Mengel, C. E., 1972, In vivo hemolysis due to hyperoxia: Role of H_2O_2 accumulation, *Aerospace Med.* **43**:943–945.

Kann, H. E., Jr., Mengel, C. E., Smith, W., and Horton, B., 1964, Oxygen toxicity and vitamin E, *Aerosp. Med.* **35**:840–844.

Karpatkin, S., and Weiss, H. J., 1972, Deficiency of glutathione peroxidase associated with high levels of reduced glutathione in Glanzmann's thrombasthenia, *New Engl. J. Med.* **287**:1062–1066.

Kaschnitz, R. M., and Coon, M. J., 1975, Drug and fatty acid hydroxylation by solubilized human liver microsomal cytochrome P-450—Phospholipid requirement, *Biochem. Pharmacol.* **24**:295–297.

Kaufman, S., and Fisher, D. B., 1974, Pterin-requiring aromatic amino acid hydroxylases, in *Molecular Mechanisms of Oxygen Activation* (O. Hayaishi, ed.), pp. 285–369, Academic Press, New York.

Kellog, E. W., and Drifovich, I., 1975, Superoxide, hydrogen peroxide and singlet oxygen in lipid peroxidation by a xanthine oxidase system, *J. Biol. Chem.* **250**:8812–8817.

King, M. M., Lai, E. K., and McCay, P. B., 1975, Singlet oxygen production associated with enzyme-catalyzed lipid peroxidation in liver microsomes, *J. Biol. Chem.* **250**:6496–6502.

Kokatnur, M. G., Bergan, J. G., and Draper, H. H., 1965, A rapid method for the preparation of peroxides from autoxidized methyl esters of fatty acids, *Anal. Biochem.* **12**:325–331.

Krinsky, N. I., 1947, Singlet excited oxygen as a mediator of the antibacterial action of leukocytes, *Science* **186**:363–365.

Labuza, T. P., 1971, Kinetics of lipid oxidation in foods, *CRC Crit. Rev. Food Technol.* **2**:355–405.

Labuza, T., Tsyuki, H., and Karel, M., 1969, Kinetics of linoleate oxidation in model systems, *J. Am. Oil Chem. Soc.* **46**:409–416.

Lankin, V. Z., 1976, Inhibition of lipid peroxidation and detoxification of lipoperoxides by protective enzyme systems (superoxide dismutase, glutathione peroxidase and glutathione reductase) during experimental neoplastic growth, *Dokl. Akad. Nauk. SSSR* **226**(3):705–708.

Lapidot, A., and Irving, C. S., 1974, The electronic structure of coordinated oxygen, in *Molecular Oxygen in Biology* (O. Hayaishi, ed.), pp. 33–80, North-Holland, Amsterdam.

Lawrence, R. A., Sunde, R. A., Schwartz, G. L., and Hoekstra, W. G., 1974, Glutathione peroxidase activity in rat lens and other tissues in relation to dietary selenium intake, *Exp. Eye Res.* **18**:563–569.

Lee, C-Y, Akera, T., and Brody, T. M., 1976, Comparative effects of chlorpromazine and its free-radical on membrane functions, *Biochem. Pharm.* **25**:1751–1756.

Levander, O. A., Morris, V. C., and Higgs, D. J., 1974, Characterization of the selenium in rat liver mitochondria as glutathione peroxidase, *Biochem. Biophys. Res. Commun.* **58**:1047–1052.

Losowsky, M. S., Kelleher, J., Walker, B. E., Davies, T., and Smith, C. L., 1972, Intake and absorption of tocopherol, *Ann. NY Acad. Sci.* **203**:212–222.

Lu, H., 1976, Liver microsomal drug-metabolizing enzyme system: Functional components and their properties, *Fed. Proc.* **35**:2460–2463.

Lyman, R. L., Cook, C. P., and Williams, M. A., 1964, Liver lipid accumulation in isoleucine-deficient rats, *J. Nutr.* **82**:432–438.

Lyman, R. L., Fosmire, M. A., Giotas, C., and Miljanich, P., 1970, Inhibition of desaturation of stearic acid in livers of rats fed ethionine, *Lipids* **5**:583–589.

Lyman, R. L., Giotas, C., Fosmire, M. A., and Miljanich, P., 1969, Effect of feeding ethionine on the distribution and metabolism of 1-^{14}C-linoleic acid in the rat, *Can. J. Biochem.* **47**:11–17.

Mahoney, L. R., 1969, Antioxidants, *Angew. Chem. Int. Ed. Engl.* **8**:547–555.

Malmsten, C., Hamberg, M., Svensson, J., and Samuelsson, B., 1975, Physiological role of an endoperoxide in human platelets: Hemostatic defect due to platelet cyclo-oxygenase deficiency, *Proc. Natl. Acad. Sci. USA* **72**:1446–1450.

Malmsten, C., Granstrom, E., and Samuelsson, B., 1976, Cyclic AMP inhibits synthesis of prostaglandin endoperoxide (PGG$_2$) in human platelets, *Biochem. Biophys. Res. Commun.* **68**:569–576.

Marbrouk, A. F., and Dugan, L. R., 1961, Kinetic investigation into glucose-, fructose- and sucrose-activated autoxidation of methyl linoleate emulsion, *J. Am. Oil Chem. Soc.* **38**:692–695.

Marklund, S., Beakman, G., and Stigbrand, T., 1976, A comparison between the common type and a rare genetic variant of human cuprozinc superoxide dismutase, *Eur. J. Biochem.* **64**:415–422.

Marks, J., 1962, Clinical appraisal of the therapeutic value of α-tocopherol, *Vit. Horm.* **20**:573–598.

Marshall, W. J., and McLean, A. E. M., 1971, A requirement for dietary lipids for induction of cytochrome P-450 by phenobarbitone in rat liver microsomal fraction, *Biochem. J.* **122**:569–573.

Matsuda, I., Oka, Y., Taniguchi, N., Furuyama, M., Kodama, S., Arashima, S., and Mitsuyama, T., 1976, Leukocyte glutathione peroxidase deficiency in a male patient with chronic granulomatous disease, *J. Pediatr.* **88**:581–583.

May, H. E., and McCay, P. B., 1968, Reduced triphosphopyridine nucleotide oxidase-catalyzed alterations of membrane phospholipids, II. Enzymic properties and stoichiometry, *J. Biol. Chem.* **243**:2296–2305.

McCay, P. B., and Poyer, J. L., 1976, Enzyme-generated free-radicals as initiators of lipid peroxidation in biological membranes, in *The Enzymes of Biological Membranes*, Vol. 4 (A. Martonosi, ed.), pp. 239–256, Plenum, New York.

McCay, P. B., Fong, K-L, King, M., Lai, E., Weddle, C., Poyer, L., and Hornbrook, K. R., 1976a, Enzyme-generated free-radicals and singlet oxygen as promoters of lipid peroxidation in cell membranes, in *Lipids* (R. Paoletti, G. Porcellati, and G. Jacini, ed.), pp. 157–168, Vol. 1, Raven Press, New York.

McCay, P. B., Gibson, D. D., Fong, K-L, and Hornbrook, K. R., 1976b, Effect of glutathione peroxidase activity on lipid peroxidation biological membranes, *Biochim. Biophy. Acta* (in press).

McCay, P. B., Poyer, J. L., Floyd, R. A., Fong, K-L, and Lai, E., 1976c, Spin-trapping of radicals produced during enzymic NADPH-dependent and CCl_4-dependent microsomal lipid peroxidation, *Fed. Proc.* **35**:1421.

McCord, J. M., Keele, B. B., Jr., and Fridovich, I., 1971, An enzyme based theory of obligate anaerobiosis: The physiological role of superoxide dismutase, *Proc. Natl. Acad. Sci. USA* **68**:1024–1027.

Menzel, D., Roehm, J. N., and Lee, S. D., 1972, Vitamin E: The biological and environmental antioxidant, *J. Agr. Food Chem.* **20**:481–486.

Mishin, V., Pokrovsky, A., and Lyakhovich, V. V., 1976, Interaction of some acceptors with superoxide anion radicals formed by the NADPH specific flavoprotein in rat liver microsomal fractions, *Biochem. J.* **154**:307–310.

Misra, H., 1974, Generation of superoxide free radical during the autoxidation of thiols, *J. Biol. Chem.* **249**:2151–2155.

Monserrat, A. J., Ghoshal, A. K., Hartroft, W. S., and Porta, E. O., 1969, Lipoperoxidation in the pathogenesis of renal necrosis in choline-deficient rats, *Am. J. Pathol.* **55**:163–189.

Monserrat, A. J., Hamilton, F., Ghoshal, A. K., Porta, E. D., and Hartroft, W. S., 1972, Lysosomes in the pathogenesis of the renal necrosis of choline-deficient rats, *Am. J. Pathol.* **68**:113–145.

Montgomery, M. R., and Holtzman, J. L., 1975, Drug induced alteration in hepatic fatty acid desaturase activity, *Biochem. Pharmacol.* **24**:1343–1347.

Moore, T., 1940, The effect of vitamin E deficiency on the vitamin A reserves of the rat, *Biochem. J.* **34**:1321–1328.

Najman, A., Lichtens, H., Buc, H., and Gorin, N. C., 1974, Glutathione peroxidase deficiency, acanthocytosis, and hemolytic anemia during cirrhosis, *Sem. Hop. (Paris)* **50**:3127–3130.

Necheles, T. F., Boles, T. A., and Allen, D. M., 1968, Erythrocyte glutathione peroxidase deficiency and hemolytic disease of the newborn infant, *J. Pediat.* **72**:319–324.

Necheles, T. F., Maldonado, N., Barquet-Chediak, A., and Allen, D. M., 1969, Homozygous erythrocyte glutathione peroxidase deficiency: Clinical and biochemical studies, *Blood* **33**:164–169.

Necheles, T. F., Steinberg, M. H., Cameron, D., 1970, Erythrocyte glutathione-peroxidase deficiency, *Br. J. Haematol.* **19**:605–612.

Newberne, P. M., Wilson, R., and Rogers, H. E., 1971, Effects of a low lipotrope diet on the response of young male rats to the pyrrolizidine alkaloid, monocrotaline, *Toxicol. Appl. Pharmacol.* **18**:397–399.

Nishikimi, M. and Machlin, L. J., Oxidation of α-tocopherol model compound by superoxide anion, *Arch. Biochem. Biophys.* **170**:684–689.

Nitowsky, H. M., Tildon, J. T., Levin, S., and Gordon, H. H., 1962, Studies of tocopherol deficiency in infants and children, VII. The effect of tocopherol on urinary, plasma, and muscle creatine, *Am. J. Clin. Nutr.* **10**:368–378.

Nordblom, G. D., White, R. E., and Coon, M. J., 1976, Studies on hydroperoxide dependent substrate hydroxylation by purified microsomal cytochrone P-450, *Arch. Biochem. Biophys.* **175**:524–533.

Norman, R. O. C., and Lindsay Smith, J. R., 1965, in *Oxidases and Related Redox Systems* (T. E. King, H. S. Mason, and M. Morrison, eds.), Vol. 1, p. 131, Wiley, New York.

Norred, W. P., and Wade, A. E., 1972, Dietary fatty acid-induced alterations of hepatic microsomal drug metabolism, *Biochem. Pharmacol.* **21**:2887–2897.

Ogura, Y., 1955, Catalase activity at high concentration of hydrogen peroxide, *Arch. Biochem. Biophys.* **57**:288–300.

Omaye, S. T., and Tappel, A. L., 1974, Effect of dietary selenium on glutathione peroxidase in the chick, *J. Nutr.* **104**:747–753.

Oppenheimer, E. H., 1956, Focal necrosis of striated muscle in infant with cystic fibrosis of pancreas and evidence of lack of absorption of fat soluble vitamins, *Bull. Johns Hopkins Hosp.* **98**:353–359.

Orrenius, S., and Ernster, L., 1971, Enzymatic properties of membrane-associated cell fractions as related to drug metabolism, in *Cell Membranes: Biological and Pathological* (G. W. Richter and D. G. Scarpelli, eds.), p. 38, Williams & Wilkins, Baltimore.

Orrenius, S., and Ernster, L., 1974, Cytochrome P-450-linked monooxygenase systems, in *Molecular Mechanisms of Oxygen Activation* (O. Hayaishi, ed.), Academic Press, New York.

Pani, P., Gravella, E., Mazzarino, C., and Burdino, E., 1972, On the mechanism of fatty liver in white phosphorus poisoned rats, *Exp. Mol. Pathol.* **16**:201–209.

Peake, I. R., and Bieri, J. G., 1971, Alpha and gamma-tocopherol in the rat: *In vitro* and *in vivo* tissue uptake and metabolism, *J. Nutr.* **101**:1615–1622.

Peake, I. R., Windmueller, H. G., and Bieri, J. G., 1972, A comparison of the intestinal absorption lymph and plasma transport and tissue uptake of α- and γ-tocopherols in the rat, *Biochim. Biophys. Acta* **260**:679–688.

Pearson, C. K., and McC-Barnes, M., 1970, Absorption of tocopherols by small intestinal loops of the rat *in vivo, Int. Z. Vitaminforsch.* **40**:19–22.

Pedersen, N. D., Whanger, P. D., and Weswig, P. H., 1975, Effect of dietary selenium depletion and repletion of glutathione peroxidase levels in rat tissues, *Nutr. Rep. Int.* **11**:429–436.

Pederson, T. C., and Aust, S. D., 1973, The role of peroxide and singlet oxygen in lipid peroxidation promoted by xanthine oxidase, *Biochem. Biophys. Res. Commun.* **52**:1071–1078.

Peeters-Joris, C., Vandervoordi, A-M, and Baudhuim, P., 1975, Subcellular localization of SOD in rat liver, *Biochem. J.* **150**:31–39.

Petkau, A., Kelly, K., Chelack, W. S., and Barefoot, C., 1976, Protective effect of superoxide dismutase on erythrocytes of x-irradiated mice, *Biochem. Biophys. Res. Commun.* **70**:456–458.

Pfeifer, P. M., and McCay, P. B., 1971, Reduced triphosphopyridine nucleotide oxidase-catalyzed alterations of membrane phospholipids, V. Use of erythrocytes to demonstrate enzyme-dependent production of a component with the properties of a free-radical, *J. Biol. Chem.* **246**:6401–6408.

Prins, H. K., Oort, M., Loos, J. A., Zürchen, C., and Beckers, T., 1966, Congenital nonspherocytic hemolytic anemia associated with glutathione deficiency of the erythrocytes: Hematologic biochemical and genetic studies, *Blood* **27**:145–166.

Prohaska, J. R., and Wells, W. W., 1975, Copper deficiency in the developing rat brain: Evidence for abnormal mitochondria, *J. Neurochem.* **25**:221–228.

Rahimtula, A., and O'Brien, P. J., 1976, The possible involvement of singlet oxygen in prostaglandin biosynthesis, *Biochem. Biophys. Res. Commun.* **70**:893–899.

222   Lloyd A. Witting

Rao, K. S., and Recknagel, R. O., 1969, Early incorporation of carbon-labeled carbon tetrachloride into rat liver particulate lipids and proteins, *Exp. Mol. Pathol.* **10**:219–228.

Recknagel, R. O., 1967, Carbon tetrachloride hepatotoxicity, *Pharmacol. Rev.* **19**:145–208.

Recknagel, R. O., and Glende, E. A., Jr., 1973, Lipid peroxidation in acute carbon tetrachloride liver injury, in *Intermediary Metabolism of the Liver* (H. Brown and D. Hardwick, eds.), p. 23, C C Thomas, Springfield, Illinois.

Recknagel, R. O., and Glende, E. A., Jr., 1973, Carbon tetrachloride hepatotoxicity: An example of lethal cleavage, *CRC Crit. Rev. Toxicol.* **2**:263–297.

Recknagel, R. O., and Ghoshal, A. K., 1966, Lipoperoxidation as a vector in carbon tetrachloride hepatotoxicity, *Lab. Invest.* **15**:132–148.

Roehm, J. N., Hadley, J. G., and Menzel, D. B., 1972, The influence of vitamin E on lung fatty acids of rats exposed to ozone, *Arch. Environ. Health* **24**:237–242.

Rosenberger, H. M., and Johnson, R. H., 1970, Factors influencing peroxide formation in diethyl ether, *Am. Lab.* **1970**(6):8–15.

Rotruck, J. T., Pope, A. L., Ganther, H. E., Swanson, N. B., Hafeman, D. G., and Hoekstra, W. G., 1973, Selenium: Biochemical role as a component of glutathione peroxidase, *Science* **179**:588–590.

Rowe, L., and Wills, E. D., 1976, The effect of dietary lipids and vitamin E on lipid peroxide formation, cytochrome P-450 and oxidative demethylation in the endoplasmic reticulum, *Biochem. Pharmacol.* **25**:175–179.

Samuel, D., and Steckel, F., 1974, The physico-chemical properties of molecular oxygen, in *Molecular Oxygen in Biology* (O. Hayaishi, ed.), pp. 1–32, North-Holland, Amsterdam.

Schwarz, K., 1948, Uber die lebertranshadigung der ratte und ihre verhultüng durch tocopherol, *Hoppe-Seyler's Z. Physiol. Chem.* **283**:106–112.

Schwarz, K., and Foltz, C. M., 1957, Selenium as an integral part of factor three against dietary necrotic liver degeneration, *J. Am. Chem. Soc.* **79**:3292–3293.

Scott, G., 1965, *Atmospheric Oxidation and Antioxidants,* Elsevier, Amsterdam.

Scott, T. W., Cook, L. J., and Mills, S. C., 1971, Protection of dietary polyunsaturated fatty acids against microbial hydrogenation in ruminants, *J. Am. Oil Chem. Soc.* **48**:358–364.

Shafferman, A., and Stein, G., 1975, Study of the biochemical redox processes by the technique of pulse radiolysis (review), *Biochim. Biophys. Acta* **416**:287–317.

Shaver, S. L., and Mason, K. E., 1951, Impaired tolerance to silver in vitamin E-deficient rats, *Anat. Rec.* **109**:382.

Shih, J. C. H., Sandholm, M., and Combs, G. F., Jr., 1976, Enzymatic analyses of tissues of selenium-deficient chicks, *Fed. Proc.* **35**:576.

Shimasaki, H., Takatori, T., Anderson, W. R., Horten, H. L., and Privett, O. S., 1976, Alteration of lung lipids in ozone exposed rats, *Biochem. Biophys. Res. Commun.* **68**:1256–1262.

Siakotos, A. N., Goebel, H. H., Patel, V., Watanabe, I., and Zeman, W., 1973, The morphogenesis and biochemical characteristics of ceroid isolated from cases of neuronal ceroid-lipofuscinosis, in *Sphingolipids, Sphingolipidoses and Allied Disorders* (B. W. Volk and S. M. Aronson, eds.), p. 53, Plenum Press, New York.

Siakotos, A. N., Koppang, N., Youmans, B. S., and Bucana, C., 1974, Blood and tissue levels of alpha-tocopherol in a disorder of lipid peroxidation: Batten's disease, *Am. J. Clin. Nutr.* **27**:1152–1157.

Siakotos, An. N., Watanabe, I., Saito, A., and Fleischer, S., 1970, Procedures for the isolation of two distinct lipopigments from human brain: Lipofuscin and ceroid, *Biochem. Med.* **4**:361–375.

Sims, P., and Glover, P. L., 1974, Epoxides in polycyclic aromatic hydrocarbon metabolism and carcinogenesis, *Adv. Cancer Res.* **20**:166–274.

Smith, P. J., Tappel, A. L., and Chow, C. K., 1974, Glutathione peroxidase activity as a function of dietary selenomethionine, *Nature (London)* **247**:392–393.

Stanier, R. Y., Doudoroff, M., and Adelberg, E. A., 1970, in *The Microbial World,* 3rd ed., pp. 663–664, Prentice-Hall, Englewood Cliffs, N.J.

Staudt, H., Lichtenberger, F., and Ullrich, V., 1974, The role of NADH in uncoupled microsomal monoxygenations, *Eur. J. Biochem.* **46**:99–106.

Steinberg, M. H., and Necheles, T. F., 1971, Erythrocyte glutathione peroxidase deficiency. Biochemical studies on the mechanisms of drug induced hemolysis, *Am. J. Med.* **50**:542–546.

Steinberg, M. H., Brauer, M. J., and Necheles, T. F., 1970, Acute hemolytic anemia associated with erythrocyte glutathione-peroxidase deficiency, *Arch. Int. Med.* **125**:302–303.

Stier, A., 1976, Lipid structure and drug metabolizing enzymes, *Biochem. Pharmacol.* **25**:109–113.

Strehler, B. L., Mark, D. D., Mildvan, A. S., and Gee, M. V., 1959, Rate and magnitude of age pigment accumulation in the human myocardium, *J. Gerontol.* **14**:430–439.

Strubel, H. W., Lu, A. YH., Heidema, J., and Coon, M. J., 1970, Phosphatidyl choline requirement in the enzymatic reduction of hemoprotein P-450 and in fatty acid, hydrocarbon and drug hydroxylate, *J. Biol. Chem.* **245**:4851–4854.

Sugioka, G., Porta, E. D., and Hartroft, W. S., 1969, Early changes in livers of rats fed choline-deficient diets at four levels of protein, *Am. J. Pathol.* **57**:431–455.

Tam, B. K., and McKay, P. B., 1970, Reduced triphosphopyridine nucleotide oxidase-catalyzed alterations of membrane phospholipids, III. Transient formation of phospholipid peroxides, *J. Biol. Chem.* **245**:2295–2300.

Tappel, A. L., 1967, Where old age begins, *Nutrition Today* **2**:2–10.

Tappel, A. L., 1968, Will antioxidant nutrients slow aging processes?, *Geriatrics* **23**(10):97–105.

Tappel, A. L., 1974, Selenium-glutathione peroxidase and vitamin E, *Am. J. Clin. Nutr.* **27**:960–965.

Tappel, A. L., 1975, Lipid peroxidation and fluorescent molecular damage to membranes, in *Pathological Aspects of Cell Membranes* (B. F. Trumpal and A. Arstila, eds.), Academic Press, New York.

Thomas, H. V., Mueller, P. K., and Lyman, R. L., 1967, Lipoperoxidation of lung lipids in rats exposed to nitrogen dioxide, *Science* **159**:532–534.

Thompson, J. N., and Scott, M. L., 1970, Impaired lipid and vitamin E absorption related to atrophy of the pancreas in selenium-deficient chicks, *J. Nutr.* **100**:797–809.

Thompson, J. N., Beare-Rogers, J. L., Erdödy, P., and Smith, D. C., 1973, Appraisal of human vitamin E requirement based on examination of individual meals and a composite Canadian diet, *Am. J. Clin. Nutr.* **26**:1349–1354.

Thurman, R. G., Marazzo, D. P., and Scholz, R., 1975, Mixed function oxidation and intermediary metabolism: Metabolic interdependencies in the liver, in *Cytochromes P-450 and 65* (D. Y. Cooper, O. Rosenthal, R. Snyder, and C. Witmer, eds.), pp. 355–367, Plenum Press, New York.

Tinsley, I. J., and Lowry, R. R., 1972, An interaction of DDT in the metabolism of essential fatty acids, *Lipids* **7**:182–185.

Uri, N., 1961a, Physico-chemical aspects of autoxidation, in *Autoxidation and Antioxidants* (W. O. Lundberg, ed.), pp. 55–106, Vol. 1, Wiley (Interscience), New York.

Uri, N., 1961b, Mechanism of antioxidation, in *Autoxidation and Antioxidants* (W. O. Lundberg, ed.), Vol. 1, pp. 133–169, Wiley (Interscience), New York.

Wade, A. E., and Norred, W. P., 1976, Effect of dietary lipid on drug-metabolizing enzymes, *Fed. Proc.* **35**:2475–2479.

Wattenberg, L. W., Loub, W. O., Lam, L. K., and Speier, J. L., 1976, Dietary constituents altering the responses to chemical carcinogens, *Fed. Proc.* **35**:1327–1331.

Weatherholz, W. M., Campbell, T. C., and Webb, R. E., 1969, Effects of dietary protein levels on the toxicity and metabolism of heptachlor, *J. Nutr.* **98**:90–94.

Webb, B. E., and Miranda, C. L., 1973, Effect of quality of dietary protein on heptachlor toxicity, *Food Cosmet. Toxicol.* **11**:63–67.

Weddle, C. C., Hornbrook, K. R., and McCay, P. B., 1977, Lipid peroxidation and alteration of membrane lipids in isolated hepatocytes exposed to carbon tetrachloride, *J. Biol. Chem.* **251**:4973–4978.

Weichselbaum, T. E., 1935, Cystine deficiency in the albino rat, *QJ Exp. Physiol.* **25**:363–367.

Weves, R., Roos, D., Weening, R. S., Vulsma, T., and Van Gelder, B. F., 1976, An EPR study of myeloperoxidase in human granulocytes, *Biochim. Biophys. Acta* **421**:328–333.

Williams, D. M., Lynch, R. E., Lee, G. R., and Cartwright, G. E., 1975, Superoxide dismutase in copper-deficient swine, *Proc. Soc. Exp. Biol. Med.* 149:534–536.

Wills, E. D., 1966, Mechanisms of lipid peroxide formation in animal tissues, *Biochem. J.* **99**:667–676.

Wills, E. D., 1969, Lipid peroxide formation in microsomes, the role of non-haem iron, *Biochem. J.* **113**:325–332.

Wills, E. D., 1972, Effects of iron overload on lipid peroxide formation and oxidative demethylation by the liver endoplasmic reticulum, *Biochem. Pharmacol.* **21**:239–247.

Wiss, O., Bunnell, R. H., and Gloor, U., 1962, Absorption and distribution of vitamin E in tissues, *Vit. Horm.* **20**:441–455.

Witting, L. A., 1965, Lipid peroxidation in vivo, *J. Am. Oil Chem. Soc.* **42**:908–913.

Witting, L. A., 1967, The effect of antioxidant-deficiency on tissue lipid composition in the rat, IV. Peroxidation and interconversion of polyunsaturated fatty acids in muscle phospholipids, *Lipids* **2**:109–113.

Witting, L. A., 1969, The oxidation of alpha-tocopherol during the autoxidation of ethyl oleate, linoleate, linolenate and arachidonate, *Arch. Biochem. Biophys.* **129**:142–151.

Witting, L. A., 1970, The interrelationship of polyunsaturated fatty acid and antioxidants in vivo, in *Progress in the Chemistry of Fats and Other Lipids* (R. T. Holman, ed.), Vol. 9, pp. 519–553, Pergamon Press, Oxford.

Witting, L. A., 1972, The role of polyunsaturated fatty acids in determining vitamin E requirement, *Ann. NY Acad. Sci.* **203**:192–198.

Witting, L. A., 1973, Fatty liver induction: Effect of ethionine on polyunsaturated fatty acid synthesis, *Biochim. Biophys. Acta* 296:271–286.

Witting, L. A., 1974, Vitamin E-polyunsaturated lipid relationships in diet and tissues, *Am. J. Clin. Nutr.* 27:952–959.

Witting, L. A., 1975, Vitamin E as a food additive, *J. Am. Oil Chem. Soc.* **52**:61–64.

Witting, L. A., and Horwitt, M. K., 1964, Effect of degree of fatty acid unsaturation in tocopherol-deficiency induced creatinuria, *J. Nutr.* **82**:19–33.

Witting, L. A., and Horwitt, M. K., 1967, The effect of antioxidant-deficiency on tissue lipid composition in the rat, I. Gastrocnemius and quadriceps muscle, *Lipids* **2**:89–96.

Witting, L. A., and Lee, L., 1975a, Recommended dietary allowance for vitamin E: relation to dietary, erythrocyte and adipose tissue linoleate, *Am. J. Clin. Nutr.* **28**:577–583.

Witting, L. A., and Lee, L., 1975b, Dietary levels of vitamin E and polyunsaturated fatty acids and plasma vitamin E, *Am. J. Clin. Nutr.* **28**:571–576.

Witting, L. A., Harvey, C. C., Century, B., and Horwitt, M. K., 1961, Dietary alterations of fatty acids of erythrocytes and mitochondria of brain and liver, *J. Lipid Res.* **2**:412–418.

Witting, L. A., Century, B., and Horwitt, M. K., 1967a, Susceptibility to dietary modification of lipids of various tissues as a factor in nutritional disease, in *Nutrition and Health* (J. Kuhnau, ed.), Vol. 1, pp. 199–202, Friedr. Vieweg, and Sohn, Braunschweig, Germany.

Witting, L. A., Harmon, E. M., and Horwitt, M. K., 1967b, Influence of dietary selenium on whole body levels of α-tocopherol, *Fed. Proc.* **26**:475.

Witting, L. A., Likhite, V. N., and Horwitt, M. K., 1967c, The effect of antioxidant-deficiency on tissue lipid composition in the rat, III. Testes, *Lipids* **2**:103–108.

Witting, L. A., Theron, J. J., and Horwitt, M. K., 1967d, The effect of antioxidant-deficiency on tissue lipid composition in the rat, II. Liver, *Lipids* **2**:97–102.

Wolfe, L. S., Rostworowski, K., and Marion, J., 1976, Endogenous formation of the prostaglandin endoperoxide metabolite, Thromboxane B_2, by brain tissue, *Biochem. Biophys. Res. Commun.* **70**:907–909.

Wynder, E. L., 1976, Nutrition and cancer, *Fed. Proc.* **35**:1309–1315.

Yamamoto, S., Yamauchi, T., and Hayaishi, O., 1972, Alkylamine-dependent amino acid oxidation by lysine monooxygenase fragmented substrate of oxygenase, *Proc. Natl. Acad. Sci. USA* **69**:3723–3726.

Yamazaki, I., 1974, Peroxidase, in *Molecular Mechanisms of Oxygen Activation* (O. Hayaishi, ed.), pp. 535–558, Academic Press, New York.

Yamazaki, I., Mason, H. S., and Piette, L. H., 1960, Identification by electron paramagnetic resonance spectroscopy of free radicals generated from substrates by peroxidase, *J. Biol. Chem.* **235**:2444–2449.

Young, J. D., Nimmo, I. A., and Hall, J. G., 1975, The relationship between GSH, GSSG and non-GSH thiol in GSH-deficient erythrocytes from Finnish Landrace and Tasmanian Merino sheep, *Biochim. Biophys. Acta* **404**:124–131.

Zannoni, V. G., and Sato, P. H., 1976, The effect of certain vitamin deficiencies on hepatic drug metabolism, *Fed. Proc.* **35**:2464–2469.

Zeman, W., and Rider, J. A., 1975, *The Dissection of a Degenerative Disease*, American Elsevier, New York.

Chapter 9

The Role of Copper and Zinc in Cholesterol Metabolism

Leslie M. Klevay

1. Introduction

Ischemic heart disease is the leading cause of death in the United States
(Anonymous, 1974a). Previously termed coronary heart disease (Anonymous,
1969), this disease causes about 35% of deaths, twice as many as are caused by
malignant neoplasms or cancer. In the United States, risk of ischemic heart
disease is higher among men than among women, among smokers of cigarettes
than among nonsmokers, among diabetics and among those with certain abnormalities of the electrocardiogram. Risk also increases with age, with blood
pressure, and with the concentration of cholesterol in serum (Insull, 1973). The
predictive utility of the concentration of cholesterol in serum or plasma (Fredrickson and Levy, 1972) and the identification of many dietary components that
can alter these concentrations have been the sources of most interest among
nutritionists in the metabolism of cholesterol.

A new hypothesis regarding the etiology of ischemic heart disease has been
proposed (Klevay, 1975c). This hypothesis, which has been developed in a series
of papers (Klevay, 1973, 1974a, b, d, 1975a–c, 1976a–c; Klevay and Forbush,
1976; Klevay *et al.*, 1975, 1976), states that a "metabolic imbalance in regard to
zinc and copper is a major factor" in the etiology of ischemic heart disease. This
metabolic imbalance is "either a relative or absolute deficiency of copper characterized by a high ratio of zinc to copper." Following a brief review of experiments which implicate chemical elements in the metabolism of cholesterol, information relating copper, zinc, and cholesterol metabolism, which has been

Leslie M. Klevay • Agricultural Research Service, Human Nutrition Laboratory, United States Department of Agriculture, Grand Forks, North Dakota 58201

found since the proposal of the zinc/copper hypothesis (Klevay, 1975c) will be summarized. Data on the effects of metallic elements on the metabolism of other lipids or on atherosclerosis have not been included.

2. Chemical Elements That Alter Cholesterol Metabolism

Twenty of the 29 elements (Fig. 1) that have been associated with the epidemiology of ischemic heart disease or with the metabolism of cholesterol or other lipids have been implicated in the metabolism of cholesterol. References to the nine other elements have been published (Klevay, 1976b).

2.1. Sodium

Meneely and Ball (1958) showed that the concentration of cholesterol in serum of rats was increased by diets high in sodium chloride. The effect was "more striking" in the rats that became hypertensive. Dahl (1960) confirmed these findings with rats and produced hypercholesterolemia in dogs fed diets containing over 2% sodium chloride. Page *et al.* (1974) reported that among six tribal groups living in the Solomon Islands, the three groups that consumed higher amounts of sodium chloride had higher concentrations of cholesterol in serum than did the three groups that consumed lower amounts.

2.2. Magnesium

Hellerstein *et al.* (1960) found an inverse relationship between dietary magnesium and serum cholesterol in male rats fed a diet containing hydrogenated cottonseed oil. No similar relationship was found when the diet contained corn oil or large amounts of cholesterol and cholic acid. Vitale *et al.* (1963) found hypercholesterolemia in Cebus monkeys deficient in magnesium.

2.3. Sulfur

Diets containing protein of low quality and large amounts of cholesterol have produced hypercholesterolemia in Cebus monkeys (Mann *et al.*, 1953), mice and rats (Fillios and Mann, 1954), and cockerels (Stamler *et al.*, 1958). This hypercholesterolemia could be decreased by the addition of methionine to the diets.

"Onion and garlic are reported to lower serum cholesterol levels and stimulate fibrinolytic activity. The active principle appears to be in the essential oil and may be an organic sulfide" (Anonymous, 1976).

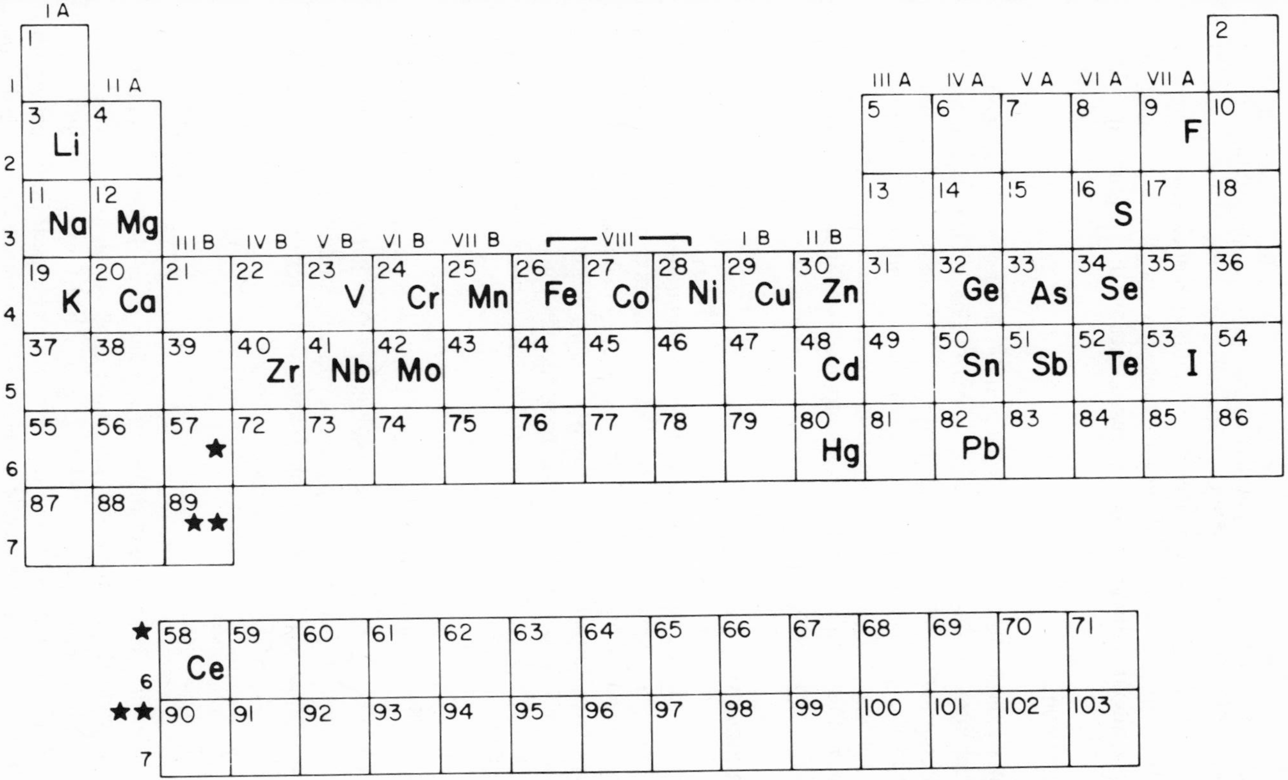

Fig. 1. Elements of ischemic heart disease. Elements identified by symbols have been implicated in the epidemiology of ischemic heart disease or the metabolism of cholesterol or other lipids. The list is inclusive rather than exclusive (from Klevay, 1976c).

2.4. Calcium

Calcium salts (usually carbonate) given orally to men and women (Yacowitz et al., 1965; Maibach, 1967, 1969; Carlson et al., 1971; Birenbaum et al., 1972; Albanese et al., 1973) have decreased the concentration of cholesterol in plasma. Iacono (1974) found the concentration of cholesterol in plasma of rabbits to be increased in calcium deficiency and decreased when the diet was high in calcium.

2.5. Vanadium

Hopkins and Mohr (1974) have reviewed the effect of vanadium deficiency on the concentration of cholesterol in plasma of chicks. Deficiency produced hypocholesterolemia at 28 days of age and hypercholesterolemia at 49 days. The latter results are consistent with the decrease in the concentration of cholesterol in serum of young men fed diammonium oxytartratovanadate (Curran et al., 1959) and are in contrast to the increase in concentration found in male rats fed vanadyl sulfate (Schroeder, 1968).

2.6. Chromium

Schroeder (1968, 1969) produced a lower concentration of cholesterol in the serum of rats of both sexes by adding trivalent chromium to drinking water. In an earlier experiment (Schroeder and Balassa, 1965), female rats given chromium had higher concentrations of cholesterol. Staub et al. (1969) decreased the rise in serum cholesterol of male rats fed an atherogenic diet with a drinking solution of chromic acetate. The results of Schroeder (1969) have been confirmed (Whanger and Weswig, 1975).

2.7. Manganese

Doisy (1973) noted hypocholesterolemia in a man fed a formula diet from which the manganese had been omitted accidently; he attributed the effects to an inability to synthesize farnesyl pyrophosphate, which resulted from manganese deficiency.

2.8. Nickel

Nielsen et al. (1975) found the concentration of cholesterol in plasma of male rats deficient in nickel to be higher than those supplemented with nickel chloride. Female rats deficient in nickel generally had lower concentrations of cholesterol in plasma than did those that were supplemented. Statistical significance was attained only in the second generation of deficiency for both males and females. These results are similar to the results of Schroeder and Balassa (1965) with chromium. In contrast, Schroeder (1968) found no effect of nickel supplementation with male rats; female rats whose diets were supplemented had significantly lower concentrations of cholesterol.

2.9. Copper and Zinc

Hypercholesterolemia was produced in male rats (Klevay, 1973) that ingested a high ratio of zinc to copper. Rats were assigned to groups matched by mean weight and were fed the diet shown in Table I. This diet is deficient in both copper and zinc; these elements were provided as drinking solutions of cupric sulfate and zinc acetate. The concentrations (μg/ml) of zinc and copper in these solutions were 10 and 2 (low ratio) or 10 and 0.25 (high ratio) or 20 and 0.50 (high ratio). Calculation of the ratio of zinc to copper ingested from both food (containing 1.8 μg zinc and 0.7 μg copper per gram of diet, respectively) and water demonstrates a low ratio of 4 and a high ratio of either 14 or 20. Rats were housed in stainless steel cages, with water bottles of soft glass and silicone stoppers. Environmental conditions were considered to be clean when the experiments were done in a specially built room (Klevay *et al.,* 1971) with a filtered air supply; conditions were considered conventional when rats were housed under identical conditions with the exception of extreme air filtration. Results are shown in Figs. 2 and 3 (Klevay, 1974d). Higher absolute amounts of copper delayed the onset of hypercholesterolemia.

2.10. Nickel, Germanium, Arsenic, Selenium, Zirconium, Niobium, Cadmium, Antimony, Tellurium, and Lead

One cannot peruse the biology of trace elements without discovering the work of the late Henry A. Schroeder. His lengthy experiments and extensive reviews provide a great amount of information. Some of his experiments have been cited (Sections 2.6. and 2.8.). In an experiment (Schroeder, 1968) in which male rats were fed 13 trace elements as various salts (anions or cations); ger-

Table I. Composition of the Diet

Ingredient	g
Sucrose	623.5
Fibrous cellulose powder	30.0
Zn- and Cu-free Jones Foster salt mix	40.0
Choline chloride	1.5
Egg white	200.0
Rat vitamin mix	5.0
Vitamin ADE mix	10.0
Corn oil	90.0

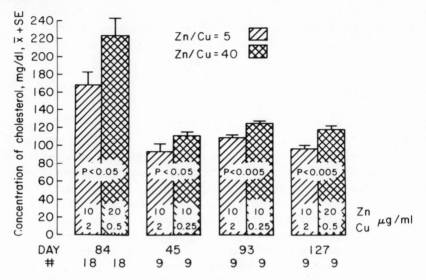

Fig. 2. Rats raised in a clean environment. The concentrations of zinc and copper in the drinking water of the animals are shown at the base of each bar. The number of animals per group and the experimental day on which statistically significant differences were first found are shown beneath the bars (from Klevay, 1974d, fig. 1).

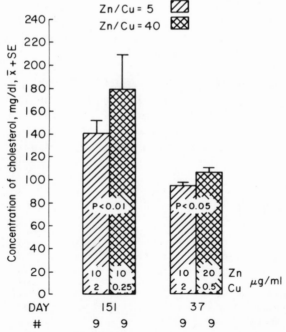

Fig. 3. Rats raised in a conventional environment. See Fig. 2 (from Klevay, 1974d, fig. 2).

manium decreased the concentration of cholesterol in plasma; nickel, niobium, tin, and lead had no effect; and vanadium, arsenic, selenium, zirconium, cadmium, antimony, and tellurium increased the concentration in comparison to rats fed 1 ppm chromium. Females responded differently—vanadium, nickel, selenium, niobium, and antimony decreased the concentration; germanium, arsenic, zirconium, cadmium, tin, tellurium, and lead had no significant effect. The difference in response between the sexes was at least partly explained by an apparently greater requirement of female rats for chromium. Rats of both sexes deficient in chromium had hypercholesterolemia. The lack of an effect of lead on cholesterolemia is in contrast to the hypercholesterolemia produced by injected lead acetate which Selye (1970) attributed to Sroczýnski *et al.*

2.11. Iodine

"In primary hypothyroidism the serum cholesterol concentration is usually increased, . . . A classic effect of thyroid hormones is to lower the concentration of cholesterol in plasma (Ingbar and Woeber, 1974)." This information has been used in the study of atherosclerosis (Friedberg, 1966).

2.12. Summary

Schroeder (1968) noted that feeding of elements of the A groups of the periodic table produced higher "levels of serum cholesterol" than did feeding of the elements of the B groups. The major exceptions to this rule are magnesium and calcium which were not tested by Schroeder, but which have been shown to lower the concentration of cholesterol in serum and plasma (Sections 2.2. and 2.4.). Considering all the elements in Fig. 1 it is noteworthy that only the noble gases and members of group III have not been associated with ischemic heart disease. Toxicity or deficiency of noble gases seem unlikely. However, is there something peculiar and important about group III, or are these elements absent from the table because they have not been tested?

3.   The Zinc/Copper Hypothesis

3.1. Background

The etiology of ischemic heart disease is presently incomprehensible; it seems reasonable to assume that the disease has a nutritional component. The study of pellagra, a comprehensible disease of nutritional origin, has been useful to me in the consideration of ischemic heart disease. This usefulness arises from my belief that analogy can be a source of knowledge (Lorenz, 1974). The study of ischemic heart disease today is similar to the study of pellagra in 1915 when the second century of search for an etiology was ending (Roe, 1973; Terris, 1964). The multiplicity of factors (Table II) which once were implicated in the origin of pellagra and filled many volumes has been simplified until a contemporary text summarizes important information on pellagra in about three pages (Follis and Van Itallie, 1974).

Before the etiology of a disease becomes thoroughly understood, i.e., comprehensible, so many apparently dissimilar observations are recorded that it is easy to conclude that risk of disease is random or that the disease has a vast number of determinants. In the latter case, the disease is considered to be of multifactorial origin (Friedberg, 1966). This term, of apparently recent vintage when used in the context of human disease, has been applied to ischemic heart disease because of the large number of factors, in addition to those enumerated by Insull (1973) (Section 1), which have been associated with risk (Table III).

It has been hypothesized (Klevay, 1975c) (section 1) that zinc and copper are major factors in the etiology of ischemic heart disease. This hypothesis is based on the belief that ischemic heart disease may not be of multifactorial origin and the hope that the concepts of ischemic heart disease may be simplified as were those of pellagra. High ratios of zinc to copper are associated with high risk and high concentrations of cholesterol in serum or plasma; low ratios of zinc to copper are associated with low risk and low concentrations. This hypothesis has resulted from the induction of hypercholesterolemia in rats by an increase in the ratio of zinc to copper ingested (Klevay, 1973) (Section 2.9.). Considering the protean nature of the factors and the number of elements (Fig. 1) that have been

Table II. Multifactorial Disease—Pellagra[a]

Host	Agent
Hartnup's disease, 145	Air pollution, 65
Heredity, 52	Bacteria, 137
Malabsorption, 138	Corn
Environment	Gluten deficiency, 56
Sewage disposal, 90	Fungal toxin, 59
Sunstroke, 150	Insects
Cures	Simulium, 85
Arsenic	Stomoxys, 90
Fowler's solution, 92	Intestinal infection, 91
Salvarsan, 93	Isoniazid, 148
Blood transfusion, 95	Millet, 148
Electricity, 97	Misery, 67
Iron, 95	Poverty
Other, 95	Taxation, 42
Quinine, 95	Depression, 128
Strychnine, 95	Protozoa, 88
	Sheep infection, 51

[a]Numbers refer to pages in Roe (1973).

Table III. Multifactorial Disease—Ischemic Heart Disease[a]

Host	Agent
Heredity	Cigarettes
Environment	Cirrhosis
Exercise	
Industrialization	Fat
The "water factor"	Amount
	Type
Cures	Fiber
Diet	Hemodialysis
Drugs	Milk formulas
Surgery	Protein, animal
	Sucrose
	Zinc/Copper

[a] Many of these concepts have been reviewed in Klevay (1975c, 1976b).

associated with risk of ischemic heart disease and the metabolism of cholesterol, the zinc/copper phenomenon can be considered of paramount importance only because it seems to fulfill Selye's triad (Selye, 1964). That is, the phenomenon is "true, generalizable and surprising at the same time."

When the original experiments were published (Klevay, 1973) an increase in the concentration of cholesterol in plasma of rats attributable to a high ratio of zinc to copper had been found in five experiments in two environments during 3 years. This reproducibility suggests that the results are true (other confirmatory work will be found in Sections 4.3–4.5.). No reference common to cholesterol, copper, and zinc had been found in Nutrition Reviews (Hegsted *et al.*, 1942–1973) at the time of original publication or since (Hegsted *et al.*, 1974 to Aug. 1976). The results are therefore surprising.

The degree to which the results are "generalizable" has been published (Klevay, 1975c). Nonnutritional phenomena which have been associated with a high ratio of zinc to copper and either high risk of ischemic heart disease or hypercholesterolemia are pregnancy, nephrotic syndrome, chronic hemodialysis and hypertension. Associated with a low ratio of zinc to copper and either low risk of mortality or hypocholesterolemia are hepatic cirrhosis, infectious hepatitis, strenuous exercise, and the availability of hard water.

3.2. Copper—Intake and Requirements

Comparison of the putative adult requirement for copper with the amount of copper thought to be contained in the average daily diet has led to the following statements. "Copper deficiency does not occur in human beings . . . (Scheinberg and Sternlieb, 1960)." "It is unlikely that copper deficiency occurs in man solely as a result of a dietary deficiency of copper (Cartwright and Wintrobe,

1964)." "Copper deficiency ... has never been reported in human adults ... (Darby *et al.*, 1973)." "Severe copper deficiency is rare in man (Anonymous, 1974b)." "Copper deficiency in man occurs only under very unusual circumstances (Sandstead, 1974)." These quotations, which summarize prevalent medical and nutritional opinion, are based on data published between 10 and 32 years ago. Excessive emphasis has been placed on mean daily intakes of copper; low individual intakes have been ignored.

Cartwright (1950) summarized several balance studies and concluded that neither net retention nor net loss of copper occurs at dietary intakes of about 2 mg/day. "The daily American diet contains about 2 to 5 mg of copper (Scheinberg and Sternlieb, 1960)." "The average dietary intake of copper is of the order of 2 to 5 mg (Cartwright and Wintrobe, 1964)." " ... Most diets provide the 2 mg/day sufficient to maintain balance (Anonymous, 1974b)." These statements are, in turn, based on four publications.

The most extensive of the experiments in these publications was that of Leverton and Binkley (1944), who measured the intakes of copper of 65 healthy young women who consumed their customary, self-chosen diets for 95 periods lasting 1 week. They also measured the intake of four women who consumed a constant diet for continuous periods of 75 to 140 days. The mean daily intakes of copper for these groups of women were 2.6 and 2.4 mg, respectively; the range of individual, daily intakes was 1.0–5.0 mg. Cartwright (1950) summarized six studies of 119 adults (including the 65 of Leverton and Binkley); mean daily intake of copper ranged from 1.1 to 8.1 mg. More recently, Cartwright *et al.* (1954) found the daily intake of copper of nine adults ranged from 1.4 to 3.8 mg. Finally, Schroeder *et al.* (1966) found the amount of copper in an institutional diet for 1 day to be 3.5 mg.

A search for data on the ratio of zinc to copper of diets in the United States (Klevay, 1975a) revealed daily amounts of copper ranging from less than 0.2 mg to 3.5 mg. These data and some of our own measurements (Klevay *et al.*, 1976) are summarized in Fig. 4. Diets were not included unless both copper and zinc had been measured on the same samples. Each diet is described by its type, e.g., general or low sodium and the place of preparation and analysis. More detailed descriptions are available (Klevay, 1975a; Klevay *et al.*, 1976). The width of the bar (2.0–2.1 mg) showing the adult requirement is based upon the beliefs of the Food and Nutrition Board of the U.S. National Academy of Sciences (Anonymous, 1974b) and the World Health Organization (Darby *et al.*, 1973). This latter estimate was stated as 30 μg/kg of body weight, an estimate confirmed by Hartley *et al.* (1974).

Although most of the diets shown in Fig. 4 were prepared upon consideration of sound nutritional principles, most of the diets did not contain the 2 mg of copper thought to be required. Some are close to this amount; most contain less than half that amount.

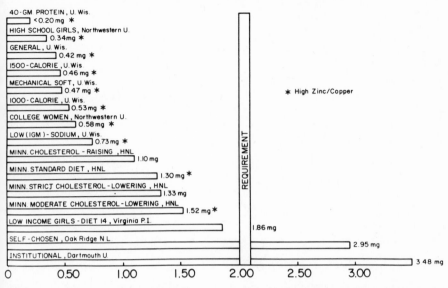

Fig. 4. Dietary copper (mg/day) diets in which both copper and zinc have been measured [data taken from Klevay, 1975a and Klevay *et al.*, (1976)]. Diets are identified by type and place of preparation and analysis [the requirement is taken from Anonymous (1974b) (NAS) and Darby *et al.* (1973) (WHO)]. Asterisks indicate diets with a ratio of zinc to copper greater than which have been shown to cause hypercholesterolemia in rats (Klevay, 1973).

3.3. Dietary Changes with Time

The diets shown in Fig. 4 were prepared for analysis in the last 10 years. These diets seem to contain less copper than those which have received more nutritional notice (Leverton and Brinkley, 1944; Cartwright, 1950; Cartwright *et al.*, 1954; Schroeder, 1966). Whether or not diets contain less copper than they did earlier is difficult to determine. By comparing the analysis of 44 foods published in 1942 and 1966, we (Klevay and Forbush, 1976) concluded that the foods probably contained less copper recently. This period of time coincides rather closely with the period for which the most uniform statistics on heart disease death rates are available in the United States (Fig. 5). During this latter period, the death rate due to arterosclerotic heart disease increased about 44%.

Others have considered the possibility that changes in average diets over long period of time have contributed to increase in ischemic heart disease. It has been suggested that consumption of fat (Rizek *et al.*, 1974) and sucrose (Yudkin, 1974) has increased and that consumption of vegetable fiber (Trowell, 1976) has decreased. In the United States the apparent increase in the consumption of fat may be an artifact of the way in which the data were collected (Call and Sanchez,

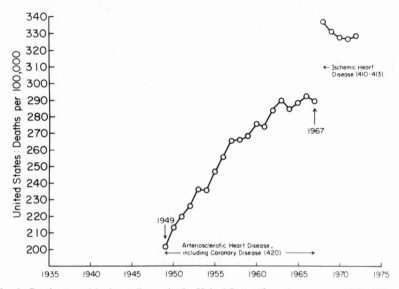

Fig. 5. Deaths caused by heart disease in the United States (from Anonymous, 1969, 1974a, and previous and subsequent editions). These are the only years for which death rates are available. The leading cause of death in the United States (30–35% of all deaths) was redefined and renamed between 1967 and 1968, the period of discontinuity of the curves.

1967). There was no increase in the availability of fat per capita between 1940 and 1965 (Call and Sanchez, 1967), a period (Fig. 5) very close to that (1949–1967) in which risk of atherosclerotic heart disease increased about 44%. The per capita availability of linoleic acid in 1972 was more than double that in 1909 (Rizek *et al.*, 1974). Even if consumption of fat has increased in the United States, only 8% of the long-term increase in risk of coronary heart disease can be explained by changes in dietary fat (Kahn, 1970).

3.4. Zinc/Copper and Dietary Fat

3.4.1. Amount of Fat

A positive association between the amount of fat and the ratio of zinc to copper of 71 foods has been found (Fig. 6); foods high in fat have high ratios of zinc to copper (and *vice versa*). The major exceptions to this relationship are beef liver, walnuts, pecans, and peanut butter, which have amounts of fat in excess of the median for 71 foods (1.1 g/100 g) and ratios of zinc to copper less than the median (11.2). Brussels sprouts, cauliflower, and tuna contain less than the median amount of fat and have ratios of zinc to copper greater than the median; these foods conform to the general distribution of the 71 foods, however. Skim milk falls somewhat outside this distribution, being low in fat and having a high ratio.

Therefore, individuals who attempt to decrease their intakes of fat will, on the average, decrease their dietary ratio of zinc to copper. Similar decreases in the dietary ratio of zinc to copper may have occurred in the therapeutic trials involving the fat hypothesis, with regard to the etiology of ischemic heart disease (Turpeninen *et al.*, 1968; Leren, 1966, 1970; Miettinen *et al.*, 1972). Although the Leren diet is described as a diet "low in animal fats and dietary cholesterol and rich in vegetable oil" (Leren, 1970) the diet actually seems to be a diet low in meat, dairy products and eggs and high in vegetables to which vegetable oil had been added (Leren, 1966). According to data in Fig. 6, these maneuvers probably decreased the ratio of zinc to copper of the diet.

3.4.2. Type of Fat

Proponents of the fat hypothesis with regard to the etiology of ischemic heart disease have placed more emphasis on type than on amount of dietary fat.

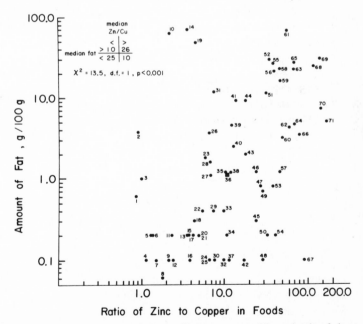

Fig. 6. Each datum point represents a food classified by amount of fat and ratio of zinc to copper. Foods of particular interest are 2, beef liver; 10, walnuts; 14, pecans; 19, peanut butter; 36, shrimp (median fat = 1.1 g/100 g, median ratio = 11.2); 50, brussels sprouts, 53, tuna; 54, cauliflower; 67, skim milk. Other foods are identified in the original reference. The miniature fourfold table contains the number of foods in each quadrant of the scatter diagram (from Klevay, 1974a).

Processed fats and oils do not contribute much zinc and copper to human diets because transition elements are removed from refined products to prevent decay (McPherson, 1976).

Four of the diets shown in Fig. 4 (labeled Minn.) are those of Anderson *et al.* (1973) arranged in order of the amount of copper contained in the meals for one day. These diets were formulated (Anderson *et al.*, 1973) to decrease the concentration of cholesterol in serum by progressive decreases in dietary fat and cholesterol and a progressive increase in polyunsaturated fats. If the diets are considered in the descending order of the concentration of cholesterol in serum of individuals consuming them, the ratios of zinc to copper were 9.4, 12.9, 11.8, and 7.4 (Klevay *et al.*, 1976) when measured by atomic absorption spectrophotometry. Ratios of zinc to copper calculated from tabulated values for copper and zinc and intakes for 1 week were 14.7, 13.1, 12.1, and 11.5, respectively (Klevay *et al.*, 1976).

Thus, although the data suggest declining trends for the concentration of cholesterol in serum and the ratio of zinc to copper of the diets, the relationship is not perfect. If preparation and analysis of these diets in other geographic areas would result in a greater similarity of the trends in cholesterol and ratio of zinc to copper, some of the effects of the "diet high in unsaturated fat in preventing complications of atherosclerosis" (Dayton *et al.*, 1969) may be explainable in terms of zinc and copper. Similar experiments could be designed to determine the relative importance of metabolic effects of dietary fats and trace elements and to determine whether or not the effects are complementary. Experiments with animals (Section 4.6.) demonstrate that effects of copper and zinc can be detected in diets of enormous variety.

3.5. Nursing

Osborn (1968) examined the coronary arteries by necroscopy of 109 persons up to the age of 20 and classified them according to the degree of reduction of luminal size due to atherosclerosis. Simultaneous classification of the individuals according to the length of time they were breast fed in infancy resulted in his concluding that breast feeding of greater than 1 month was associated with less arterial pathology than was artificial feeding. The results were statistically highly significant ($\chi^2 = 16.1$, df $= 4$, $p < 0.005$) (Klevay, 1975b).

Until recently, most artificial feeding of infants has been done with formulas based on cow's milk. As there is a positive correlation ($r = 0.354$, $P < 0.02$) between the ratio of zinc to copper of cow's milk available in 47 cities of the United States and the mortality rate attributable to coronary heart disease in these cities (Klevay, 1975b), the difference in ratio of zinc to copper of human and of cow's milk (6 and 38, respectively) may be important. The correlation between the ratio of zinc to copper of cow's milk and the mortality rate caused by coronary heart disease accounts for 12.5% of the variation mortality and represents the only epidemiologic association between a chemical characteristic of food or diet and risk of coronary heart disease which has been found in the United States. No significant correlation between either copper or zinc in milk and mortality rate was found.

4.   Effects of Zinc and Copper on Cholesterol Metabolism

4.1.   Hypercholesterolemia Due to Ascorbic Acid

Recently it has become fashionable to ingest large amounts of ascorbic acid (vitamin C) in the hope of decreasing the frequency or severity of upper respiratory infections collectively known as "colds." Ascorbic acid inhibits the absorption of copper from the intestinal tracts of animals (Evans, 1973).

Rats were fed a diet (Table I) supplemented with copper and zinc, with or without ascorbic acid. The concentration of cholesterol in the plasma of rats fed ascorbic acid was 129 mg/dl, as compared to a control concentration of 109 ($p <$ 0.02) (Klevay, 1976a). Although copper and zinc were not measured in this experiment, Gipp *et al.* (1974) recorded without comment an apparent increase in the ratio of zinc to copper in the livers of pigs fed extra ascorbic acid. The amount of ascorbic acid used in this experiment (Klevay, 1976a) corresponds to between 82 and 630 mg in excess of that consumed from dietary sources, an amount which is often exceeded by those in search of respiratory benefit. These data are consonant with data on humans tested under reasonable dietary control (Anderson *et al.*, 1958; Hodges *et al.*, 1971); many other experiments on this topic have been difficult to interpret (for review see Klevay, 1976a).

4.2.   Hypocholesterolemia Due to Phytic Acid

Trowell (1976) has hypothesized that populations whose diet is rich in fiber have a lower mortality from ischemic heart disease than do those that have a Western-type diet. A low intake of dietary fiber may be associated more closely with a high risk of ischemic heart disease than is a high intake of fat (Klevay, 1974c, 1976c). A protective factor in diets high in fiber may be phytic acid (*myo*-inositol hexaphosphate) (Klevay, 1975c; Trowell, 1976), which, in theory (Klevay, 1975c), should lower the ratio of zinc to copper absorbed from the intestinal tract and thereby prevent the hypercholesterolemia associated with the absorption of a high ratio. Fiber may adversely affect the absorption of zinc. Reinhold *et al.* (1976) have shown that in some individuals, high-fiber Iranian bread produced a greater impairment of the absorption of zinc than could be explained by the amount of phytate it contains.

Rats fed sodium phytate (Klevay *et al.*, 1975) had lower (17–20%, $p <$ 0.025) concentrations of cholesterol in plasma than rats fed control diets. The copper and zinc in the hair of the animals fed phytate were increased 17% and decreased 33%, respectively ($p < 0.02$); these measurements can be used to assess nutriture (for review see Klevay, 1970a,b). Absorption of ^{65}Zn from the intestinal tract was impaired by phytate; no effect on the absorption of ^{64}Cu could be demonstrated.

4.3. Zinc/Copper Effects

Rats fed a diet deficient in vitamin A and based on 68% cornstarch, 22% casein, and 6% cottonseed oil by Erdman and Lachance (1974) had a significantly ($p < 0.01$) smaller degree of hypercholesterolemia when cholesterol was added to the diet if the salt mixture of the diet had a low ratio of zinc to copper than if the salt mixture had a high ratio of zinc to copper.

Barboriak et al. (1958) reported that in rats fed a diet containing all known nutrients based on 56.6% lard, 20.0% casein, and 14.0% dextrin neither notable arterial pathology nor uniform hypercholesterolemia was shown to develop in a period of about 2 years. Inspection of this diet in the light of personal experience regarding probable content of zinc and copper resulted in the conclusion that the ratio of zinc to copper was low. The experiment was repeated (Klevay, 1974b) with the diet modified with regard to zinc and copper. The concentrations of cholesterol in the plasma of rats fed a high ratio of zinc to copper (20.4) and a low ratio of zinc to copper (4.6) were 142 and 116 mg/dl, respectively ($p < 0.001$).

Mice fed trimmed sirloin of beef became hypercholesterolemic as compared to mice fed beef liver–sirloin (1 : 3, by weight) (Jacob et al., 1975). A mixture of liver and muscle has a lower ratio of zinc to copper than does muscle alone (Klevay, 1974a). The concentrations of cholesterol in plasma of mice fed the high and low ratios of zinc to copper were 121 and 96 mg/dl, respectively ($p < 0.01$).

Tsai and Evans (1976) fed rats a diet based on 35% egg yolk, 13% soy protein, 20% cornstarch, and 20% sucrose and found that cholesterol in hearts decreased as dietary copper was increased. Although the probability that the results were caused by chance ($p < 0.10$) was somewhat greater than usually accepted as being significant ($p < 0.05$) they interpreted the data as ''supporting the Zn/Cu ratio theory.''

Walravens et al. (1976) have given the mothers of newborn infants enough formulae containing either 1.8 or 5.8 mg zinc/liter to supply the needs of their infants for 6 months. The infants were fed the formulas beginning 4–6 days after birth; at that time, the ratios of zinc to copper of the two diets were either 5 or 17. The mothers decided when to add other food items to the diets. After 6 months, when the diets of the two groups probably were almost identical because of the depletion of the formulas and the introduction of other foods, the mean concentrations of cholesterol in plasma of the two groups were almost identical. The group consuming the formula with the higher ratio of zinc to copper had a higher concentration of cholesterol (127 vs. 115 mg/dl) 3 months after birth. Although the difference was not statistically significant, the concentrations should not be assumed to be identical. Presumably there was still a difference in the ratio of zinc to copper of the two diets at that time. This experiment cannot be considered a test of the zinc/copper hypothesis.

4.4. Zinc Deficiency

Burch et al. (1975) produced zinc deficiency in pigs by feeding them a diet based on 68% glucose, 15% soy protein, and 10% corn oil. One group received the zinc-deficient diet; a second group received an equal amount of diet supplemented with an adequate amount of zinc. A third group received the supplemented diet *ad libitum*. The concentrations of cholesterol in the sera of the three groups were 75, 118, and 140 mg/dl, respectively. The difference between the first and second groups was highly significant ($p < 0.005$). About one-third of the difference between the first and third groups could be attributed to the restricted feeding.

Patel *et al.* (1975) found the concentration of cholesterol in serum of rats deficient in zinc (50 mg/dl) to be 44% of that of the control value (115 mg/dl) ($p < 0.01$). The diet was based on 53% glucose, 26% egg white, and 10% corn oil. No animals were fed restrictively, but comparison with the work of Burch *et al.* (1975) indicates that the effect probably was not due to decreased food intake.

4.5. Copper Deficiency

Allen and Klevay (1976) produced hypercholesterolemia (107 mg/dl) in rats deficient in copper compared to pair-fed animals (65 mg/dl) ($p < 0.001$). The copper in the diets of deficient and control groups was 0.57 and 5.0 μg/g of diet, respectively. The diet was based on 51% sucrose, 20% egg white, 20% fat (cottonseed oil, coconut oil, corn oil/95 : 95 : 10, by weight).

4.6. Summary

The effects of ascorbic and phytic acids on the concentration of cholesterol in plasma are the result of alterations in copper and zinc metabolism. Ingestion of large amounts of ascorbic acid to prevent "colds" may have undesired consequences. Phytic acid may be a protective factor in diets high in fiber.

Whenever one investigates a new phenomenon, no matter how reproducibly it appears, one must wonder whether it is an artifact of the experimental system. Copper and zinc have been connected with the metabolism of cholesterol of rats, mice and pigs fed many different diets. Sucrose, egg white, and corn oil (Klevay, 1973); cornstarch, casein, and cottonseed oil (Erdman and Lachance, 1974);

lard, casein, and dextrin (Klevay, 1974b); beef liver–sirloin (Jacob *et al.*, 1975); egg yolk, soy protein, cornstarch, sucrose (Tsai and Evans, 1976); glucose, soy protein, corn oil (Burch *et al.*, 1975); glucose, egg white, and corn oil (Patel *et al.*, 1975) have been used. One can conclude from this mélange that the effect on cholesterol metabolism was produced by an alteration in the metabolism of copper and zinc and is not by some other dietary component closely associated with copper or zinc. Similarly, an effect of copper and zinc on cholesterol metabolism can be detected in spite of variations in the amount and type of fat, the presence or absence of cholesterol or plant sterols, the use of monosaccharides, disaccharides, or starch, or plant or animal proteins in experimental diets.

The degree of cholesterolemia which can be produced in rats by the manipulation of copper and zinc is quite high. The highest values to date for means, individual values and increases over a control value are 223 mg/dl (Klevay, 1973), 413 mg/dl (Klevay, 1973), and 65% (Allen and Klevay, 1976).

5.   Factors That Alter Zinc and Copper Metabolism

5.1.   The "Water Factor"

Sharrett and Feinlieb (1975) have collected the evidence regarding the lesser risk of ischemic heart disease where hard drinking water is available. I believe that further investigation of this "Factor" should be experimental rather than epidemiologic. In the reports of the United States Geological Survey (Durfor and Becker, 1964) hardness of water is reported as calcium carbonate, indicating a close association between hardness and calcium. The correlation coefficient (Crawford *et al.*, 1968; Klevay, 1975c) between hardness and calcium is about 0.96. The decrease in concentration of cholesterol in plasma of people whose diets have been supplemented with calcium has been noted (Section 2.4.).

It was hypothesized (Klevay, 1975c) that higher amounts of dietary calcium would produce a lower ratio of zinc to copper in liver. Because liver is the major site for the synthesis (Dietschy and Wilson, 1970) and degradation (Haslewood, 1963) of cholesterol, it is assumed that alterations in copper and zinc metabolism affect cholesterol metabolism there. Pond and Walker (1975) have shown that rats fed diets high in calcium have almost twice as much copper in their livers as rats fed a diet low in calcium; the ratio of zinc to copper of the livers of rats fed more calcium was probably lower, as the amounts of zinc in liver were similar in the two groups.

5.2.   Acrodermatitis Enteropathica

Acrodermatitis chronica enteropathica is "a chronic, often fatal disease of unknown origin, possibly transmitted as a recessive trait." "It is characterized by a vesiculopustulobullous eruption symmetrically arranged on the extremities and around the mouth and anal areas, by loss of hair and by diarrhea'' (Burgoon, 1975). The disease has been treated empirically with diiodohydroxyquin. Recently, complete remission of symptoms has been obtained by treatment with zinc sulfate (Moynahan, 1974; Portnoy and Molokhia, 1974; Neldner and Hambidge, 1975). It has been noted that many of the signs of the disease are similar to those produced in animals by zinc deficiency (Moynahan, 1974; Neldner and Hambidge, 1975).

Low concentrations of cholesterol in plasma, generally 70–145 mg/dl, and in one instance 207 mg/dl, have been found (Zaidman *et al.*, 1971; Kayden and Cox, 1973; Neldner *et al.*, 1974). These findings have been attributed to abetalipoproteinemia (Zaidman *et al.*, 1971) or hypobetalipoproteinemia (Neldner *et al.*, 1974). These observations may be related to the hypocholesterolemia found in zinc-deficient pigs and rats (Section 4.4). No data for cholesterol in plasma of patients with acrodematitis enteropathica who have been treated with zinc have been found.

6. Conclusions

6.1. Copper—Dietary Amounts and Requirements

As shown in Fig. 4, of the 15 diets summarized, only two contain the 2 mg of copper thought to be the daily requirement (Section 3.2). Eight of the diets contained less than 1.0 mg and five diets contained less than 0.5 mg. Assuming the analyses from other laboratories are correct—our measurements are in error by less than 2% (Klevay *et al.*, 1976)—many diets made from common foods are inadequate in copper.

6.2. Copper—Recommended Dietary Allowance

Although the metabolism of copper in women is somewhat different than in men (Klevay, 1970b; Evans, 1973), I believe data are not adequate to determine whether sexual differences in metabolism necessitate differences in recommendations. Assuming the requirement for copper is about 2 mg, a recommended allowance should exceed this amount. Butler and Daniel (1973), whose data suggest a requirement of slightly more than 2 mg, recommend an allowance of 4.5–5.0 mg. Dietary analyses imply that this amount may be difficult to attain. Data on the availability of copper for intestinal absorption will be useful in determining whether this amount is necessary.

6.3. Dietary Ratio of Zinc to Copper

Definition of an optimal dietary ratio of zinc to copper must be determined by experimentation. Several observations can be used for guidance. The ratio of zinc to copper of human milk is approximately 6, in contrast to that of cow's milk, which is about 38 (Klevay, 1975b). Considering the apparent protective effect of nursing (Osborn, 1968) on coronary atherosclerosis (Section 3.5), a ratio of 6 must be considered desirable for infants and a ratio of 38 must be considered too high. Whether a ratio of 6 will prove desirable for adults is unknown. The Recommended Dietary Allowance (Anonymous, 1974b) for adults for zinc is 15 mg; no recommendation is made for copper (Anonymous, 1974b). If a recommended allowance for copper should exceed 2 mg (Section 6.2) the ratio for adults would be less than 7.5.

Hypercholesterolemia was produced in rats by a ratio of 14 to 20 (Klevay, 1973, 1974d) compared to a control ratio of less than 4. A ratio of 14, which must be considered high for a rat, is not necessarily high for a human. As nutritional requirements vary among species, so may optimal ratios vary among species.

The median ratio of zinc to copper of the diets summarized in Fig. 4 is 17 (Klevay, 1975a, Klevay *et al.*, 1976). The ratios ranged from 3 to greater than 38. Common diets may have ratios of zinc to copper higher than that which produced hypercholesterolemia in rats.

The amounts of zinc in the diets summarized in Fig. 4 either were close to the Recommended Dietary Allowance for adults (Anonymous, 1974b) or exceeded (Klevay, 1975a, Klevay *et al.*, 1976) that 15 mg. Sandstead (1973) has found some diets in the United States that appear to be deficient in zinc. Similarly, Hambidge *et al.* (1972, 1976) have found some short children who may have been deficient in zinc. Attempts to decrease the ratio of zinc to copper of diets probably should be based on an increase in copper. Whether absolute or relative deficiency of copper is of primary importance in the zinc/copper phenomenon may be determined by further research.

6.4. Summary

Either the adult daily requirement for copper is not 2 mg, or many diets available in the United States do not contain the required amount. Some diets in the United States have ratios of zinc to copper that seem too high. If diets with low ratios of zinc to copper are desired, a decrease in ratio should be based upon increases in dietary copper.

Note Added in Proof

Several abstracts of work to be presented at the 61st annual meeting of the Federation of American Societies for Experimental Biology, which are germane to this topic, have been published. O'Neal *et al.* (1977) found that the ratio of zinc to copper of hair from a Missouri population increased gradually as the concentration of cholesterol in serum of men increased or as the concentration of triglycerides in serum of women increased. The relationship between the ratio of zinc to copper of hair and the concentration of triglycerides of men or the concentration of cholesterol of women was not as close. These authors concluded that dietary ratios of zinc to copper, as reflected by hair values, may be a factor in hyperlipidemia and coronary heart disease.

Three groups reported analyses of human diets containing amounts of copper lower than the 2 mg of copper thought to be the daily requirement (Fig. 4). We (Klevay *et al.*, 1977) found a mean daily amount of copper of 0.76 mg for five hospital diets. Wolf *et al.* (1977) reported a mean daily intake of 1.01 mg; these diets were self-selected, noninstitutional diets. Brown *et al.* (1977) studied hospital diets and found that the daily amount of copper in the vegetarian diet (1.10 mg) exceeded that in the general diet (0.90 mg), which exceeded that (0.51 mg) in the diet served patients with renal disease. If vegetarian diets are generally found to contain more copper than those containing meat, this observation may be related to the lesser risk of ischemic heart disease associated with the habitual consumption of vegetarian diets (Trowell, 1976), as beef is a poor source of dietary copper (Darby *et al.*, 1973).

If the small amount of copper found in the diets of patients with renal disease is representative of renal diets in general, this finding may a provide partial explanation of the increased risk of coronary heart disease experienced by these patients (for references, see Klevay, 1975c). This increased risk has been attributed to a loss of bone, which results in relatively large amounts of zinc being released to the rest of the body and to the infusion of relatively large amounts of zinc during hemodialysis (Klevay, 1975c).

Lei (1977) attributed hypercholesterolemia in rats fed a diet low in copper and high in zinc to a shift of cholesterol from liver to serum or to a decrease in the degradation of cholesterol. Looney *et al.* (1977), in what seems to have been a similar experiment, found that a diet high in zinc and low in copper produced a greater hypercholesterolemia than did a diet low in copper.

References

Albanese, A. A., Edelson, A. H., Woodhull, M. L., Lorenze, E. J., Jr., Wein, E. H., and Orto, L. A., 1973, Effects of a calcium supplement on serum cholesterol, calcium, phosphorus and bone density of "normal, healthy" elderly females, *Nutr. Rep. Int.* **8**:119.

Allen, K. G. D., and Klevay, L. M., 1976, Hypercholesterolemia in rats caused by copper deficiency, *J. Nutr.* **106**(7):XXII.

Anderson, J. T., Grande, F., and Keys, A., 1958, Dietary ascorbic acid and serum cholesterol, *Fed. Proc.* **17**:468.

Anderson, J. T., Grande, F., and Keys, A., 1973, Cholesterol-lowering diets, *J. Am. Dietet. Assoc.* **62**:133.

Anon., 1969, Vital Statistics of the United States 1967. Mortality, Vol. 2, Part A, Table 1–7, Washington, D.C., U.S. Department of Health, Education and Welfare.

Anon., 1974a, Vital Statistics of the United States 1970. Mortality, Vol. 2, Part A, Table 1-7, Washington, D.C., U.S. Department of Health, Education and Welfare.

Anon., 1974b, Recommended Dietary Allowances, Food and Nutrition Board, National Research Council, National Academy of Sciences, Washington, D.C., pp. 92–96.

Anon., 1976, The herbs and the heart, *Nutr. Rev.* **34**:43.

Barboriak, J. J., Krehl, W. A., Cowgill, G. R., and Whedon, A. D., 1958, Influence of high-fat diets on growth and development of obesity in the albino rat, *J. Nutr.* **64**:241.

Birenbaum, M. L., Fleischman, A. I., and Raichelson, R. I., 1972, Long term human studies on the lipid effects of oral calcium, *Lipids* **7**:202.

Brown, E. D., Howard, M. P., and Smith, J. C., Jr., 1977, The copper content of regular, vegetarian, and renal diets, *Fed. Proc.*, **36**:1122.

Burch, R. E., Williams, R. V., Hahn, H. K. J., Jetton, M. M., and Sullivan, J. F., 1975, Serum and tissue enzyme activity and trace-element content in response to zinc deficiency in the pig, *Clin. Chem.* **21**:568.

Burgoon, C. F., Jr., 1975, The skin, in *Nelson's Textbook of Pediatrics* (V. C. Vaughan III, R. J. McKay, and W. E. Nelson, eds.), 10th ed., pp. 1514–1568, W. B. Saunders, Philadelphia.

Butler, L. C., and Daniel, J. M., 1973, Copper metabolism in young women fed two levels of copper and two protein sources, *Am. J. Clin. Nutr.* **26**:744.

Call, D. L., and Sanchez, A. M., 1967, Trends in fat disappearance in the United States 1909–65, *J. Nutr.* **93** (Suppl. 1) Part II, 1.

Carlson, L. A., Olsson, A. G., Orö, L., and Rössner, S., 1971, Effects of oral calcium upon serum cholesterol and triglycerides in patients with hyperlipidemia, *Atherosclerosis* **14**:391.

Cartwright, G. E., 1950, Copper metabolism in human subjects, in *Copper Metabolism* (W. D. McElroy and B. Glass, eds.), pp. 274–314, Johns Hopkins Press, Baltimore.

Cartwright, G. E., and Wintrobe, M. M., 1964, The question of copper deficiency in man, *Am. J. Clin. Nutr.* **15**:94.

Cartwright, G. E., Hodges, R. E., Gubler, C. J., Mahoney, J. P., Daum, K., Wintrobe, M. M., and Bean, W. B., 1954, Studies on copper metabolism. XIII. Hepatolenticular degeneration, *J. Clin. Invest.* **33**:1487.

Crawford, M. D., Gardner, M. J., and Morris, J. N., 1968, Mortality and hardness of local water supplies, *Lancet* **1**:827.

Curren, G. L., Azarnoff, D. L., and Bolinger, R. E., 1959, Effect of cholesterol synthesis inhibition in normocholesteremic young men, *J. Clin. Invest.* **38**:1251.

Dahl, L. K., 1960, Effects of chronic excess salt feeding. Elevation of plasma cholesterol in rats and dogs, *J. Exp. Med.* **112**:635.

Darby, W. J., Friberg, L., Mills, C. F., Parizek, J., Ramalingaswami, V., Ronaghy, H. A., Sandstead, H. H., Underwood, E. J., Rao, N. M., Par, R., Bengoa, J. M., DeMaeyer, E. M., and Mertz, W., 1973, Trace Elements in Human Nutrition, Report of a WHO expert committee. World Health Organization Technical Report Series, No. 532 World Health Organization, Geneva, pp. 15–19.

Dayton, S., Pearce, M. L., Hashimoto, S., Dixon, W. J., and Tomiyasu, U., 1969, A controlled clinical trial of a diet high in unsaturated fat, *Circulation* **39, 40** (Suppl. II):1.

Doisy, E. A., Jr., 1973, Micronutrient controls on biosynthesis of clotting proteins and cholesterol, in *Trace Substances in Environmental Health. VI* (D. D. Hemphill, ed.), pp. 193–199, Univ. of Missouri Press, Columbia.

Dietschy, J. M., and Wilson, J. D., 1970, Regulation of cholesterol metabolism, *New Engl. J. Med.* **282**:1128.

Durfor, C. N., and Becker, E., 1964, Public water supplies of the 100 largest cities in the United States, 1962, Washington, D.C., U.S. Geological Survey, Water-Supply Paper 1812.

Erdman, J. W., Jr., and Lachance, P. S., 1974, Effect of salt mixture and cholesterol upon rat serum and liver zinc, vitamin A, and cholesterol, *Nutr. Rep. Int.* **9**:319.

Evans, G. W., 1973, Copper homeostasis in the mammalian system, *Physiol. Rev.* **53**:535.

Fillios, L. C., and Mann, G. F., 1954, Influence of sulfur amino acid deficiency on cholesterol metabolism, *Metabolism* **3**:16.

Follis, R. H., and Van Itallie, T. B., 1974, Pellagra, in *Harrison's Principles of Internal Medicine* (M. M. Wintrobe, G. W. Thorn, R. D. Adams, E. Braunwald, K. J. Isselbacher, and R. G. Petersdorf, eds.), pp. 427–430, McGraw-Hill, New York.

Fredrickson, D. S., and Levy, R. I., 1972, Familial hyperlipoproteinemia, in *The Metabolic Basis of Inherited Disease* (J. B. Stanbury, J. B. Wyngaarden, and D. S. Fredrickson, eds.), 3rd ed., pp. 545–614. McGraw-Hill, New York.

Friedberg, C. K., 1966, *Diseases of the Heart, 3rd ed.*, pp. 652, 669, W. B. Saunders, Philadelphia.

Gipp, W. F., Pond, W. G., Kallfelz, F. A., Tasker, J. B., Van Campen, D. R., Krook, L., and Visek, W. V., 1974, Effect of dietary copper, iron, and ascorbic acid levels on hematology, blood and tissue copper, iron and zinc concentrations and ^{64}Cu and ^{59}Fe metabolism in young pigs, *J. Nutr.* **104**:532.

Hambidge, K. M., Hambidge, C., Jacobs, M., and Baum, J. D., 1972, Low levels of zinc in hair, anorexia, poor growth, and hypogeusia in children, *Pediat. Res.* **6**:868.

Hambidge, K. M., Walravens, P. A., Brown, R. M., Webster, J., White, S., Anthony, M., and Roth, M. L., 1976, Zinc nutrition of preschool children in the Denver Head Start program, *Am. J. Clin. Nutr.* **29**:734.

Hartley, T. F., Dawson, J. B., and Hodgkinson, A., 1974, Simultaneous measurement of Na, K, Ca, Mg, Cu, and Zn balances in man, *Clin. Chim. Acta* **52**:321.

Haslewood, G. A. D., 1963, Bile salts, in *Sterols, Bile Acids and Steroids* (M. Florkin and E. H. Stotz, eds.), Vol. 10, pp. 23–31, Elsevier, Amsterdam.

Hegsted, D. M., *et al.*, 1942–1976, *Nutr. Rev.* **1–34**(8).

Hellerstein, E. E., Nakamura, M., Hegsted, D. M., and Vitale, J. J., 1960, Studies on the interrelationships between dietary magnesium, quality and quantity of fat, hypercholesterolemia and lipidosis, *J. Nutr.* **71**:339.

Hodges, R. E., Hood, J., Canham, J. E., Sauberlich, H. E., and Baker, E. M., 1971, Clinical manifestations of ascorbic acid deficiency in man, *Am. J. Clin. Nutr.* **24**:432.

Hopkins, L. L., and Mohr, H. E., 1974, Vanadium as an essential nutrient, *Fed. Proc.* **33**:1773.

Iacono, J. M., 1974, Effect of varying the dietary level of calcium on plasma and tissue lipids of rabbits, *J. Nutr.* **104**:1165.

Ingbar, S. H., and Woeber, K. A., 1974, The thyroid gland, in *Textbook of Endocrinology* (R. H. Williams, ed.), 5th ed., pp. 95–232, W.B. Saunders, Philadelphia.

Insull, W., Jr., 1973, *Coronary Risk Handbook,* p. 3, American Heart Association, New York.

Jacob, R. A., Klevay, L. M., and Thacker, E. J., 1975, Hypercholesterolemia due to meat anemia, *Fed Proc.* **34**:899.

Kahn, H. A., 1970, Change in serum cholesterol associated with changes in the United States civilian diet, 1909–1965, *Am. J. Clin. Nutr.* **23**:879.

Kayden, H. J., and Cox, R. P., 1973, Evidence for normal metabolism and interconversions of unusual fatty acids in acrodermatitis enteropathica, *J. Pediat.* **83**:993.

Klevay, L. M., 1970a, Hair as a biopsy material. I. Assessment of zinc nutriture. *Am. J. Clin. Nutr.* **23**:284.

Klevay, L. M., 1970b, Hair as a biopsy material. II. Assessment of copper nutriture. *Am. J. Clin. Nutr.* **23**:1194.

Klevay, L. M., 1973, Hypercholesterolemia in rats produced by an increase in the ratio of zinc to copper ingested, *Am. J. Clin. Nutr.* **26**:1060.

Klevay, L. M., 1974a, An association between the amount of fat and the ratio of zinc to copper in 71 foods: Inferences about the epidemiology of coronary heart disease, *Nutr. Rep. Int.* **9**:393.

Klevay, L. M., 1974b, Lard diets (56.6%) don't produce hypercholesterolemia in rats unless zinc/copper metabolism is altered, *Clin. Res.* **22**:629A.

Klevay, L. M., 1974c, Coronary heart disease and dietary fiber, *Am. J. Clin. Nutr.* **27**:1202.

Klevay, L. M., 1974d, Interactions among dietary copper, zinc, and the metabolism of cholesterol and phospholipids, in *Trace Element Metabolism in Animals* (W. G. Hoekstra, J. W. Suttie, H. E. Ganther, and W. Mertz, eds.), Vol. 2, pp. 553–556, University Park Press, Baltimore.

Klevay, L. M., 1975a, The ratio of zinc to copper of diets in the United States, *Nutr. Rep. Int.* **11**:237.

Klevay, L. M., 1975b, The ratio of zinc to copper in milk and mortality due to coronary heart disease: An association, in *Trace Substances in Environmental Health. VIII* (D. D. Hemphill, ed.), p. 9, Univ. of Missouri Press, Columbia.

Klevay, L. M., 1975c, Coronary heart disease: The zinc/copper hypothesis, *Am. J. Clin. Nutr.* **28**:764.

Klevay, L. M., 1976a, Hypercholesterolemia due to ascorbic acid, *Proc. Soc. Exp. Biol. Med.* **151**:579.

Klevay, L. M., 1976b, Elements of ischemic heart disease, *Perspect. Biol. Med.* **20**:186.

Klevay, L. M., 1976c, Ischemic heart disease: The fiber hypothesis, *Proceedings of the Miles Symposium, Nutrition Society of Canada* (in press).

Klevay, L. M., and Forbush, J., 1976, Copper metabolism and the epidemiology of coronary heart disease, *Nutr. Rep. Int.* **14**:221.

Klevay, L. M., Evans, G. W., and Sandstead, H. H., 1975, Zinc/copper hypercholesterolemia: The effect of sodium phytate, *Clin. Res.* **23**:460A.

Klevay, L. M., Petering, H. G., and Stemmer, K. L., 1971, A controlled environment for trace metal experiments on animals, *Environ. Sci. Technol.* **5**:1196.

Klevay, L. M., Vo-Khactu, K. P., and Jacob, R. A., 1976, The ratio of zinc to copper of cholesterol-lowering diets, in *Trace Substances in Environmental Health. IX* (D. D. Hemphill, ed.), pp. 131–138, Univ. of Missouri Press, Columbia.

Klevay, L. M., Reck, S., and Barcome, D. F., 1977, United States diets and the copper requirement, *Fed. Proc.,* **36**:1175.

Lei, K. Y., 1977, Dietary copper deficiency: Effects on cholesterol metabolism in the rat. *Fed. Proc.* **36**:1151.

Leren, P., 1966, The effect of plasma cholesterol lowering diet in male survivors of myocardial infarction, *Acta Med. Scand. Suppl.* **446**:1.

Leren, P., 1970, The Oslo diet-heart study, eleven-year report, *Circulation* **42**:935.

Leverton, R. M., and Binkley, E. S., 1944, The copper metabolism and requirement of young women, *J. Nutr.* **27**:43.

Looney, M. A., Lei, K. Y., and Kilgore, L. T., 1977, Effects of dietary fiber, zinc, and copper on serum and liver cholesterol levels in the rat, *Fed. Proc.,* **36**:1134.

Lorenz, K. Z., 1974, Analogy as a source of knowledge, *Science* **185**:229.

Maibach, E., 1967, Die Beeinflussung des Gesamtcholesterins der β-Lipoproteide und Gesamtlipide des Serums durch orale und parenterale Calciumzufuhr, *Schweiz med. Woch.* **97**:418.

Maibach, E., 1969, Uber die Beeinflussung des Gesamtcholesterins und der Gesamtlipide im Serum des Menschen durch orale Kalziumzufuhr, *Wien. med. Woch.* **118**:1059.

Mann, G. V., Andrus, S. B., McNally, A., and Stare, F. J., 1953, Experimental atherosclerosis in Cebus monkeys, *J. Exp. Med.* **98**:195.

McPherson, W. J., 1976, Nickel Catalyst Scavenging Process, Technical Data Sheet No. 2625, Chicago, Illinois, Wurster and Sanger.

Meneely, G. R., and Ball, C. O. T., 1958, Experimental epidemiology of chronic sodium chloride toxicity and the protective effect of potassium chloride, *Am. J. Med.* **25**:713.

Miettinen, M., Turpeinen, O., Karvonen, M. J., Elosuo, R., and Paavilainen, E., 1972, Effect of cholesterol-lowering diet on mortality from coronary heart-disease and other causes, *Lancet* **2**:835.

Moynahan, E. J., 1974, Acrodermatitis enteropathica: a lethal inherited human zinc-deficiency disorder, *Lancet* **2**:399.

Neldner, K. H., Hagler, L., Wise, W. R., Stifel, F. B., Lufkin, E. G., and Herman, R. H., 1974, Acrodermatitis enteropathica, *Arch. Dermatol.* **110**:711.

Neldner, K. H., and Hambidge, K. M., 1975, Zinc therapy of acrodermatitis enteropathica, *New Engl. J. Med.* **292**:879.

Nielsen, F. H., Myron, D. R., Givand, S. H., Zimmerman, T. J., and Ollerich, D. A., 1975, Nickel deficiency in rats, *J. Nutr.* **105**:1620.

O'Neal, R. M., Abrahams, O. G., Paulsen, D. S., Lorah, E. J., Eklund, D. L., and Dowdy, R. P., 1977, The relationships of hair zinc to copper ratios with serum cholesterol and triglyceride levels, *Fed. Proc.*, **36**:1122.

Osborn, G. R., 1968, Stages in development of coronary disease observed from 1500 young subjects. Relationship of hypotension and infant feeding to aetiology, in *Le rôle de la paroi artérielle dans l'athérogénèse*, (M. J. Lenègre, L. Scebat and J. Renais, Eds.), pp. 93–139, Centre National de la Recherche Scientifique, Paris.

Page, L. B., Damon, A., Moellering, R. C., 1974, Antecedents of cardiovascular disease in six Solomon Islands societies, *Circulation* **49**:1132.

Patel, P. B., Chung, R. A., and Lu, J. Y., 1975, Effect of zinc deficiency on serum and liver cholesterol in the female rat, *Nutr. Rep. Int.* **12**:205.

Pond, W. G., and Walker, E. F., Jr., 1975, Effect of dietary Ca and Cd level of pregnant rats on reproduction and on dam and progeny tissue mineral concentrations, *Proc. Soc. Exp. Biol. Med.* **148**:665.

Portnoy, B., and Molokhia, M., 1974, Zinc in acrodermatitis enteropathica, *Lancet* **2**:663.

Reinhold, J. G., Faradji, B., Abadi, P., and Ismail-Beigi, F., 1976, Binding of zinc to fiber and other solids of whole meal bread, in *Trace Elements in Human Health and Disease* (A. S. Prasad and D. Oberleas, eds.), Vol. 1, pp. 163–180, Academic Press, New York.

Rizek, R. L., Friend, B., and Page, L., 1974, Fat in today's food supply—Level of use and sources, *J. Am. Oil Chem. Soc.* **51**:244.

Roe, D. A., 1973, *A Plague of Corn,* Cornell Univ. Press, Ithaca, N.Y.

Sandstead, H. H., 1973, Zinc nutrition in the United States, *Am. J. Clin. Nutr.* **26**:1251.

Sandstead, H. H., 1974, Macroelement and trace element deficiencies, in *Harrison's Principles of Internal Medicine* (M. M. Wintrobe, G. W. Thorn, R. D. Adams, E. Braunwald, K. J. Isselbacher, and R. G. Petersdorf, eds.), pp. 441–443, McGraw-Hill, New York.

Scheinberg, I. H., and Sternlieb, I., 1960, Copper metabolism, *Pharmacol. Rev.* **12**:355.

Schroeder, H. A., 1968, Serum cholesterol levels in rats fed thirteen trace elements, *J. Nutr.* **94**:475.

Schroeder, H. A., 1969, Serum cholesterol and glucose levels in rats fed refined and less refined sugars and chromium, *J. Nutr.* **97**:237.

Schroeder, H. A., and Balassa, J. J., 1965, Influence of chromium, cadmium, and lead on rat aortic lipids and circulating cholesterol, *Am. J. Physiol* **209**:433.

Schroeder, H. A., Nason, A. P., Tipton, I. H., and Balassa, J. J., 1966, Essential trace metals in man: Copper, *J. Chron. Dis.* **19**:1007.

Selye, H., 1964, *From Dream to Discovery,* p. 104, McGraw-Hill, New York.

Selye, H., 1970, *Experimental Cardiovascular Diseases,* p. 308. Springer-Verlag, New York.

Sharrett, A. R., and Feinleib, M., 1975, Water constituents and trace elements in relation to cardiovascular disease, *Prev. Med.* **4**:20.

Stamler, J., Pick, R., and Katz, L. M., 1958, Effects of dietary proteins, methionine and vitamins on plasma lipids and atherogenesis in cholesterol fed cockerels, *Circ. Res.* **6**:442.

Staub, H. W., Reussner, G., and Thiessen, R., Jr., 1969, Serum cholesterol reduction by chromium in hypercholesterolemic rats, *Science* **166**:746.

Terris, M., ed., 1964, *Goldberger on Pellagra,* Louisiana State Univ. Press, Baton Rouge.

Trowell, H., 1976, Definition of dietary fiber and hypotheses that it is a protective factor in certain diseases, *Am. J. Clin. Nutr.* **29**:417.

Tsai, C. M. E., and Evans, J. L., 1976, Influence of dietary ascorbic acid and copper on tissue trace elements, cholesterol, and hemoglobin, in *Trace Substances in Environmental Health. IX* (D. D. Hemphill, ed.), pp. 441–449, Univ. of Missouri Press, Columbia.

Turpeinen, O., Miettinen, M., Karvonen, M. J., Roione, P., Pekkarinen, M., Lehtosuo, E. J., and

Alivirten, P., 1968, Dietary prevention of coronary heart disease: Long-term experiment, *Am. J. Clin. Nutr.* **21**:225.

Vitale, J. J., Velez, H., Guzman, C., and Correa, P., 1963, Magnesium deficiency in the Cebus monkey, *Circ. Res.* **12**:642.

Walravens, P. A., and Hambidge, K. M., 1976, Growth of infants fed a zinc supplemented formula, *Am. J. Clin. Nutr.* **29**:1114.

Whanger, P. D., and Weswig, P. H., 1975, Effects of selenium, chromium, and antioxidants on growth, eye cataracts, plasma cholesterol, and blood glucose in selenium deficient, vitamin E supplemented rats, *Nutr. Rep. Int.* **12**:345.

Wolf, W. R., Holden, J., and Green, F. E., 1977, Daily intake of zinc and copper from self-selected diets, *Fed. Proc.,* **36**:1175.

Yacowitz, H., Fleischman, A. I., and Bierenbaum, M. L., 1965, Effects of oral calcium upon serum lipids in man, *Br. Med. J.* **1**:1352.

Yudkin, J., 1974, Sugar and coronary disease, in *Controversy in Internal Medicine. II* (F. J. Ingelfinger, R. V. Ebert, M. Finland, and A. S. Relman, eds.), pp. 199–207, W. B. Saunders, Philadelphia.

Zaidman, J. L., Julsary, A., Kook, A. I., Szeinberg, A., Wallis, K., and Azizi, E., 1971, Abetalipoproteinemia in acrodermatitis enteropathica, *New Engl. J. Med.* **284**:1387.

Chapter 10

Relationship between Nutrition and Aging

Charles H. Barrows and Gertrude C. Kokkonen

1. Introduction

One of the most critical needs in nutritional research at the present time is a knowledge of the levels of intake of various nutrients at specific intervals in the life cycle which will optimize physical and mental development, physiological performance during adulthood, and the retention of health and vigor in senescence. In the past, emphasis had been placed primarily on the establishment of the nutritional requirements of young growing animals. Few efforts have been made to determine whether changes occur in nutritional requirements following growth cessation. This is unfortunate because many age changes in physiological functions may result in increased nutritional needs in later life. In addition, it is generally accepted that intakes moderately above the recommended allowances are optimal for the well-being of an organism. However, a number of studies carried out on animals have demonstrated that longevity was increased when intakes of certain nutrients were lower than the recommended allowances. Therefore, an attempt is made here to review the pertinent literature on (1) the effect of age on nutritional requirements following the cessation of growth and (2) the effect of nutrition on lifespan. This information may provide useful knowledge for the optimal nutrition of the aged and for an understanding of the basic mechanisms of biological aging. Studies concerned with the first problem area

Charles H. Barrows and Gertrude C. Kokkonen • Laboratory of Cellular and Comparative Physiology, Gerontology Research Center, National Institute on Aging, National Institutes of Health, PHS, U.S. Department of Health, Education and Welfare, Bethesda, Maryland and Baltimore City Hospital, Baltimore, Maryland 21224.

have been carried out principally in human subjects, whereas those with the second area, in animals.

2.   Physiological and Biochemical Changes with Age

Only those physiological and biochemical changes with age which may influence nutrient requirements or aid in an interpretation of the effects of nutrition on lifespan will be discussed here. Among those changes which seem important are basal metabolic rates, physical activities, synthesis of tissue proteins, and organ functions which influence digestion and absorption of nutrients.

2.1.   Basal Metabolic Rate and Physical Activity

The two most important bodily functions that control caloric intake are basal metabolic rate and physical activity. Basal metabolic rate based on the surface area of humans has consistently been reported to decrease with age. Furthermore, Shock et al. (1963) have demonstrated that this decline results from a loss of cells as measured by antipyrine space (Fig. 1). However, from a nutritional point of view, it matters little whether the BMR/cell decreases or there is a loss of cells with age. The data indicate that fewer dietary calories must be provided to maintain basal metabolic rate in the older individual. Unfortunately, similar quantitative data on large segments of the population relating age to total daily energy expenditure are not presently available. Therefore, it is not possible at this time to recommend age-associated adjustments in total caloric intake.

2.2.   Synthesis of Tissue Proteins

A variety of theories of aging suggest the formation of transcriptional and/or translational errors during protein synthesis which may contribute to the aging process. Post-transcriptional changes could result from such a myriad of causes that, at present, such findings are most difficult to interpret. However, imperfections in transcription may be manifested in two ways. In the first, the synthesis of the message may be so impaired that the coded protein cannot be synthesized. In the second, the message may be essentially intact so that the synthesized protein could be identified as such but exhibits some type of imperfection. There is ample evidence that enzyme synthesis is reduced during aging. However, more important is the observation that only approximately 30% of the enzymes reported are affected (Finch, 1972). This may indicate selective parts of the genetic code are damaged. Although unequivocal proof that aging results in transcriptional imperfections of the second type is not available, alterations in tissue proteins associated with senescence have been summarized in a recent review by Comfort (1974).

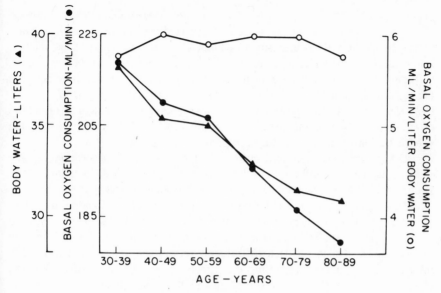

Fig. 1. Effect of age on the basal oxygen consumption and body water of men (Shock, 1976).

2.3. Digestion and Absorption

Although there have been many reports regarding changes with age in the gastrointestinal tract, it is difficult at this time to evaluate the influence of these changes on digestion and absorption. Nevertheless, it seems appropriate to present these findings. Among the factors that may result in agewise decreases in dietary intake are dental insufficiencies and age-related losses in the senses of smell and taste (Lemming et al., 1973). The most common age change associated with the stomach is achlorhydria. For example, after the age of 60, more than 25% of subjects tested have histamine-fast achlohydria (Lemming et al., 1973).

Age-related changes in the activity of digestive enzymes have also been reported. For example, it has been shown that resting and stimulated secretion of saliva is significantly lower in volume and amylase content in old people. Other studies reported that the activity of pepsin and hydrochloric acid in fasting gastric juice, and of amylase, trypsin, and lipase in fasting duodenal juice showed a significant depression in the aged. On the other hand, pancreatic secretion stimulation minimized these age differences (Necheles et al., 1942). Another intestinal enzyme that has recently been investigated and may be of importance in the nutrition of the aged is lactase. Its deficiency results in a lactose intolerance which is associated with marked abdominal distress. Studies indicate that certain segments of the population may be very susceptible to this insufficiency. For example, 81% of blacks, but only 12% of whites, were found to have abnormal lactose tolerance tests in a study of 166 hospitalized male patients (Bayless *et al.,* 1975). In addition, it has been suggested that a lactase deficit may be age related. For example, its incidence is negligible at birth but rises to 31% in adults (Littman, 1966). Indeed, the prevalence of the deficiency seems to be between the ages of 20 and 50 years (Cuatrecasas *et al.,* 1965; Dunphy *et al.,* 1965). However, there has been no systematic study of this enzyme system as a function of age in adult populations. Therefore, at this time, care should be taken in the administration of milk in feeding programs and care of the aged.

Other variables that may influence digestion and/or absorption are the motility and cellular replication of the gastrointestinal tract, and intestinal microflora. Cineradiologic studies have shown diminished or absent esophageal and intestinal peristalsis in most elderly subjects (Strauss, 1971). In addition, Fry *et al.* (1961) have reported an age-associated diminished rate of cellular replication in the duodenum and jejunum of mice by autoradiographic techniques. Intestinal microflora have been reported to influence nutritional states. However, little information is available regarding the effects of age on intestinal microflora. Gorbach *et al.* (1967) have reported quantitative and qualitative data of the microflora of the feces of 70 normal individuals, aged 20–100 years. The intestines of elderly subjects were found to contain fewer anaerobic lactobacilli and more coliforms and fungi than in younger persons. The microbial counts of staphylococci, streptococci, and aerobic lactobacilli were found to be unchanged. Unfortunately, no information is available regarding how these changes may relate to nutrient availability.

In spite of these apparent age-associated deficits, studies presently available do not indicate a marked impairment in the absorption of various nutrients in elderly subjects. For example, Chinn *et al.* (1956) reported no decrements in the rate of digestion of ^{131}I-labeled albumin in 12 subjects 72–88 years of age. Similarly, Watkin *et al.* (1955) and Bogdonoff *et al.* (1953), using the nitrogen-balance technique, found no decrease in protein absorption with age. Yiengst and Shock (1949) administered 100,000 U vitamin A orally to subjects between 40 and 89 years of age and observed marked elevations in the plasma levels of the vitamin in all age groups. There were no indications of a change with age in the rate of absorption of the vitamin following age 60. These data do not indicate marked malabsorption of these selected nutrients in older subjects.

3.   Effect of Age on Nutritional Status of Humans

Two types of studies have been used to assess the nutritional status of humans as a function of age: (1) national surveys of large segments of a population and (2) studies on limited numbers of subjects as part of specific laboratory investigations. There are many variables that make such assessments difficult.

This is especially so in the United States due to the marked differences in genetic background, social environment, and economic status of the population. Furthermore, the great selection of foods available and the determination of their nutrient content complicates the problem even more. Other difficulties arise because of the various ways of assessing nutritional status; namely, records of dietary intake; plasma, blood, or tissue contents of the nutrient; urinary or fecal excretion under various intakes; and the measurement of a biochemical system in which the nutrient plays a role.

3.1. Dietary Intake

3.1.1. National Surveys

In order to assess the nutritional status of the general population in the United States, the Department of Agriculture has obtained information at intervals since 1936 on nutrient intakes of large numbers of individuals. The most recent study was in 1965 (U.S. Department of Agriculture, 1968) based on 14,500 persons from 6174 households. The survey did not include those living in institutions and rooming houses and thus omitted many of the aged who were ill or disabled. The results showed that the average nutrient intakes per day for men aged 55 and over were adequate, except for calcium which declined with age and averaged 84% of the recommended dietary allowances (RDAs; National Academy of Sciences—National Research Council, 1968), whereas the intakes of thiamine, riboflavin, and calcium of women in this age groups were 87, 84, and 64%, respectively of the RDA. When individuals with low incomes were evaluated separately, it was found that dietary adequacy (based on RDAs) declined with income. Of those with incomes after taxes less than $3000, 63% had inadequate diets.

The Ten-State Nutrition Survey conducted by the U.S. Department of Health, Education and Welfare (Center for Disease Control, 1972) was designed to assess the nutritional status of certain groups considered to be at risk for undernutrition, such as poverty groups, migrant workers, Spanish-speaking people in the southwest United States, inner-city residents and individuals in industrial states who had migrated from the south in the previous 10–20 years. These groups were selected from districts with average incomes in the lowest quartile according to the 1960 census. The assessment involved a series of clinical and biochemical measurements and a dietary evaluation. The clinical assessments used included skeletal weight, obesity, and dental evaluations. It was concluded that persons 60 years of age and older consumed far less food than was needed to meet the nutrient standards for their age, sex, and weight. In addition to a low caloric intake, other limiting nutrients included protein, iron, and vitamin A. Although the differences were minor, the Ten-State Survey tended to report lower intakes than the USDA study due to the low economic level of the group assessed. After age 50, skeletal weight generally decreased with age, reflecting the loss and thinning of bone. Low-income groups had lower skeletal weights than high-income groups. Obesity was more prevalent in higher than lower-income groups. The percentages in females in the 45–55-year-old groups were 50% for Blacks and 40% for Whites. These percentages declined markedly with age to about 20–25% in both races by age 75–85. Men of both races had lower incidences of obesity as compared to females and there were no age-associated patterns. The major dental problem in adults was periodontal disease. The incidence of periodontal disease increased with age to over 90% in nearly every subgroup of the survey population by the age of 65–75 years. However, there were no correlations between dental disorders and several selected biochemical measures (serum albumin, plasma vitamin A, and serum vitamin C). Taken as a whole, the clinical assessments did not indicate a high incidence of severe malnutrition in the older subjects examined. Similarly, the biochemical tests for nutritional status did not suggest marked age-associated nutritional deficiencies.

At present, the findings of the first Health and Nutrition Examination Survey (HANES) for the U.S. population in 1971–1972 are only available in a preliminary report (National Center for Health Statistics, 1974). The sample studied represented civilian, noninstitutionalized persons 1–74 years of age. The design allows for detailed analysis of the data for the total population, as well as for those groups considered to be at high risk for malnutrition—the poor, preschool children, women of childbearing age, and the elderly. The preliminary results include data on 10,126 persons, which represent 72.8% response of the individuals selected for sampling. Because it was assumed that the main evidence of malnutrition in this population would be early subclinical symptoms, the measures of nutritional status used were those considered early risk signals. These included (1) dietary intake data (recall of the previous 24-h consumption of calories, protein, calcium, iron, and vitamins A and C), (2) biochemical tests (hematocrit, hemoglobin, serum iron, percent transferrin saturation, total protein, albumin, and vitamin A), (3) clinical signs of malnutrition, and (4) anthropometric measurements. The results indicated that, among persons over 60 years of age with incomes above the poverty level (Orshansky; 1968), 16% of the white and 18% of the black population consumed less than 1000 cal/day. In those with incomes lower than the poverty level, these percentages rose to 27% and 36%, respectively. The intake of protein as well as of calories in this age group was also related to income in both races; however, protein intakes per 1000 cal showed no variation with race or income. Calcium intakes were less than the standards for 37% of all persons over 60. The intakes of vitamin A were below standards in 52–62% and consumption of vitamin C was low in 39–59% in all adults in this age group. The only biochemical indications of nutritional problems among the elderly in this study were the high percentages of blacks aged 60 years and over with low values for hemoglobin (29.6%) and hematocrit (41.7%).

However, iron deficiency was not considered the cause, because the majority of this group did not have low levels of serum iron or percentage of transferrin saturation. The biochemical tests have not been completed, nor are the clinical assessments available from this study. The final, still unpublished report will include data on serum folate, vitamin C, magnesium, cholesterol and total iron-binding capacity and urinary creatinine, thiamine, riboflavin, and iodine.

These surveys do not indicate consistent evidence of poor nutritional status or of marked deficiencies in nutrient intake among older members of the general population in the United States. However, significant percentages of many of the groups studied consumed less than recommended amounts of certain nutrients, especially of protein, calcium, ascorbic acid, and vitamin A. One of the most consistent findings was that low intakes were more likely to occur if income was low. The same conclusion about the relationship between poverty and diets containing less than the RDA's was reached by Watkin (1968), who indicated that nutrient intake or nutritional status of the elderly was more related to health and poverty than to age *per se*.

3.1.2. Laboratory Studies

In order to evaluate the influence of age on dietary intake with minimal effects due to economic factors, a study was carried out by McGandy *et al.* (1966) on a group of apparently healthy, highly educated, successful men engaged in or retired from professional and managerial occupations. Seven-day dietary histories were obtained on a group of 250 individuals between the ages of 23 and 99 years who resided in the Baltimore–Washington area. The results showed a marked decline with age in total daily calories consumed (Fig. 2). As would be expected, there were also progressive decreases with age in the intakes of iron, thiamine, calcium, and vitamin A (Fig. 2), as well as riboflavin and niacin (Fig. 3). However, except for calcium, the National Research Council's suggested allowances were met by the great majority of subjects. Efforts were made to identify the causes of the age-related decrease in dietary calories. The data on the effect of age on daily caloric intakes were analyzed as shown in Fig. 4. Figure 4A shows the age decrement in total calories (a decline of 12.4 calories per day per year) whereas Figure 4B shows the basal metabolism of the same subjects, which fell by 5.23 calories per day per year. This difference, which amounts to 7.6 calories per day per year, (Fig. 4C) must be related to the reduction in calories required for other purposes including physical activity.

Interviews were conducted with 167 of the subjects to obtain an estimate of their physical activities. Table I shows the age-wise decrease in the energy expenditure required for physical activity. In addition to this estimate, the energy needed for activity was also calculated for these individuals as the difference between the total caloric intake and the basal metabolism. The age regressions in energy expended for activity estimated by these two methods were in excellent

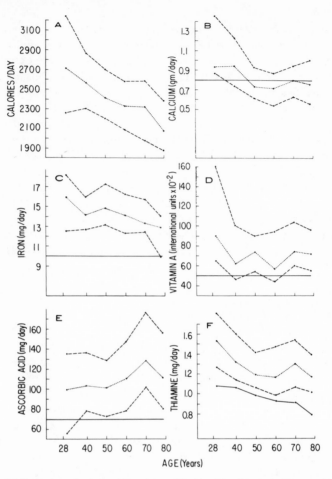

Fig. 2. Total daily intakes of calories (A), calcium (B), iron (C), vitamin A (D), ascorbic acid (E), and thiamine (F) in men of different ages. The medians are represented by the dotted lines and the first and third quartiles by the dashed lines. Solid lines represent National Research Council recommended allowances (McGandy *et al.*, 1966).

agreement (−8.3 calories per day per year for both). It should be noted that this value is very similar to that obtained for the total group by the difference method (−7.6 calories per day per year). Thus, it seems that the age decrement in total dietary caloric intake can be accounted for by the age-associated decrease in basal metabolism plus the decline in physical activity.

Therefore, nutritional status, estimated in a population in which economic factors were minimized, indicates that age is accompanied by a decrease in dietary caloric intake. This decrement can be essentially accounted for by decreases in basal metabolic caloric needs as well as calories necessary for energy expenditures for physical activity.

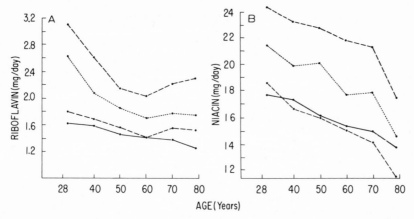

Fig. 3. Total daily intakes of riboflavin (A) and niacin (B) in men of different ages. The medians are represented by the dotted lines and the first and third quartiles by the dashed lines. Solid lines represent National Research Council recommended allowances (McGandy *et al.*, 1966).

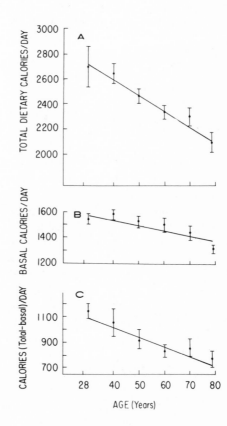

Fig. 4. Mean total daily caloric intakes (A), basal metabolic rates (B), energy expenditures (C), and energy expenditures per unit of body weight. Vertical bars represent standard errors of the means. Correlation coefficients for regressions on age of total calories; basal calories, and total basal calories were -0.374, -0.374, and -0.231, respectively, and were all statistically significant ($p < 0.01$) (McGandy, *et al.*, 1966).

Table I. Energy Expenditure Required for Reported
Activities in Men of Different Ages[a]

Age group (years)	n	Energy expenditure (kcal/day)	
		Mean	Standard deviation
20–34	13	1175	307
35–44	32	1166	333
45–54	41	982	280
55–64	34	950	269
65–74	36	928	239
75–99	13	640	245

[a] McGandy et al. (1966).

3.2. Age Changes in Nutritional Requirements

There are a small number of laboratory studies that provide information regarding the effects of age on nutritional requirements. It should be pointed out that some of the variables which make nutritional status difficult to establish also apply to nutritional requirements.

3.2.1. Vitamins

Data are available for both men and women on the plasma and blood levels of total carotenes, α- and β-carotenes, vitamin A (Kirk and Chieffi, 1948), thiamine (Kirk and Chieffi, 1949), DPN (Kirk, 1954), total ascorbic acid (Kirk and Chieffi, 1953), and α-tocopherol (Chieffi and Kirk, 1951). The results were calculated as linear regressions but the only statistically significant age-associated difference was a decrease in the blood ascorbic acid level in men. Morgan and associates conducted studies on 250 men and 280 women between the ages of 50 and 80, and failed to demonstrate any effect of age on the serum content of vitamin A, carotene (Gillum et al., 1955), or ascorbic acid (Morgan et al., 1955). Similar results were obtained in another study by Brewer et al. (1956). In addition, no age-wise changes in thiamine or riboflavin in serum were found by Horwitt (1953) who employed the depletion–repletion technique on a small number of subjects (6 with a mean age of 71 years and 5 with a mean age of 36 years). After a control period of 2 years on diets containing 4 mg thiamine and 3.1 mg riboflavin per day, the intakes of both vitamins were decreased for 3 months (to 0.2 mg and 0.75 mg, respectively) and then increased during a 90-day repletion period (to 6.4 mg and 2.1 mg, respectively). There were no age differences in either initial serum levels, the rates of depletion or repletion, or the levels attained following the periods of depletion or repletion.

In contrast, there is evidence that there are age-wise differences in certain other vitamin levels. For example, Ranke et al. (1960) estimated the pyridoxal phosphate (PLP) levels by the activity of serum glutamic-oxalacetic acid trans-aminase (SGOT), an enzyme system that requires PLP for its activity. The data demonstrated marked age decrements (21% for men and 30% for women) in the activities of the enzyme and, therefore, suggested age dependent decreases in serum PLP levels. In addition, more recently in a study carried out on 617 community-dwelling subjects, Rose et al. (1974) found that the average PLP of men not taking a vitamin supplement ($n = 414$) was 12.3 ± 0.3 SD ng/ml, with 25% of the values below 7.5 ng/ml. There was a statistically significant decrease in plasma PLP with age of 0.9 ng/ml per decade. On the other hand, no significant age-related regression was observed in subjects taking supplements. Similarly age-dependent decrements in vitamin B_{12} levels have been reported by Gaffney et al. (1957).

In the first experiment, 144 subjects were studied. Of these, 89 were ambulant, apparently healthy males selected from the Infirmary Division at Baltimore City Hospitals. The remaining 55 subjects were physicians and other male staff members of the participating research institutions. The results indicated an age-wise regression in vitamin B_{12} content of serum with age. As the diets of the younger subjects were different from those of the older individuals who lived in the institutional environment, the study was repeated on 97 apparently healthy male volunteer subjects who were inmates of a state penal institution in the same geographic region. All had been in the institution for at least 5 years prior to the test and represented the same age range as in the first study. The results were essentially the same in both studies, approximately a 35% decrement in vitamin B_{12} levels over the age span of 25–70 years, although the variance was reduced in the second study. These data indicate that deficiency states among the elderly are found only for selected vitamins. However, there is no evidence to suggest a general increase in vitamin requirements with age.

3.2.2. Proteins and Amino Acids

Data presently available on the effects of age on protein requirements were obtained by the nitrogen-balance technique. Unfortunately, the results of these studies are not in complete agreement, as shown by the data (Table II) of Kountz et al. (1947, 1951), Ohlson et al. (1948), and Schulze (1955). A similar lack of agreement on the effect of age is found when the requirements for the amino acids are considered. For example, Tuttle et al. (1957) fed five men aged 52–68 a diet that contained amino acids in amounts that exceeded Rose's minimal recommended allowance for 25-year-old male subjects (Table III). When fed either this diet, or one that contained twice the amount of tryptophan, all subjects were in negative nitrogen balance. These data indicate an increased requirement for one or more of the essential amino acids with age. In contrast, in a study carried

Table II. The Effect of Age on Protein Requirements

Investigator	Subjects (n)	Age (years)	Sex	Estimated requirements (g/kg/day)
Kountz et al. (1947)	27	41–86	M	>1.0
Kountz et al. (1951)	4	69–76	M	0.7
Schulze (1955)	4	61–79	M	0.5
Ohlson et al. (1948)	8	50–75	F	0.9 (0.7–1.3)

out by Watts et al. (1964), 6 black men, 65–85 years of age were fed semipurified diets containing essential amino acids in ratios corresponding to the FAO references protein and the pattern of milk protein (Table III). Although the amounts of the essential amino acids fed these subjects were actually lower than those fed the subjects of Tuttle, both diets were adequate for maintaining nitrogen equilibria in all subjects. A comparison of the other variables failed to explain the lack of agreement between these two studies. For example, the caloric intakes of the subjects of Tuttle (30 to 41 cal/kg) were similar to those of Watts (28–44 cal/kg). The dietary periods used by Tuttle were 4–6 days, while those of Watts were 7–12 days. The greatest difference may be found in the total

Table III. Comparison of Daily Intakes of Amino Acids in Various Studies with the Minimal Requirements of Rose[a]

Essential amino acid	Rose's minimum for 25-year male (g)	Tuttle's AA I (g)	FAO pattern (360 mg Trypt) Amount fed (g)	FAO pattern (360 mg Trypt) Rose's minimum (%)	Milk pattern (500 mg S-AA) Amount fed (g)	Milk pattern (500 mg S-AA) Rose's minimum (%)
Isoleucine	0.70	1.28	1.08	154	1.01	144
Leucine	1.10	1.72	1.22	111	1.54	140
Lysine	0.80	1.39	1.08	135	1.25	156
Methionine	1.10	1.12[b]	1.20	109	0.58	53
Phenylalanine	1.10	1.56[c]	1.44	131	1.58	144
Threonine	0.50	0.77	0.72	144	0.72	144
Tryptophan	0.25	0.38	0.36	144	0.24	96
Valine	0.80	1.29	1.08	135	1.10	138
Histidine	—	0.43	—	—	—	—
Amino acid N	0.74	1.11	0.96	—	0.95	—

[a] Watts et al. (1964).
[b] Calculated from published data: 0.56 g methionine + 0.45 g cystine.
[c] Calculated from published data: 0.96 g phenylalanine + 0.60 g tyrosine.

nitrogen intake, which, in the study of Tuttle, was 7 g, whereas that of Watts was 10 g. However, later studies by Tuttle *et al.* (1959) indicated that increasing the total nitrogen intake from 7 to 15 g resulted in an even greater negative nitrogen balance in these older subjects. The interpretation of such nitrogen-balance studies is made difficult due to the variations in quality of protein used, caloric intake, lack of simultaneous study of young subjects, and the observation that subjects can be maintained at nitrogen equilibrium at various levels of nitrogen intake. This latter point has been expressed by Hegsted (1952), who also pointed out that an individual on self-selected diets would show a nitrogen balance varying from positive balance through equilibrium to negative balance, depending on changing factors of stress or food intake, or both. Horwitt (1953) reported that 31 individuals who were in positive nitrogen balance when fed a diet containing 11 g nitrogen per day, immediately went into negative balance when the intake was reduced to 6.5 g/day. However, when this same low intake was continued over a 3-month period, these individuals returned to a positive nitrogen balance. No differences between the young and old subjects were found in these studies. Thus, at present, there is no indication that age affects the protein and amino acid requirements of humans.

3.2.3. Calcium

Most of the studies to establish calcium requirements as a function of age failed to include young subjects. These studies (Table IV) indicate that the calcium intakes necessary to establish equilibrium exceed the RDA of 800 mg/day. This suggests that old individuals require higher calcium intakes. However, in the study of Ohlson *et al.* (1952) (Table IV), where the subjects ranged from 30 to 89 years of age, no correlation with age and calcium requirements was observed. It should be pointed out that the possibility of establishing calcium equilibrium at various levels of intake make these data difficult to interpret. This latter point is evident from the data of the following studies. Malm (1958) reported that, regardless of age, prisoners who were in calcium equilibria on intakes of 937 mg/day could maintain calcium balance on intakes estimated to be as low as 440 mg/day. In addition, Hegsted *et al.* (1952) determined that Peru-

Table IV. The Effect of Age on Calcium Requirements

Investigators	Subjects (*n*)	Age	Sex	Estimated requirements (mg/day)
Bogdonoff *et al.* (1953)	7	66–83	M	850
Ackermann *et al.* (1953)	8	70–88	M	1020
Ohlson *et al.* (1952)	135	30–89	F	840
Roberts *et al.* (1948)	9	52–74	F	929
Ackermann *et al.* (1953a)	8	48–83	F	920

vian prisoners could maintain calcium equilibria on mean intakes of only 216 mg/day. Therefore, age does not appear at present to have a marked effect on the requirements for calcium.

3.3.   Nutritional Deficiencies

3.3.1.   Physiological Impairments

Unfortunately, there is little evidence presently available to correlate age-associated nutritional deficiency states with clinical findings, physiological functions, or biochemical changes. Chieffi and Kirk (1949) in a study on 106 subjects aged 16–99 years found no significant correlations between vitamin A levels in serum and (1) dark-adaptation time, (2) the number of epithelial cells excreted daily in the urine, or (3) the percentage of keratinized cells in the urinary sediment. However, the frequency of dryness of the skin, conjunctivitis, and the percent of keratinized cells in conjunctival smears were higher in subjects with low vitamin A plasma levels (1–5 μg%) than in those with high levels (26–60 μg%). Nevertheless, the differences between the groups were not marked and could not be considered clinically useful. Gillum et al. (1955) attempted to relate the frequency of gingivitis in subjects over 50 years to serum levels of ascorbic acid. However, they found that the incidence of the disease was essentially the same over the range of serum ascorbic acid levels from 0 to 1.1 + mg%. In addition, although thickening of the bulbar conjunctiva was noted in 94% of the subjects, this condition was not marked in individuals whose serum vitamin A levels were low. In a more recent study, Davis et al. (1966) attempted to find biochemical and hematological changes attributable to low serum levels of vitamin B_{12}. However, in this study on 275 subjects aged 49 to 89 years, there were no significant correlations between the serum vitamin B_{12} level and (1) serum lactic acid concentration, (2) serum lactic dehydrogenase activity, or (3) hematocrit. Taken as a whole, these data indicate that vitamin deficiency states, if they exist in some old subjects, are generally not severe enough to be manifested by clinical or biochemical changes.

The importance of calcium nutrition in the elderly is best indicated by its relationship to the increased incidence of osteoporosis with age. The high frequency of the disease among the elderly was shown by Gitman and Kamholtz (1965) who performed routine X-ray examinations of the dorsal lumbar spine on all admissions to a large geriatric facility. They found that in a group of 933 females, the incidence of osteoporosis increased linearly from approximately 50% between the ages of 65 to 70 to 90% in women over 90 years. Unfortunately, the reports of attempts to relate calcium intake to the incidence of osteoporosis are not in complete agreement. Lutwak (1963) and Nordin (1961) reported that the average intakes of calcium of subjects with osteoporosis are statistically lower than those of normal subjects. On the other hand, Garn et al.

(1967) found no correlation between the intake of calcium and the cortical thickness of the second metacarpal in 382 subjects ranging in age from 25 to 85 years. Likewise, Smith and Frame (1965) in a radiographic survey of 2000 women, 45 years of age or older, found no significant differences in the calcium intake of the subjects with high or low vertebral densities or vertebral compressions. The difficulty in correlating calcium intake with the incidence or severity of osteoporosis has been recently discussed by Garn (1970), who pointed out that at least four variables of dietary origin (reduced calcium intake, altered calcium to phosphorus ratio, decreased protein intake, or a change in acid–base balance) could contribute to bone loss in the adult. Therefore, at the present time, the specific relationship between nutritional status with respect to calcium and osteoporosis cannot be defined.

3.3.2. Reversal of Deficiencies by Supplementation

There are a number of examples that indicate that nutritional deficiency states associated with age can be corrected simply by supplementation with the specific nutrient. Kirk and Chieffi (1953a) administered 100 mg of ascorbic acid daily to 19 old subjects whose serum ascorbic acid levels were low (approximately 0.25 mg%). In 16 of the 19 subjects, the blood levels of the vitamin rose immediately to values approximately 1.2 mg% and were maintained at these high levels during the 12-week supplementation period. Following the withdrawal of the supplement, the levels dropped very rapidly to those originally observed. Similar results were obtained by Davis et al. (1965) when 20 μg of vitamin B_{12} per day were administered orally to 40 elderly men whose initial serum vitamin B_{12} levels were low, i.e., 150 μg/ml (Fig. 5). During supplementation, the plasma level rose to that seen in young subjects (300 $\mu\mu$g/ml). When the supplement was withdrawn, the serum levels of the vitamin returned to the original low values. Similarly, Chernish et al. (1957) demonstrated that values comparable to those of young individuals may be attained by old subjects following the oral administration of vitamin B_{12}. Finally, Ranke et al. (1960) showed that the activities of serum glutamico-xalacetic transaminase in old subjects, which was taken as an index of pyridoxal phosphate levels, could be raised to those of young individuals by the oral administration of vitamin B_6.

In view of the uncertainty regarding the relationship between calcium intake and osteoporosis, as previously discussed, it is not surprising that a lack of agreement exists regarding the therapeutic effects of increased calcium intake. Bogdonoff et al. (1953), Lutwak and Whedon (1963), Nordin (1962), and Harrison et al. (1961) have shown that an increase in dietary calcium resulted in positive calcium balance in older individuals. However, in the report by Harrison et al. (1961), the senile osteoporotics in the study group, although they maintained positive calcium balance on high intakes of calcium, i.e., 40 mg/kg body weight per day, showed no increased bone density estimated radiographically. In

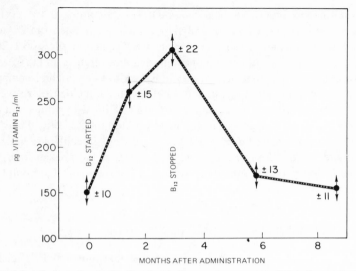

Fig. 5. Mean serum vitamin B_{12} levels of 40 men (average age 87) before and after the oral administration of 20 μg vitamin B_{12}. Vertical lines represent standard error for each mean value (Davis *et al.*, 1965).

contrast, Schmid (1962) and Nordin and Smith (1964) did find increases in radiographic density in osteoporotics given dietary calcium supplements. Therefore, at the present time, there is still controversy regarding the beneficial effects of increased calcium intake in older individuals.

3.3.3. Frequency

Some estimates of the frequency of nutritional deficiencies among the aged are found in studies reported by Brewer *et al.* (1956) and by Chinn (1956). In the latter study, the nutritional status of approximately 500 elderly patients admitted to a hospital for long term illnesses over a period of three years was assessed. The data showed that only 35 (7%) had significant primary nutritional problems. Of these, 15 were undernourished, whereas the remaining 20 had a problem of obesity. Similar data were obtained by Brewer on 107 subjects who were admitted to county institutions for the aged in Michigan. Nutritional assessment was made on the basis of the concentration of hemoglobin and plasma levels of ascorbic acid, vitamin A, and carotene. Only 5–10% of the residents could be considered in a poor nutritional state with respect to vitamin A and ascorbic acid. Brin *et al.* (1965) examined 234 elderly subjects whose average age was 71.0 ± 8.9 years (mean ± SD). They measured hematocrits and evaluated nutritional status with respect to ascorbic acid, vitamin A and carotene, riboflavin, and thiamine. The latter was estimated on the basis of urinary excretion as well as erythrocyte transketolase activity. They concluded that 5% of the men and 13% of the women had hematocrits in the deficient range according to the Interdepartmental Committee on Nutrition for National Defense(ICNND) (1963) criteria. Plasma ascorbic acid levels were low in only 8% of the total population. Plasma vitamin A and carotene levels were in the acceptable to high range and urinary riboflavin excretion values showed no deficiencies. Thiamine deficiency was indicated for 18% of the population if the standard of ICNND (1963) was used and for 21% on the basis of Pearson's criteria (1962). However, a biochemical defect, on the basis of the erythrocyte transketolase data, was evident in only 6% of the group. The authors concluded that this ambulant, well, aged, surveyed population was fairly well nourished. Therefore, on the basis of data presently available, the frequency of serious nutritional problems among the aged is estimated to be approximately 5–10%.

4. Effect of High Vitamin Intakes on Older People

On the basis of the low frequencies of nutritional deficiencies and the complete lack of information on the effect of continued long-term vitamin therapy in older people, it seems unwise to propose mass vitamin and other nutrient supplements to the aged at this time. It seems more reasonable to administer vitamin therapy on the merits of individual cases. There are studies, however, which suggest a beneficial effect of consuming vitamins in quantities greater than the recommended daily allowances. Chope (1951, 1954) studied 577 individuals over fifty years of age in the San Mateo County of California. The dietary information secured in 1948–1949 was compared with the mortality data obtained in 1952. The data suggest a relationship exists between mortality and the intake of vitamin A, C, and niacin.

For example, among the individuals consuming less than 5000 IU of vitamin A per day, a mortality rate of 13.9% was observed. For those subjects whose intake was above 5000 IU/day, the mortality was 5.4%. For people who consumed diets that provided less than 50 mg vitamin C per day the mortality was 18.5%. On the other hand, for these subjects whose intake was greater than 50 mg/day the mortality was 4.5%. A reduction in mortality with increasing niacin intake was apparent but not as marked as that for vitamins A and C.

Similar data for vitamin C were observed by Schlenker (1976). Dietary records were obtained in 1948 on 100 women in Michigan. Between 1948 and 1972, 60 of the women died. The mean intake of vitamin C in 1948 was 51 and 73 mg for those who died and those who survived until 1972, respectively. Interpretation of these findings is made difficult by the fact that in 1948, the ages of the women who died prior to 1972 averaged 67.4 yrs while those who survived averaged 52.1 yrs. In addition, the dietary records did not include estimates of supplementary vitamins.

Both of the previous studies suggest beneficial effects on longevity as a result of high vitamin intakes. Nevertheless, the retrospective nature of these reports necessitates further investigation. Such studies should include dietary records as well as the supplementation of varying, but known, amounts of vitamins. Owing to the positive response of the elderly to attention, it is important that the effects of placebo also be evaluated.

5. Food Additives

Harman (1956) proposed that aging may be caused in part by the deleterious side effects of free radicals produced in the normal course of metabolism. He proposed that raising the concentrations of compounds such as cysteine and other chemicals capable of reacting rapidly with free radicals, would tend to slow down the aging process and thus lead to an extension of the normal lifespan. In the first study to test this proposal (Harman, 1957), weanling AKR male mice and C_3H female mice, which develop spontaneous lymphatic leukemia and mammary carcinoma, respectively, were used. The experimental animals were offered *ad libitum* a powdered diet to which was added the compounds to be tested. The results indicated that AKR mice fed diets that contained either cysteine hydrochloride (1.0%), 2-mercaptoethylamine hydrochloride (MEA : 1.0%), or 2,2'-diaminodiethyl disulfide dihydrochloride (0.5%) had half-survival times of approximately 10 months, which were about 20% greater than that of the controls. However, the differences were not statistically significant. Ascorbic acid (2.0%) and 2-mercaptoethanol (0.5%) had no effect. None of these five compounds influenced the lifespan of C_3H mice. In the second study (Harman, 1961), these same strains as well as Swiss male animals were employed. The mice were fed *ad libitum* a pelleted mouse diet to which the compounds to be tested were added before pelleting. The results indicated that none of the antioxidants studied, 2-mercaptoethylamine hydrochloride (1%), 2,2'-diaminodiethyl disulfide (1%), hydroxylamine hydrochloride (1 and 2%), and cysteine hydrochloride (1%) prolonged the lifespan of Swiss male mice. Cysteine hydrochloride (1%) and hydroxylamine hydrochloride (2%) extended the half-survival time of AKR male mice from 9.6 to 11.0 months, respectively, an increase of about 15%. MEA (1%) prolonged the half-survival time of C_3H female mice from 14.5 to 18.3 months (plus 26%), whereas hydroxylamine hydrochloride (1%) produced a slight prolongation (plus 7%). However, again, none of these differences was statistically significant. In the third study of this series (Harman, 1968), weanling LAF male mice were used. The control animals were offered *ad libitum* either a powdered commercial diet (Rockland, Teklad, Inc.) or a synthetic diet (20% casein, 68% sucrose, 5% corn oil, and adequate amounts of vitamins and minerals). The experimental mice were also fed *ad libitum* and received one of these diets to which were added compounds to be tested (MEA 0.5%, 1.0%, and BHT

0.25% and 0.5%). The addition of MEA (0.5%, 1%) to the commercial diet failed to increase the lifespan of the mice. The addition of cysteine hydrochloride, propyl gallate, 2,6 di-ter-butyl hydroquinone or hydroxylamine hydrochloride to the synthetic diet also failed to increase the mean lifespan. However, when MEA (1%) was added to the synthetic diet, lifespan increased by 12%. The inclusion of BHT at the level of 0.25% and 0.50% increased longevity by 17.6% and 44.6%, respectively. One major difficulty in interpreting the results of this study is the fact that mean lifespan of the control mice fed the commercial diet was 20.0 months, whereas that of the control animals fed the synthetic diet was 14.5 months. The authors pointed out that this finding may be due to the higher frequency of amyloidosis in the animals fed the synthetic control diet (60%) as compared to that of those offered the commercial diet (20%). Furthermore, the lifespan of the treated synthetic diet groups was less than that of the mice fed the commercial diet.

Kohn (1971) carried out a series of three experiments to test the effect of antioxidants (MEA and BHT) on the lifespan of C57BL/6J female mice. The compounds were added to Wayne Mouse Breeder Blox, which served as the control diet. In the first and second studies, each group was composed of 50–100 female breeders. In the third study which tested only BHT, weanling female mice were used. The MEA was maintained at a level of 1% of the diet in both the first and second studies. The other antioxidant (BHT) caused early weight losses and death when offered at the 1% level. It was therefore varied between 0.2% and 0.5% throughout the lifespan of the animals in the second and third studies. The results of these experiments indicated that when the control animals had a 50% survival time of 121 weeks and a maximum lifespan of 148 weeks, the antioxidants were without effect. When the survival of the controls was suboptimal, the agents increased the lifespan, but not beyond optimal control values. It was concluded by Kohn (1971) that antioxidants do not inhibit aging but increase lifespan by beneficially affecting some harmful environmental or nutritional factors.

Tappel *et al.* (1973) carried out studies on the effects of three diets on various criteria related to aging in 9 month old retired breeder CDI male mice. One diet served as a control and contained normal amounts of antioxidants such as vitamin E. Two test diets contained supplements of antioxidants and related nutrients including vitamin E, BHT, selenium, ascorbic acid, and methionine. Age-related variables measured were (1) accumulation of fluorescent products in testes, heart, brain, skeletal muscle, and kidney, (2) rates of calcium uptake in muscle microsomes, (3) the percent solids in urine, (4) cage and treadmill activities, and (5) mortality rates. The results showed no significant differences in mortality among the various groups. Animals fed the diet supplemented with high levels of antioxidants had significantly lower amounts of fluorescent products in the testes than the other mice. No other age-related indices were affected by diet.

Oeriu and Vochitu (1965) proposed that aging is associated with increases in the concentrations of cystine and oxidized glutathione as well as decreases in the activities of thiolic enzymes. Furthermore, compounds which contain sulfhydryl groups, such as cysteine and N-formylcysteine, have been shown to exert a favorable action on various age-related enzymatic changes. Therefore, the effect of treatment with compounds that contain sulfhydryl groups initiated at 13 months on the subsequent lifespan of male mice, female rats, and female guinea pigs was investigated. The animals were fed natural-product diets *ad libitum*. They were given 21 subcutaneous injections on alternate days of either cysteine (30 mg/kg), a combination of thiazolidincarboxylic acid (30 mg/kg) and folic acid (0.75 mg/kg) or saline. Control (saline-injected) male mice experienced a 50% survival of 14.2 months, whereas that of animals injected with cysteine or the combination of thiazolidincarboxylic acid and folic acid was 18.0 and 15.0 months, respectively. Fifty percent of the female guinea pigs injected with saline survived to 18.3 months, whereas those treated with cysteine and the other compounds had half-survival times of 20.2 and 21.0 months, respectively. Neither treatment had an effect on the 50% survival (16 months) of female rats. Unfortunately, no statistical analysis of the data was presented. Furthermore, the strains of the various species were not given so that the survival curves cannot be compared with other published data. However, the lifespans of the animals in this study seemed short, even among the treated animals.

These studies do not offer convincing statistical evidence that the addition of various antioxidants increases the lifespan of animals. A number of inconsistencies are apparent within a given laboratory and among laboratories. In many instances, the mean lifespans of control animals have been considerably lower than that reported by other laboratories. Another criticism which may be made is the lack of information regarding dietary intakes. For example, in the studies of Harman (1968) and of Kohn (1971) low body weights were observed in the experimental groups that may indicate reduced dietary intakes. Although final proof of the effectiveness of dietary antioxidants on lifespan is lacking, the data available are impelling and the problem area should be further explored.

6. Effect of Dietary Restriction on Aging

Dietary restriction has been shown to increase the lifespan of laboratory animals. In general, dietary restriction has been brought about by (1) reducing the daily intake of a nutritionally adequate diet (one which supports maximal growth); (2) intermittently feeding a nutritionally adequate diet (e.g., feeding every second, third, or fourth day); and (3) feeding *ad libitum* a diet containing insufficient amounts of proteins to support maximal growth. The results of these experiments have been divided into whether dietary restriction was imposed on young growing animals or on adult organisms.

6.1. Lifespan

6.1.1. Young Growing Animals

Increased lifespan associated with underfeeding has been reported in the following animal model systems: *Tokophyra* (Fig. 6), *Campanularia flexuosa* (Fig. 7), *Daphnia* (Fig. 8), rotifers (Table V), *Drosophila* (Loeb and Northrup, 1917), and fish (Fig. 9). In addition, a number of laboratory experiments have been carried out on rodents. McCay *et al.* (1935, 1939, 1943) carried out a series of three studies that supported the observation that nutritional deprivation increases lifespan. Essentially, McCay and associates fed an adequate diet *ad libitum* to control animals and only enough food to maintain the body weight of the animals in the retarded groups. However, the authors indicated that when any of the retarded group appeared to be failing from the deficiency of calories, the entire group was supplemented until they gained 10 g. At the end of the periods of 300, 500, 700, and 1100 days, the various groups of retarded animals were fed *ad libitum*. The results indicated that animals subjected to dietary restriction lived longer than those fed *ad libitum*. Since these early studies, the increased lifespan associated with underfeeding has been reported in rats by Berg and Simms (1960) (Table VI), Ross (1959) (Table VII), Leveille (1972) (Fig. 10), and Riesen *et al.* (1947), and in mice by Leto *et al.* (1976a) (Fig. 11).

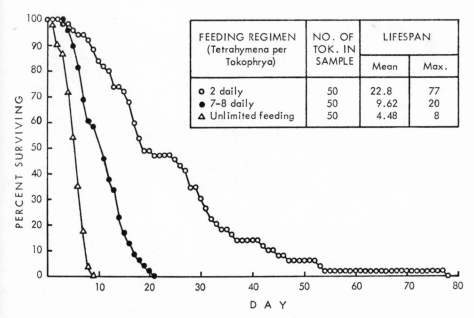

FEEDING REGIMEN (Tetrahymena per Tokophrya)	NO. OF TOK. IN SAMPLE	LIFESPAN	
		Mean	Max.
o 2 daily	50	22.8	77
• 7–8 daily	50	9.62	20
△ Unlimited feeding	50	4.48	8

Fig. 6. The percent survivorship of *Tokophrya lemnarum* (MacKeen and Mitchell, 1975).

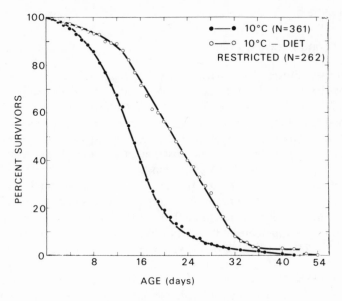

Fig. 7. The percent survivorship of *Campanularia flexuosa* fed artemia daily or every third day (Brock, 1975).

6.1.2. Adult Animals

It has been generally believed that nutritional manipulations which increase lifespan had to be imposed during early growth. This concept originated as a result of the early work of Minot (Comfort, 1974) postulating that senescence follows the cessation of growth. In addition, McCay *et al.* (1935, 1939) showed that increased lifespan of rats was associated with growth retardation. Furthermore, Lansing (1948) indicated that aging in the rotifer involves a cytoplasmic factor the appearance of which coincides with the cessation of growth. However, more recently, studies have indicated that dietary restriction imposed in adult life was effective in increasing lifespan. These data are shown in Fig. 12 and Tables VIII–XIII.

Thus, it is apparent that the life expectancy of adult animals can be increased by dietary manipulations. However, these experimental data are not in agreement regarding the effectiveness of various methods of imposing dietary restriction. In addition, Kopec (1928) and David *et al.* (1971) have shown that dietary restriction in adult *Drosophila* was ineffective. Similarly, Barrows and Roeder (1965) did not demonstrate an increase in lifespan in 13- or 19-month-old male adult Wistar rats whose daily dietary intake was reduced 50%. Thus, it is

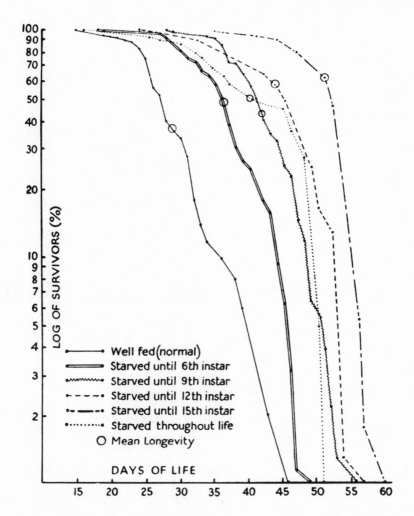

Fig. 8. Effect of restricted food upon the survivorship of *Daphnia longispina* (Ingle *et al.*, 1937).

apparent that further studies must be carried out to define effective ways of consistently increasing the lifespan of adult organisms.

6.2. Diseases

Although the incidence of many diseases increases with age, the relationship between disease and aging remains unknown. However, data presented in Figs. 13–14 and Tables XIV–XVI clearly indicate that dietary restriction which

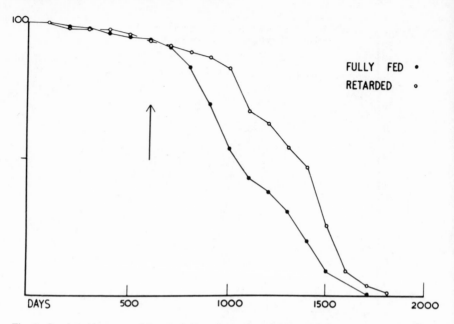

Fig. 9. Survivorship curve of female *Lebistes reticulatus* fed live *Tubifex* worms weekly (●) or biweekly (○). Arrow indicates realimentation of the restricted fish. The animals were maintained at 23°C (Comfort, 1963).

Table V. Effect of Nutrition on Lifespan of Rotifers[a]

		Mean lifespan (days)		
		Diet[b]		
Experiment		I	II	III
1	Mn	35.7	43.8	58.6
	σMn	±2.1	±3.0	±1.9
2	Mn	36.0	45.6	56.5
	σMn	±1.2	±2.5	±2.2
3	Mn	29.0	46.2	49.1
	σMn	±2.8	±3.2	±2.2
Mean	Mn	34.0	45.3	54.7
	σMn	±1.1	±1.7	±1.3

[a] Fanestil and Barrows (1965).
[b] (I) Algae and fresh pond water daily; (II) fresh pond water daily; and (III) fresh pond water—Mon., Wed., and Fri.

Table VI. The Effect of Reduced Dietary Intake
on Lifespan of Male Sprague-Dawley Rats[a]

Diet [b]	n	% Survivorship at 799 days	Maximum body weight (g)
Ad lib	50	48	448
33% Restriction	48	87	342
46% Restriction	76	81	275

[a] Berg and Simms (1960).
[b] Rockland "D-free" pellets.

Table VII. The Effect of Dietary Intakes and Protein Levels on Lifespan
of Male Sprague-Dawley Rats

Diets	Dietary intake				
	Comm.	A	B	C	D
			Unrestricted		
n	150	25	25	25	25
Casein (%)	23	30.0	50.8	8.0	21.6
Caloric value (kcal/g)	3.1	4.1	4.2	4.1	4.2
Food intake (g/day)	25.0	17.4	18.8	15.0	19.6
Max. body weight (g)	610		(not available)		
Mean lifespan (days)	730	305	595	825	600
Max. lifespan (days)	1072	347	810	1251	895
			Restricted		
n		150	60	150	135
Casein (%)		30.0	50.8	8.0	21.6
Caloric Value (kcal/g)		4.1	4.2	4.1	4.2
Food intake (g/day)		14.3	8.5	14.3	5.3
Max. body weight (g)		420	287	390	162
Mean lifespan (days)		904	935	818	929
Max. lifespan (days)		1322	1480	1287	1638

[a] Ross (1959, 1961, 1969).

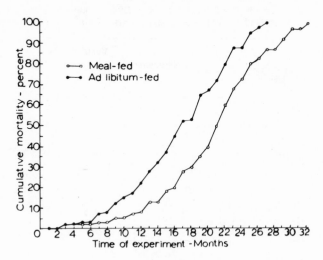

Fig. 10. Cumulative mortality for male Sprague-Dawley rats offered food for periods of 2 hr (O———O, meal-fed) or 24 h (●———●, *ad libitum*-fed) daily (Leveille, 1972).

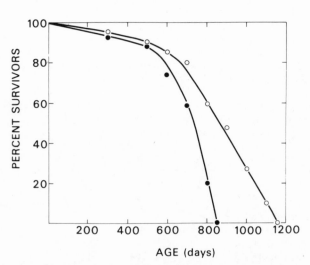

Fig. 11. Survivorship curve of female $C_{57}BL/6J$ mice fed *ad libitum* either a 26% casein diet (●) or 4% casein diet (O); the mean lifespans and SEMs were 685 ± 22.8 and 852 ± 27.4 days, respectively (Leto *et al.*, 1976a).

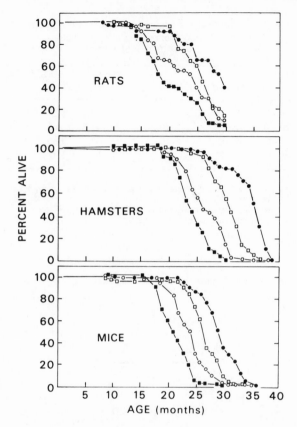

Fig. 12. Effect of various dietary regimens on the survivorship of rats, hamsters, and mice. Group 1 (■) was fed *ad libitum* throughout life; Group 2 (○) was fed one-half the amount of food consumed by Group 1 throughout their life; Group 3 (□) was fed *ad libitum* until 1 year of age and then restricted thereafter; Group 4 (●) was restricted until 1 year of age and then fed *ad libitum* thereafter (Stuchlikova *et al.*, 1975).

Table VIII. The Effect of Intermittent Feeding[a] on Lifespan of Male Wistar Rats[b]

Diet[c]	n	50% Survivorship (weeks)
Ad lib[d]	25	135
Ad lib-restricted[e]	30	156
Restricted-*ad lib*[e]	30	152
Restricted[d]	25	172

[a]Fed every other day. [b]Beauchene (personal communication). [c]Wayne Lab Blox Diet. [d]Throughout lifespan. [e]Dietary regimen changed at 1 year of age.

Table IX. The Effect of Changes in Nutrition Following
Cessation of Egg Production on the Lifespan
of Rotifer (Philodina acuticornis)[a]

n	Interval A-C[b]	Interval C-E[b]	Mean lifespan ± SEM (days)
30	I[c]	I	33.4 ± 1.27
28	I	III	41.8 ± 2.62
22	III	III	53.4 ± 3.65
29	III	I	57.7 ± 1.13

[a] Fanestil and Barrows, 1965
[b] A, day hatched; C, end of egg production (I: 13 days; III: 35 days);
E, death.
[c] (I) Algae and fresh pond water daily; (III) fresh pond water—Mon.,
Wed., and Fri.; animals maintained at 25°C.

Table X. Daily Dietary Allotments and Mortality Risk after 300 Days of Age

Diet (% casein)	Level of allotment (%)	Total food (g)	Allotments			Total (kcal)	Mortality index (× 100)
			Casein (g)	Sucrose (g)	Corn oil (g)		
Commercial[b]	100	25.0				85.5	105
	80	20.0				68.4	106
	70	17.5				59.9	83
	60	15.0				51.3	79
A (30.0%)	100	18.0	5.40	10.98	0.90	73.6	5550
	90	16.2	4.86	9.88	0.81	66.3	1180
	80	14.4	4.32	8.78	0.72	58.9	1940
	71.5	12.9	3.86	7.85	0.64	52.6	723
B (50.9%)	100	19.0	9.66	6.44	1.61	78.9	178
	78	14.8	7.53	5.02	1.25	61.4	122
	55.9	10.6	5.40	3.60	0.90	44.1	115
	40	7.6	3.86	2.58	0.64	31.6	675
C (8.0%)	100	15.0	1.20	12.45	0.75	61.4	2103
	96	14.4	1.15	11.95	0.72	58.9	2882
	86.8	13.0	1.04	10.81	0.65	53.3	3195
	79.3	11.9	0.95	9.88	0.60	48.7	3195
D (21.6%)	100	20.0	4.32	10.81	2.70	84.8	200
	80.7	16.1	3.49	8.72	2.18	68.4	126
	59.6	11.9	2.58	6.44	1.61	50.6	99
	52	10.4	2.25	5.62	1.41	44.1	68

[a] Ross (1967).
[b] Purina Lab. Chow, 23% protein.

Table XI The Effect of Reduced Dietary Intake on the
Mean Survival Time of Sprague-Dawley Rats[a]

| | Mean survival time (days) | |
Diet[b]	Males[c]	Females[c]
Ad lib	706	756
20% Restriction	856	872
40% Restriction	924	872
Ad lib for 12 weeks, 20% Restriction thereafter	801	871
Ad lib for 12 weeks, 40% Restriction thereafter	927	943
20% Restriction for 12 weeks, *Ad lib* thereafter	723	788
40% Restriction for 12 weeks, *Ad lib* thereafter	782	805

[a]Nolen (1972).
[b]Natural products diet: lipid 18.5%; protein 23%; ash 6.2%; 4.4 kcal/g.
[c]$N = 50$ weanling animals.

Table XII. The Effect of Various Dietary Regimens
on Lifespan of Female Rats[a]

Dietary regimen[b]	n	Life expectancy $\pm$ SE (days)	Significant differences
A	10	763 ± 94	$p < 0.001$
B	10	980 ± 50	$p < 0.001$
C	10	828 ± 73	
D	10	282 ± 40	
E	10	< 100	

[a]Miller and Payne (1968).
[b]A, stock diet throughout life; B, stock diet for 120 days, then 20% stock diet and 80% starch; C, 30% stock diet and 70% starch throughout life; D, 20% stock diet and 80% starch throughout life; E, protein free diet. All diets were fed *ad libitum*.

Table XIII. The Effect of Dietary Protein Levels
on the Survival of 16-Month Female Wistar Rats[a]

Dietary protein levels (%)	n	Survival (weeks)
24	44	29.5 ± 2.28[b]
12	44	37.0 ± 2.00[c]
8	44	30.0 ± 2.30
4	44	31.6 ± 1.70

[a]Barrows and Kokkonen (1975).
[b]Mean ± SEM.
[c]$p = 0.001$.

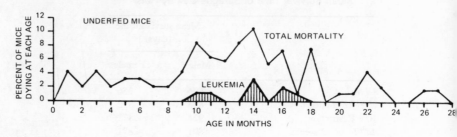

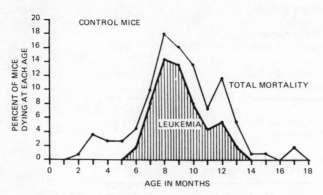

Fig. 13. Effect of underfeeding on the mortality and incidence of leukemia in Ak mice. The underfed (47 males, 47 females) mice were offered 1.5 g of Wayne Fox Food Blox daily; controls (52 males, 59 females) were given the same diet *ad libitum* (Saxton *et al.*, 1944).

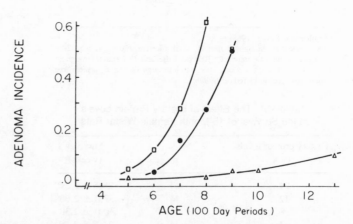

Fig. 14. Influence of dietary regimen on the incidence of adenomas in male, COBS (Charles River) rats. (□) Rats fed *ad libitum* throughout postweaning life; △ rats fed a restricted amount of diet throughout postweaning life; (●) rats fed a restricted amount of diet 21–70 days of age and then fed *ad libitum*. Composition of diet: casein, 22.0%; sucrose, 58.5%; Mazola oil, 13.5%; salt mixture (USP XII), 6.0%; vitamins and trace elements (Ross and Bras, 1971).

Table XIV. The Effect of Dietary Restriction
on the Incidence of Spontaneous Mammary Carcinoma
and the Survival of C_3H Mice[a]

	Restricted	Unrestricted
n	44	51
Caloric intake/day[b]	8.4	11.5
Protein intake/day[b]	0.64	0.65
Max. body weight (g)	15.5	32.0
% Survival at 16 months	57.0	29.0
Cumulative tumors[c] (%) at 16 months	0	63.0

[a]Visscher et al. (1942).
[b]After 100 days of age.
[c]Spontaneous mammary carcinoma.

Table XV. The Effect of Dietary Restriction on the Incidence
of Three Major Diseases in Male Sprague-Dawley Rats[a]

		% Incidence[b]		
	n	Glomerular-nephritis	Periarteritis	Myocardial degeneration
Unrestricted	24	100	63	96
33% Restricted	42	36	17	28
46% Restricted	38	13	3	24

[a]Berg and Simms (1960).
[b]At 800 days of age.

Table XVI. Progressive Glomerulonephrosis Index of Male
Sprague-Dawley Rats Fed Semisynthetic Diets[a]

	Number of cases		
Dietary groups[b]	Expected	Observed	Disease index[c]
A	186.4	46	24.7
B	88.2	1	1.1
C	152.5	16	10.5
D	10.5	4	1.9
Commercial	—	—	—

[a]Bras, 1969.
[b]Rats maintained on commercial diet were fed *ad libitum*, while the intakes of animals fed diet A, B, C, or D were restricted (see Table X).
[c]Computed from rats dying from natural death only. Disease index expressed as percentage (computed as number of actual against expected cases). Expected cases equals disease rate at each age period of "control" population times exposure of experimental population. A value of the index of less than 100 indicates a beneficial effect of the experimental diet over that of the "control" diet (commercial).

increases lifespan delays the onset of a variety of diseases. However, the data are not consistent regarding the relationship between dietary restriction and the total incidence of disease. Furthermore, there are no indications as to the mechanisms responsible for this delay in their onset.

6.3. Biochemical and Physiological Variables

In an effort to establish the biological mechanisms responsible for the increased lifespan associated with dietary restriction, comparisons have been made in various biochemical and physiological variables in animals whose lifespan was increased by dietary manipulations. These data are presented in the figures and tables described below.

Figure 15 shows that animals whose lifespan have been increased by low protein feeding (4%) have a lower rectal temperature than do those fed the control diet (24% protein). In addition, the low body temperatures of these mice were associated with an increased oxygen consumption (Fig. 16). Because of the lack of agreement on the effects of these variables on lifespan (Leto *et al.*, 1976a), these data cannot contribute at present to our knowledge of the biological mechanism responsible for increased lifespan.

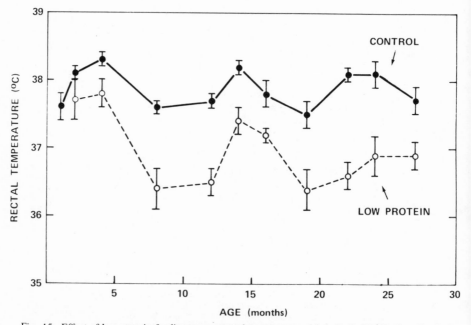

Fig. 15. Effect of low-protein feeding age on rectal temperature of female $C_{57}BL/6J$ mice. Vertical bars represent SEM. The mean lifespan and SEM of the animals fed either the low-protein or control diet was 852 ± 27.4 days and 685 ± 22.8 days, respectively (Leto *et al.*, 1976a).

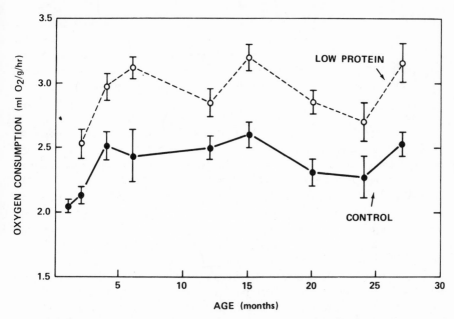

Fig. 16. Effect of low-protein feeding on oxygen consumption of C$_{57}$BL/6J female mice. Vertical bars represent SEM. The mean lifespan and SEM of the animals fed either the low-protein or control diet was 852 ± 27.4 days and 685 ± 22.8 days, respectively (Leto *et al.*, 1976a).

There have been two reports on the effects of long-term dietary restriction on immunologic functions. In one study, BALB/c male mice were fed *ad libitum* a diet that contained either 24 or 4% protein derived from casein (Stoltzner, 1976). In another study, Gerbase-DeLima *et al.* (1975) fed C57BL/6J female mice a nutritionally adequate diet (21.6% casein) *ad libitum* or every other day. Stoltzner reported that there were no differences in mitogenic response at any age between restricted and nonrestricted animals when lymphocytes were cultured with various mitogens. The data of Gerbase-DeLima, however, indicate marked differences in these immunological indices especially after middle age, 52–55 weeks. A diminished primary antibody response to sheep red blood cells was also reported by Stoltzner. On the other hand, the data of Gerbase-DeLima indicate this immunological function is markedly greater in middle-aged, restricted animals than in controls. Gerbase-DeLima infers that the immunologic system of restricted animals matures slower and is maintained longer than in control animals. These apparent discrepancies between the two reports may be explained by differences in strain, sex and/or dietary manipulation.

It was proposed that dietary restriction increases longevity by decreasing protein and RNA syntheses and thereby genetic informational transfer during the total lifespan (Barrows, 1971, 1972). On the basis of the following data, it was assumed that enzymatic activities could be used as expressions of genetic programming and that these biochemical indices are intimately associated with physiological ones, at least with egg production in the rotifer. The enzymatic activities were considered adequate expressions of program on the basis that under all conditions the following always occurred: (1) the patterns of age change in the enzymatic activities were similar, (2) the maximum levels of activity were the same, and (3) age-dependent decreases in the ratio of two enzymes, namely, malate dehydrogenase (MDH) to lactate dehydrogenase (LDH), always occurred (Figs. 17–19). Similar data have been reported by Ross (1959) for the enzymes adenosine triphosphatase and alkaline phosphatase in the livers of rats (Fig. 20). Rotifers (Table XVII) and *Daphni* (Table XVIII) apparently are programmed to produce equal numbers of eggs. Most important is that the age-associated changes in these variables occurred at later ages in animals whose lifespan was increased by dietary restriction. The data seem to support the concept that there was a program for the total lifespan of the organism and those nutritional conditions which altered the length of life did so merely by altering the rate of occurrence of specific events.

A delay in genetic informational transfer would be advantageous were aging the expression of deleterious genes in late life. However, there is no evidence to support this. Furthermore, later studies in which enzymatic activities had been based on DNA (Figs. 21–23) or numbers of hepatocytes (Fig. 24) do not support the concept of delayed genetic informational transfer; but rather they suggest a

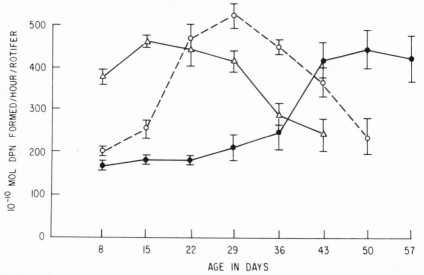

Fig. 17. Effect of nutrition on malic dehydrogenase activity in rotifers (*Philodina acuticornis*). (△) Diet I (algae and fresh pond water daily; mean lifespan = 34.0 ± 1.1); (○) Diet II (fresh pond water daily; mean lifespan = 45.3 ± 1.7); and (●) Diet III (fresh pond water—Mon., Wed., and Fri.; mean lifespan = 54.7 ± 1.3). Vertical bars represent SEM (Fanestil and Barrows, 1965).

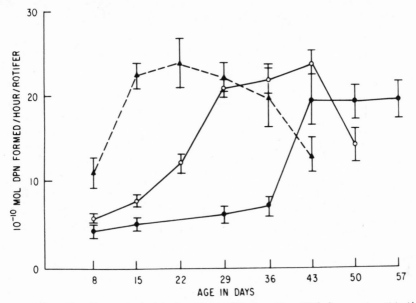

Fig. 18. Effect of nutrition on lactic dehydrogenase activity in rotifers (*Philodina acuticornis*). (▲) Diet I (algae and fresh pond water daily; mean lifespan = 34.0 ± 1.1); (○) Diet II (fresh pond water daily; mean lifespan = 45.3 ± 1.7); and (●) Diet III (fresh pond water—Mon., Wed., and Fri.; mean lifespan = 54.7 ± 1.3). Vertical bars represent SEM (Fanestil and Barrows, 1965).

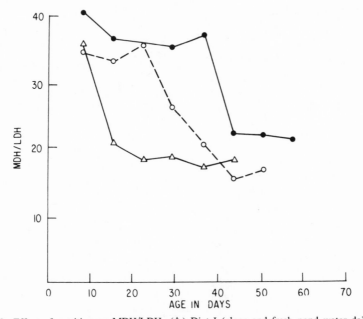

Fig. 19. Effect of nutrition on MDH/LDH. (△) Diet I (algae and fresh pond water daily; mean lifespan = 34.0 ± 1.1); (○) Diet II (fresh pond water daily; mean lifespan = 45.3 ± 1.7; (●) Diet III (fresh pond water—Mon., Wed., and Fri.; mean lifespan = 54.7 ± 1.3) (Fanestil and Barrows, 1965).

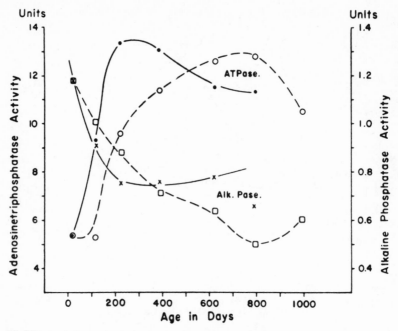

Fig. 20. Effect of diet and age on the activity of hepatic adenosinetriphosphatase and alkaline phosphatase in male Sprague-Dawley rats. Enzymatic activities are expressed as activity per milligram weight of tissue. Rats maintained on commercial diet *ad libitum:* (x) alkaline phosphatase activity; (●) adenosinetriphosphatase activity. Rats whose daily food allotment of Diet C was restricted (see Table X): (□) alkaline phosphatase activity; (○) adenosinetriphosphatase activity (Ross, 1959).

Table XVII. Effect of Diet on Lifespan and Fecundity in Rotifers

Experiment		Mean lifespan (days) Diet[a]			End of eggs (days) Diet			Mean number of eggs (days) Diet		
		I	II	III	I	II	III	I	II	III
1	Mn	35.7	43.8	58.6	12.8	19.2	27.7	40.0	40.0	44.0
	σMn	±2.1	±3.0	±1.9	±0.2	±0.3	±0.5	±1.4	±1.6	±1.0
2	Mn	36.0	45.6	56.5	11.9	19.5	25.3	28.0	36.0	38.0
	σMn	±1.2	±2.5	±2.2	±0.1	±0.3	±0.3	±2.0	±1.3	±1.3
3	Mn	29.0	46.2	49.1	11.9	29.3	38.1	47.0	39.0	33.0
	σMn	±2.8	±3.2	±2.2	±0.2	±0.9	±1.1	±1.1	±2.5	±1.5
Mean	Mn	34.0	45.3	54.7	12.1	21.7	29.8	40.9	37.6	37.9
	σMn	±1.1	±1.7	±1.3	±0.1	±0.3	±0.3	±1.0	±1.0	±0.7

[a]Fanestil and Barrows (1965).

Table XVIII. The Effect of Nutrition on Reproduction
and Lifespan of *Daphnia Longispina*[a]

Nutrition	Number of young	Time of reproduction	Lifespan
Normal	270	24	31
Starved	254	32	39
9th instar			
13th instar	230	36	44

[a]Ingle *et al.* (1937).

reduced use of the genetic code throughout lifespan. Therefore, another proposal has been offered, i.e., that dietary restriction reduces the use of the genetic code and thereby minimizes genetic imperfections as they occur in later life (Barrows and Kokkonen, 1975).

7. Summary

Nutritional surveys fail to consistently provide evidence that a poor nutritional state exists generally among members of the aging population in the United States. However, significant numbers of many of the groups studied consumed less than the recommended RDA of certain nutrients. These included protein, calcium, ascorbic acid, and vitamin A. One of the most consistent findings in the National Surveys was that low dietary intakes were associated with poor health and low income. Laboratory studies in which the economic variable was minimized indicated that age was accompanied by a decrease in the intake of all nutrients. The decrement in total caloric intake was approximated by an age-associated decrease in basal metabolism as well as a decline in physical activity. A number of carefully conducted laboratory studies in which only small numbers of subjects participated substantiate the findings of the National Surveys regarding the poor nutritional status in limited numbers of old individuals. Furthermore, they fail to provide strong evidence to indicate that age influences the vitamin requirements in humans. In general, little correlation has been observed between low plasma levels of various vitamins and physiological impairments associated with their deficiencies. Low plasma levels of vitamins when they occur in old individuals can be increased by the administration of the particular vitamin. The frequency of serious nutritional problems among the aged is estimated to be approximately 5–10%.

Dietary restriction has been shown to increase the lifespan of a variety of species. Recent studies have shown that the beneficial effects of dietary restriction can be brought about when underfeeding is initiated in adult as well as young

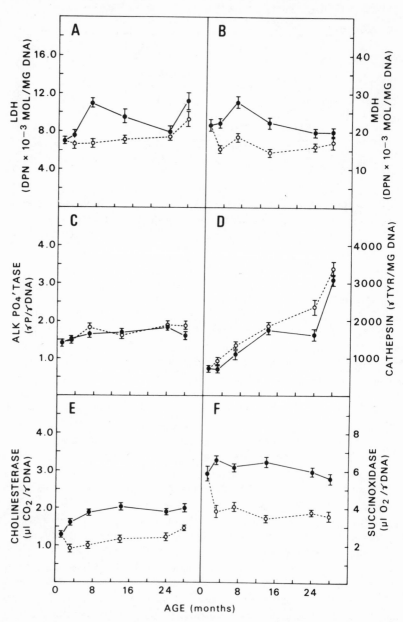

Fig. 21. Effect of age and diet on the enzymatic activities of liver of female $C_{57}BL/6J$ mice fed (●) 26% casein diet or (○) 4% casein diet. Vertical bars represent SEM. The mean lifespan and SEM of the animals fed either the low-protein or control diet was 852 ± 27.4 days and 685 ± 22.8 days, respectively (Leto *et al.*, 1976).

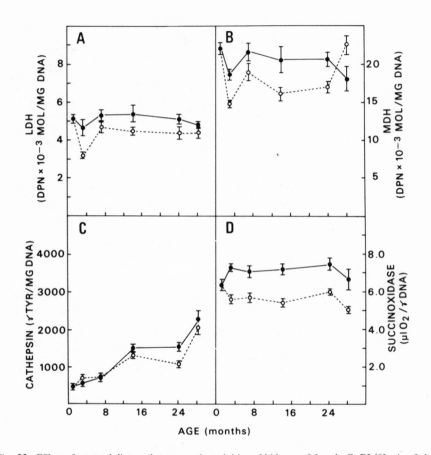

Fig. 22. Effect of age and diet on the enzymatic activities of kidneys of female $C_{57}BL/6J$ mice fed (●) 26% casein diet or (○) 4% casein diet. Vertical bars represent SEM. The mean lifespan and SEM of the animals fed either the low-protein or control diet was 852 ± 27.4 days and 685 ± 22.8 days, respectively (Leto *et al.*, 1976).

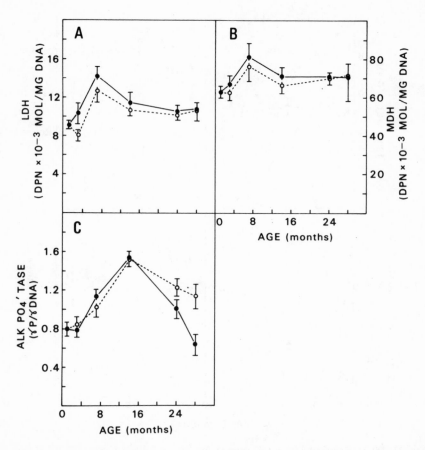

Fig. 23. Effect of age and diet on the enzymatic activities of hearts of female $C_{57}BL/6J$ mice fed (●) 26% casein diet or (○) 4% casein diet. Vertical bars represent SEM. The mean lifespan and SEM of the animals fed either the low protein or control diet was 852 ± 27.4 days and 685 ± 22.8 days, respectively (Leto et al., 1976).

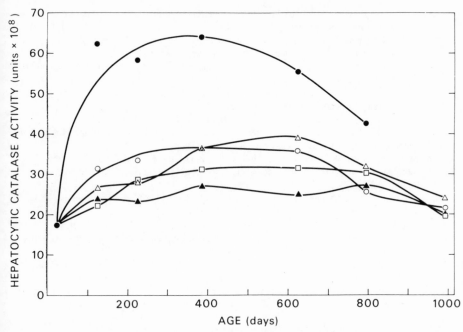

Fig. 24. Effect of diet and age on the activity of hepatic catalase in male Sprague-Dawley rats. Rats maintained on commercial diet *ad libitum* (●); rats whose daily food allotment was restricted (see Table IX): (○) Diet A; (△) Diet B; (□) Diet C; (▲) Diet D (Ross, 1969).

growing animals. Dietary restriction has been shown to delay the onset of a variety of diseases although its relationship to total incidence has not been established. Comparisons of age-associated changes in physiological variables in normal as well as underfed animals failed to establish the biological mechanism responsible for the increased lifespan. It has been proposed that dietary restriction reduces protein synthesis and increases lifespan by retarding genetic information transfer during early life, and reducing the use of the genetic code and thereby minimizing genetic imperfections as they may occur during late life.

References

Ackermann, P. G., and Toro, G., 1953, Calcium and phosphorus balance in elderly men, *J. Gerontol.* **8**:289.

Ackermann, P. G., and Toro, G., 1953a, Effect of added vitamin D on the calcium balance in elderly males, *J. Gerontol.* **8**:451.

Barrows, C. H., 1971, The challenge—Mechanisms of biological aging, *Gerontologist* **11**:5.

Barrows, C. H., 1972, Nutrition, aging and genetic program, *Am. J. Clin. Nutr.* **25**:829.

Barrows, C. H., and Kokkonen, G. C., 1975, Protein synthesis, development, growth and life span, *Growth* **39**:525.

Barrows, C. H., Jr., and Roeder, L. M., 1965, The effect of reduced dietary intake on enzymatic activities and life span of rats, *J. Gerontol.* **20**:69.

Bayless, T. M., Rothfeld, B., Massa, C., Wise, L., Paige, D., and Bedine, M. S., 1975, Lactose and milk intolerance: Clinical implications, *New Engl. J. Med.* **292**:1156.

Berg, B. N., and Simms, H. S., 1960, Nutrition and longevity in the rat. II. Longevity and onset of disease with different levels of food intake. *J. Nutr.* **71**:255.

Bogdonoff, M. D., Shock, N. W., and Nichols, M. P., 1953, Calcium, phosphorus, nitrogen, and potassium balance studies in the aged male, *J. Gerontol.* **8**:272.

Bras, G., 1969, Age-associated kidney lesions in the rat, *J. Infect. Dis.* **120**:131.

Brewer, W. D., Furnivall, M. E., Wagoner, A., Lee, J., Alsop, B., and Ohlson, M.A., 1956, Nutritional status of the aged in Michigan, *J. Am. Dietet. Assoc.* **32**:810.

Brin, M., Dibble, M. V., Peel, A., McMullen, E., Bourquin, A., and Chen, N., 1965, Some preliminary findings on the nutritional status of the aged in Onondaga County, New York, *Am. J. Clin. Nutr.* **17**:240.

Brock, M. A., 1976, Laboratory of Cellular and Comparative Physiology, Gerontology Research Center, Baltimore, Md., personal communication.

Center for Disease Control, 1972, Ten-State Nutrition Survey, 1968–1970, DHEW Publ. No. (HSM) 72-8134, Health Services and Mental Health Administration, Government Printing Office, Washington, D.C.

Chernish, S. M., Helmer, O. M., Fauts, P. J., and Hohlstaedt, K. G., 1957, The effect of intrinsic factor on the absorption of vitamin B_{12} in older people, *Am. J. Clin. Nutr.* **5**:651.

Chieffi, M., and Kirk, J. E., 1949, Vitamin studies in middle-aged and old individuals. II. Correlation between vitamin A plasma content and certain clinical and laboratory findings, *J. Nutr.* **37**:67.

Chieffi, M., and Kirk, J. E., 1951, Vitamin studies in middle-aged and old individuals. VI. Tocopheral plasma concentrations, *J. Gerontol.* **6**:17.

Chinn, A. B., 1956, Some problems of nutrition in the aged, *J. Am. Med. Assoc.* **162**:1511.

Chinn, A. B., Lavik, P. S., and Cameron, D. B., 1956, Measurement of protein digestion and absorption in aged persons by a test meal of I^{131}-labeled protein, *J. Gerontol.* **11**:151.

Chope, H. D., 1954, Relation to nutrition of health in aging persons. A four-year followup of a study in San Mateo County, *Calif. State J. Med.* **81**:335.

Chope, H. D., and Dray, S., 1951, The nutritional status of the aging: public health aspects, *Calif. Med.* **74**:105.

Comfort, A., 1963, Effect of delayed and resumed growth on the longevity of a fish (*lebistes recticulatus, Peters*) in captivity, *Gerontologia* **8**:150.

Comfort, A., 1974, The position of aging studies, *Mech. Ageing Devel.* **3**:1.

Cuatrecasas, P., Lockwood, D. H., and Caldwell, J. R., 1965, Lactase deficiency in the adult, *Lancet* **1**:14.

David, J., Van Herremege, J., and Fouillet, P., 1971, Quantitative underfeeding of *Drosophila;* effects on adult longevity and fecundity, *Exp. Gerontol.* **6**:249.

Davis, R. L., Lawton, A. H., Prouty, R., and Chow, B. F., 1965, The absorption of oral vitamin B_{12} in an aged population, *J. Gerontol.* **20**:169.

Davis, R. L., Lawton, A. H., Barrows, C. H., Jr., and Hargen, S., 1966, Serum lactate and lactic dehydrogenase levels of aging males, *J. Gerontol.* **21**:571.

Dunphy, J. V., Littman, A., Hammond, J. B., Forstner, G., Dahlqvist, A., and Crane, R. K., 1965, Intestinal lactase deficit in adults, *Gastroenterology* **49**:12.

Finch, C. E., 1972, Enzyme activities, gene function and ageing in mammals (review). *Exp. Gerontol.* **7**:53.

Fry, R. J. M., Lesher, S., and Kohn, H. I., 1961, Age effect on cell-transit time in mouse jejunal epithelium, *Am. J. Physiol.* **201**:213.

Gaffney, G. W., Horonick, A., Okuda, K., Meier, P., Chow, B. F., and Shock, N. W., 1957,

Vitamin B$_{12}$ serum concentrations in 528 apparently healthy human subjects of ages 12–94, *J. Gerontol.* **12**:32.

Garn, S. M., 1970, *The Earlier Gain and the Later Loss of Cortical Bone in Nutritional Perspective,* C C Thomas, Springfield, Ill.

Garn, S. M., Rohmann, C. G., and Wagner, B., 1967, Bone loss as a general phenomenon in men, *Fed. Proc.* **26**:1729.

Gerbase-DeLima, M., Lui, R. K., Cheney, K. E., Mickey, R., and Walford, R. L., 1975, Immune function and survival in a long-lived mouse strain subjected to undernutrition, *Gerontologia* **21**:184.

Gillum, H. L., Morgan, A. F., and Sailer, F., 1955, Nutritional status of the ageing. V. Vitamin A and carotene, *J. Nutr.* **55**:655.

Gitman, L., and Kamholtz, T., 1965, Incidence of radiographic osteoporosis in a large series of aged individuals, *J. Gerontol.* **20**:32.

Gorbach, S. L., Nahas, L., Lerner, P. I., and Weinstein, L., 1967, Studies of intestinal microflora. I. Effects of diet, age, and periodic sampling on numbers of fecal microorganisms in man, *Gastroenterology* **53**:845.

Harman, D., 1956, Aging: A theory based on free radical and radiation chemistry, *J. Gerontol.* **11**:298.

Harman, D., 1957, Prolongation of the normal life span by radiation protection chemicals, *J. Gerontol.* **12**:257.

Harman, D., 1961, Prolongation of the normal life span and inhibition of spontaneous cancer by antioxidants, *J. Gerontol.* **16**:247.

Harman, D., 1968, Free radical theory of aging: Effect of free radical reaction inhibitors on the mortality rate of male LAF mice, *J. Gerontol.* **23**:476.

Harrison, M., Fraser, R., and Mullan, B., 1961, Calcium metabolism in osteoporosis. Acute and long-term responses to increased calcium intake. *Lancet* **1**:1015.

Hegsted, D. M., 1952, False estimates of adult requirements, *Nutr. Rev.* **10**:257.

Hegsted, D. M., Moscoso, I., and Collazos, C., 1952, A study of the minimum calcium requirements of adult men, *J. Nutr.* **46**:181.

Horwitt, M., 1953, Dietary requirements of the aged, *J. Am. Dietet. Assoc.* **29**:443.

Ingle, L., Wood, T. R., and Banta, A. M., 1937, A study of longevity, growth, reproduction and heart rate in *Daphnia langispina* as influenced by limitations in quantity of food, *J. Exp. Zool.* **76**:325.

Inter-departmental Committee on Nutrition for National Defense, 1963, *Manual for Nutrition Surveys,* 2nd ed., National Institutes of Health, Bethesda, Md.

Kirk, J. E., 1954, Blood and urine vitamin levels in the aged, in *Symposium on Problems of Gerontology,* pp. 73–94, National Vitamin Foundation, New York.

Kirk, J. E., and Chieffi, M., 1948, Vitamin studies in middle-aged and old individuals. I. The vitamin A, total carotene and $\alpha + \beta$ carotene concentrations in plasma, *J. Nutr.* **36**:315.

Kirk, J. E., and Chieffi, M., 1949, Vitamin studies in middle-aged and old individuals. III. Thiamine and pyruvic acid blood concentrations. *J. Nutr.* **38**:353.

Kirk, J. E., and Chieffi, M., 1953, Vitamin studies in middle-aged and old individuals. XI. The concentration of total ascorbic acid in whole blood, *J. Gerontol.* **8**:301.

Kirk, J. E., and Chieffi, M., 1953a, Vitamin studies in middle-aged and old individuals. XII. Hypovitaminemia C. Effect of ascorbic acid administration on the blood ascorbic acid concentration, *J. Gerontol.* **8**:305.

Kohn, R. R., 1971, Effect of antioxidants on life-span of C57BL mice, *J. Gerontol.* **26**:378.

Kopec, S., 1928, On the influence of intermittent starvation on the longevity of the imaginal stage of *Drosophila melanogaster, Br. J. Exp. Biol.* **5**:204, 1928.

Kountz, W. B., Hofstatter, L., and Ackermann, P., 1947, Nitrogen balance studies in elderly people, *Geriatrics* **2**:173.

Kountz, W. B., Hofstatter, L., and Ackermann, P. G., 1951, Nitrogen balance studies in four elderly men, *J. Gerontol.* **6**:20.

Lansing, A., 1948, Evidence of aging as a consequence of growth cessation, *Proc. Natl. Acad. Sci., USA* **34**:304.

Lemming, J. T., Webster, S. P. G., and Dymock, I. W., 1973, Gastrointestinal system, in *Textbook of Geriatric Medicine and Gerontology* (J. C. Brocklehurst, ed.), pp. 321–346, Churchill Livingstone, Edinburgh.

Leto, S., Kokkonen, G. C., and Barrows, C. H., 1976a, Dietary protein, life-span, and biochemical variables in female mice, *J. Gerontol.* **31**:144.

Leto, S., Kokkonen, G. C., and Barrows, C. H., 1976b, Dietary protein, life-span, and physiological variables in female mice, *J. Gerontol.* **31**:149.

Leveille, G. A., 1972, The long-term effects of meal-eating on lipogenesis, enzyme activity, and longevity in the rat, *J. Nutr.* **102**:549.

Littman, A., 1966, Isolated lactase deficit in the adult: A present view, *J. Am. Med. Assoc.* **195**:954.

Loeb, J., and Northrop, J. H., 1917, On the influence of food and temperature upon the duration of life, *J. Biol. Chem.* **32**:103.

Lutwak, L., 1963, Osteoporosis. A disorder of nutrition, *NY State J. Med.* **63**:590.

Lutwak, L., and Whedon, C. D., 1963, Osteoporosis, in *Disease-a-Month*, pp. 1–39 (April), Year-Book Publishers, Chicago.

McCay, C. M., Crowell, M. F., and Maynard, L. A., 1935, The effect of retarded growth upon the length of life span and upon the ultimate body size, *J. Nutr.* **10**:63.

McCay, C. M., Maynard, L. A., Sperling, G., and Barnes, L. L., 1939, Retarded growth, life span, ultimate body size and age changes in the albino rat after feeding diets restricted in calories, *J. Nutr.* **18**:1.

McCay, C. M., Sperling, G., and Barnes, L. L., 1943, Growth, ageing, chronic diseases, and life span in rats, *Arch. Biochem.* **2**:469.

McGandy, R. B., Barrows, C. H., Jr., Spanias, A., Meredith, A., Stone, J. L., and Norris, A. H., 1966, Nutrient intakes and energy expenditure in men of different ages, *J. Gerontol.* **21**:581.

MacKeen, P. C., and Mitchell, R. B., 1976, Biology Department, Penn. State Univ., University Park, Penna., personal communication.

Malm, O. J., 1958, *Calcium Requirements and Adaptation in Adult Men,* Oslo Univ. Press, Oslo.

Miller, D. S., and Payne, P. R., 1968, Longevity and protein intake, *Exp. Gerontol.* **3**:231.

Morgan, A. F., Gillum, H. L., and Williams, R. I., 1955, Nutritional status of the aging. III. Serum ascorbic acid and intake, *J. Nutr.* **55**:431.

National Academy of Sciences—National Research Council, 1968, Recommended Dietary Allowances, National Research Council Publication 1964, 7th ed., National Research Council, Washington, D. C.

National Center for Health Statistics, 1974, First Health and Nutrition Examination Survey, United States, 1971–1972, DHEW Publication No. (HRA) 74-1219-1, Health Services Administration, Government Printing Office, Washington, D.C.

Necheles, H., Plotke, F., and Meyer, J., 1942, Studies on old age. V. Active pancreatic secretion in the aged, *Am. J. Digest. Dis.* **9**:157.

Nolen, G. A., 1972, Effect of various restricted dietary regimens on the growth, health and longevity of albino rats, *J. Nutr.* **102**:1477.

Nordin, B. E. C., 1961, The pathogenesis of osteoporosis, *Lancet* **1**:1011.

Nordin, B. E. C., 1962, Calcium balance and calcium requirements in spinal osteoporosis, *Am. J. Clin. Nutr.* **10**:384.

Nordin, B. E. C., and Smith, D. A., 1964, The treatment of osteoporosis, *Triangle* **6**:273.

Oeriu, S., and Vochitu, E., 1965, The effect of the administration of compounds which contain sulfhydryl groups on the survival rates of mice, rats, and guinea pigs, *J. Gerontol.* **20**:417.

Ohlson, M. A., Brewer, W. D., Cederquist, D. C., Jackson, L., Brown, E. G., and Roberts, P. H., 1948, Studies of protein requirements of women, *J. Am. Dietet. Assoc.* **24**:744.

Ohlson, M. A., Brewer, W. D., Jackson, L., Swanson, P. P., Roberts, P. H., Mangel, M., Leverton, R. M., Chaloupka, M., Gram, M. R., Reynolds, M. S., and Lutz, R., 1952, Intakes and retention of nitrogen, calcium and phosphorus by 136 women between 30 and 85 years of age, *Fed. Proc.* **11**:775.

Orshansky, M., 1968, The shape of poverty in 1966, *Soc. Sec. Bull.* **31**:3.

Pearson, W. N., 1962, Biochemical appraisal of nutritional status in man, *Am. J. Clin. Nutr.* **11**:462.

Ranke, E., Tauber, S. A., Horonick, A., Ranke, B., Goodhart, R. S., and Chow, B. F., 1960, Vitamin B_6 deficiency in the aged, *J. Gerontol.* **15**:41.

Reisen, W. H., Herbst, E. J., Walliker, C., and Elvehjem, C. A., 1947, The effect of restricted caloric intake on the longevity of rats, *Am. J. Physiol.* **148**:614.

Roberts, P. H., Kerr, C. H., and Ohlson, M. A., 1948, Nutritional status of older women. Nitrogen, calcium, phosphorus retentions of nine women, *J. Am. Dietet. Assoc.* **24**:292.

Rose, C. S., György, P., Butler, M., Andres, R., Norris, A. H., Shock, N. W., Tobin, J., Brin, M., and Spiegel, H., 1976, Age differences in vitamin B_6 status of 617 men, *Am. J. Clin. Nutr.* **29**:847.

Ross, M. H., 1969, Aging, nutrition and hepatic enzyme activity patterns in the rat, *J. Nutr.* **97**:565.

Ross, M. H., 1959, Protein, calories and life expectancy, *Fed. Proc.* **18**:1190.

Ross, M. H., 1961, Length of life and nutrition in the rat, *J. Nutr.* **75**:197.

Ross, M. H., 1967, Life expectancy modification by change in dietary regimen of the mature rat, in *Proceedings of the 7th International Congress of Nutrition,* Vol. 5, pp. 35–38.

Ross, M. H., and Bras, G., 1971, Lasting influence of early caloric restriction on prevalence of neoplasms in the rat, *J. Natl. Cancer Inst.* **47**:1095.

Saxton, J. A., Jr., Boon, M. C., and Furth, J., 1944, Observations on the inhibition of development of spontaneous leukemia in mice by underfeeding, *Cancer Res.* **4**:401.

Schlenker, E. D., 1976, The nutritional status of older women, Ph.D. thesis, Michigan State University, East Lansing.

Schmid, O., 1962, *Die Ernährung des Menschen über 50 Jahre,* Paracelsus, Stuttgart.

Schulze, W., 1955, Protein metabolism and requirement in old age, in *Old Age in the Modern World,* pp. 122–127, E. & S. Livingstone Ltd., London.

Shock, N. W., 1976, Gerontology Research Center, Baltimore, Maryland, personal communication.

Shock, N. W., Watkin, D. M., Yiengst, M. J., Norris, A. H., Gaffney, G. W., Gregerman, R. I., and Falzone, J. A., 1963, Age differences in the water content of the body as related to basal oxygen consumption in males, *J. Gerontol.* **18**:1.

Smith, R. W., Jr., and Frame, B., 1965, Concurrent axial and appendicular osteoporosis. Its relation to calcium retention, *New Engl. J. Med.* **273**:73.

Stoltzner, G., 1976, Diet restriction, longevity, and immunity in aging mice, *Proceedings of the 29th Annual Meeting, Gerontological Society, New York,* p. 43, Abst.

Straus, B., 1971, Disorders of the digestive system, in *Clinical Geriatrics* (I. Rossman, ed.), pp. 81–202, J. B. Lippincott, Philadelphia.

Stuchlikova, E., Juricova-Horakova, M., and Deyl, Z., 1975, New aspects of the dietary effect of life prolongation in rodents. What is the role of obesity in aging, *Exp. Gerontol.* **10**:141.

Tappel, A., Fletcher, G., and Deamer, D., 1973, Effect of antioxidants and nutrients on lipid peroxidation fluorescent products and aging parameters in the mouse, *J. Gerontol.* **28**:415.

Tuttle, S. G., Swenseid, M. E., Mulcare, D., Griffith, W. H., and Bassett, S. H., 1957, Study of the essential amino acid requirements of men over fifty, *Metab. Clin. Exp.* **6**:564.

Tuttle, S. G., Swenseid, M. E., Mulcare, D., Griffith, W. H., and Bassett, S. H., 1959, Study of essential amino acid requirements of men over fifty, *Metab., Clin. Exp.* **8**:61.

U.S. Department of Agriculture, Agriculture Research Service (By Consumer and Food Economics Division), 1968, Consumption of Households in the United States, Spring 1965, Household Food Consumption Survey 1965–66, Report No. 1, Government Printing Office, Washington, D. C., 212 pp.

Visscher, M. B., Ball, Z. B., Barnes, R. H., and Sivertsen, I., 1942, The influence of caloric restriction upon the incidence of spontaneous mammary carcinoma in mice, *Surgery* **11**:48.

Watkin, D. M., 1968, Nutritional problems today in the elderly in the United States, in *Vitamins in the Elderly* (A. N. Exton-Smith and D. L. Scott, eds.), pp. 66–77, John Wright and Sons, Ltd., Bristol, England.

Watkin, D. M., Parsons, J. M., Yiengst, M. J., and Shock, N. W., 1955, Metabolism in the aged; the effect of stanolone on the retention of nitrogen, potassium, phosphorus, and calcium and on the urinary excretion of 17-keto, 11-oxy, and 17-hydroxy steroids in eight elderly men on high and low protein diets, *J. Gerontol.* **10**:268.

Watts, J. H., Mann, A. N., Bradley, L., and Thompson, D. J., 1964, Nitrogen balance of men over 65 fed the FAO and milk patterns of essential amino acids, *J. Gerontol.* **19**:370.

Yiengst, M. J., and Shock, N. W., 1949, Effect of oral administration of vitamin A on plasma levels of vitamin A and carotene in aged males, *J. Gerontol.* **4**:205.

Chapter 11

Amino Acid Nutrition of the Chick

David H. Baker

1. Amino Acid Reference Diets

The earliest records of chick studies with crystalline amino acid diets involved
work by Grau and Almquist (1944), Almquist and Grau (1944), and Hegsted
(1944). Acceptable food intake, however, was a problem in all these studies. In
1950, H. M. Scott and co-workers at the University of Illinois began a meticu-
lous search for a pattern of amino acids that would permit near-optimum growth.
The first reference standard was reported by Glista *et al.* (1951). However the
chicks had to be force-fed to obtain even slow rates of growth.

During the next decade, Scott and his graduate students worked on several
modifications of the Glista Standard, but it was not until an amino acid mixture
simulating that of the fat-free chick carcass (Price *et al.*, 1953) was fed that
chicks would gain weight acceptably when fed *ad libitum* (Klain *et al.*, 1960).
Shortly after Klain and co-workers published their results, Adkins and co-
workers (1962) at Wisconsin developed an amino acid reference diet that also
seemed to perform reasonably well. From that point until 1965, numerous short-
term assays were carried out in Scott's laboratory in an attempt to improve chick
performance still further (Scott, 1972). What became known as the Dean Standard (Table I) was finally arrived at; this diet permitted growth equal in magnitude to that possible with the best practical diets (Dean and Scott, 1965).
Moreover, glutamic acid was shown to provide nitrogen for dispensable amino
acid biosynthesis as efficiently as any mixture of dispensable amino acids.

It soon became obvious that whereas the Dean Standard would permit
optimal chick performance, several amino acids were in excess of the requirement. Hence, an amino acid mixture lower in nitrogen was sought. However,

David H. Baker • Department of Animal Science, University of Illinois, Urbana, Illinois 61801.

Table I. Complete Amino Acid Mixtures for Chicks[a]

Amino acid	Dean (1965)[b]	RS (1969)[c]	MRS (1973)[d]
Arginine	1.10	1.00	0.95
Histidine	0.30	0.30	0.33[e]
Lysine	1.12	0.95	0.91
Tyrosine[f]	0.63	0.45	0.45
Tryptophan	0.23	0.15	0.15
Phenylalanine	0.68	0.50	0.50
Methionine	0.55	0.35	0.35
Cystine[g]	0.35	0.35	0.35
Threonine	0.65	0.65	0.65
Leucine	1.20	1.20	1.00
Isoleucine	0.80	0.60	0.60
Valine	0.82	0.82	0.69
Proline	1.00	0.20	0.40
Glycine	1.60	1.20	0.60
Glutamic acid	12.00	10.00	12.00

[a] Values expressed as a percent of diet.
[b] Dean and Scott (1965): "Dean Standard."
[c] Sugahara et al. (1969): "Reference Standard."
[d] Sasse and Baker (1973): "Modified Reference Standard."
[e] The published level of histidine in MRS is 0.30%, but recent data from the University of Illinois suggest 0.33% as the necessary percentage for maximum growth.
[f] Phenylalanine can meet the requirement for tyrosine with 100% molar efficiency.
[g] Methionine can meet the requirement for cystine with 100% molar efficiency, but because the molecular weight is different, methionine is only 80% efficient on a weight or concentration basis in meeting the same need.

because the interrelationships of each amino acid (i.e., amino acid balance) had been established as being highly sensitive and easily upset (Sugahara et al., 1969), caution was exercised in making any revision. What became known as the Reference Standard (Huston and Scott, 1968; Sugahara et al., 1969) was developed in the late 1960s by Scott et al. at the Illinois Station. Growth rate, however, was 10–15% slower, but nitrogen utilization 10–15% greater than had been the case with the Dean Standard. Subsequent work by Sasse and Baker (1973a) brought about the most recent revision to what has become known as the Modified Reference Standard (Table I), which performs on a par with the Dean Standard with regard to growth potential and results in nitrogen utilization (gain per gram of nitrogen intake) superior to any of the chick standards developed.

Although it is clear that the Modified Reference Standard has no amino acid deficiencies, it does indeed have some slight excesses. For example, sulfur amino acids (methionine + cysteine), aromatic amino acids (phenylalanine + tyrosine), and threonine and tryptophan are in slight excess of the minimum requirement for maximum growth in the chick. Revision, however, would probably affect the requirement for several other amino acids. It was therefore considered advisable to maintain an admitted excess of these amino acids. Certainly, it should also be kept in mind that the levels of each amino acid in Modified Reference Standard are only a "requirement" when defined in terms of the dietary concentration of both metabolizable energy and nitrogen.

Other amino acid standards have been proposed (Dobson et al., 1964; Hewitt and Lewis, 1972a,b; Woodham and Deans, 1975). Although intact proteins were used, the final pattern of amino acids arrived at was in rather close agreement with that of the Modified Reference Standard. Moreover, excellent rate and efficiency of weight gain were noted.

There are several other important features of a crystalline amino acid diet. Certainly, the physical aspects of the diet cannot be ignored. Because it is necessary to use a nonreducing carbohydrate (i.e., cornstarch or dextrin), fat is needed to give body to the diet—at least 5% fat. Energy retention, however, is maximized at a dietary concentration of 15% fat (Velu and Baker, 1974). Individual vitamin and mineral mixtures must also be developed. Because the diet is devoid of natural ingredients, all known vitamins and minerals must be considered for possible inclusion. Hence, because long-term assays were shown to result in selenium deficiency, this trace element was recently added to the Illinois mineral mixture.* Essentiality of tin, vanadium, chromium, nickel, and silicon was tested recently, but no response was noted for the addition of any one or all of these elements to the Modified Reference Standard (Baker and Molitoris, 1975).

Several unexplained examples may be cited as evidence that crystalline amino acid reference diets, regardless of how well balanced, still pose problems. Sustained growth rate over an extended period on the purified diet is therefore difficult to accomplish, even though the amino acid diet nearly maximizes gain of young broiler chicks over any two-week assay period between hatching and 9 weeks posthatching. Furthermore, gain per unit of metabolizable energy intake is considerably lower with the purified diet than with a practical diet based on natural ingredients. Finally, laying hens absolutely refuse to lay, perhaps because of poor diet intake, when fed the crystalline amino acid diet.

2. Criteria of Adequacy

Generally, maximum rate and efficiency of weight gain over a 7- to 14-day assay period have been used as response criteria in setting requirements. The correlation between weight gain and protein retention is very high in young, rapidly growing chicks (Velu et al., 1971). Other response parameters have been

*Selenium is a natural contaminent of crystalline methionine and cystine to an extent that a "selenium-free diet" must be fed for at least 5 weeks before depressed growth and biochemical lesions will appear.

evaluated, but growth rate and gain/feed ratio still appear to be the most accurate. The so-called plasma method, whereby levels of the free amino acid under study are evaluated as a function of dietary intake of that amino acid may be used, but there is high variability in these studies (Morrison *et al.*, 1961; Zimmerman and Scott, 1965; Mitchell *et al.*, 1968; Stockland *et al.*, 1970). The rate at which ^{14}C-labeled amino acids are oxidized has also been used to assess requirements (Brookes *et al.*, 1972; Ishibashi, 1973; Neale and Waterlow, 1974). Indirect methods, such as the plasma method and the amino acid oxidation method, are most suitable for large animal species, for which a long-term growth assay would be both impractical and costly. These techniques are illustrated in Fig. 1. Also, such indirect methods may yield inaccurate estimates of the requirements for amino acids that may be released (when deficient) from endogenous reserves. Examples of such amino acids are histidine (muscle carnosine and blood hemoglobin are labile sources) and cystine, hence sulfur amino acids (considerable cystine may be released from endogenous glutathione).

3.   Factors That Affect Dietary Requirements

It is extremely difficult to set forth amino acid requirements for a given species because they are influenced by a multitude of factors. Expressing re-

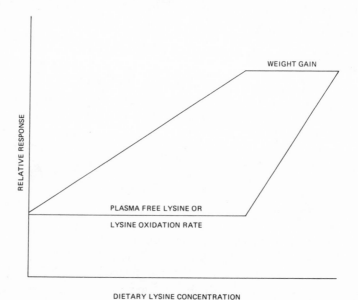

Fig. 1. Relationship between chick weight gain and (a) the concentration of free lysine in blood plasma as well as (b) the rate of U-^{14}C-lysine catabolism to $^{14}CO_2$. Above the requirement, lysine accumulates in the plasma and oxidation to CO_2 increases markedly.

quirements on a dietary concentration basis (percent of diet or percent of calories) is helpful for some species, because strain differences as well as bird variability within a given population can be dealt with. Therefore, at a given age, fast-growing chicks within a population (or fast-growing strains of chickens) have been shown to require roughly the same percentage of amino acids in their diet as do slow-growing birds (Griminger and Scott, 1959), with the absolute requirements (in grams per day) of the larger birds obviously being greater. The larger birds simply meet their daily requirement by consuming more feed. Recent work with turkeys, however, has suggested that fast-growing, large breeds may require a greater percentage of amino acids than do slow-growing breeds (Potter, 1976).

The nature of the diet used to establish amino acid requirements can have profound effects on the requirement for indispensable amino acids. Thus, protein level, energy level and type of diet (i.e., practical, semipurified with intact protein or purified with only amino acids furnishing dietary nitrogen) must be carefully considered. It would seem readily apparent that requirements established with crystalline amino acid diets, regardless of crude protein or energy level, would be lower than those arrived at by supplementing amino acid-deficient intact proteins with their limiting amino acid. The principal explanation for this difference is that a portion of the limiting amino acid present in the intact protein-containing diet is not 100% available (i.e., digestible and absorbable), thereby necessitating a greater concentration of *total* amino acid to achieve optimal performance. Crystalline amino acids, on the other hand, are 100% available.

3.1. Protein Level

The first clear-cut demonstration that the requirement for an indispensable amino acid for chicks is influenced by dietary protein level was made by Grau (1948), who fed chicks 5, 10, 20, and 30% crude protein from sesame meal. This study clearly demonstrated that the lysine requirement increased as the protein level was increased from 5% to 30%. Subsequent reports confirmed this observation. Whereas the increase in the requirement expressed as a percent of the diet appears to be linear between deficient and adequate levels of dietary nitrogen, it becomes curvilinear between adequacy and superadequacy (Boomgaardt and Baker, 1971, 1973c; Baker *et al.*, 1975).

Because the requirement for a limiting amino acid is affected by dietary protein level, it is often difficult to overcome an amino acid deficiency in an intact protein by feeding more of the amino acid-deficient protein (Wethli *et al.*, 1975; Baker and Easter, 1976). Voluntary food intake is generally reduced when excessive levels of protein are fed in an attempt to meet an amino acid requirement. Thus, although the absolute requirement (grams per day) is probably unchanged, because of the lower food intake, a greater concentration of the limiting amino acid must be present in the high-protein diet to provide the same intake of that amino acid. Because of these considerations, formulation of chick rations to a given level of one or more indispensable amino acids with no regard to dietary protein level will often result in poor performance.

Perhaps one of the better illustrations of the practical implications of the effect of protein level on the requirement for an indispensable amino acid is found in a demonstration experiment conducted at the University of Illinois (Baker and Easter, 1976). As shown in Table II, the control diet, except for necessary vitamin, mineral, and methionine fortification, consisted of corn and soybean meal and was formulated to meet the chick's minimal lysine requirement of 1.10% of the diet. To meet this restriction, the dietary protein level had to be 20.3%. When corn and sesame meal were used to achieve this same protein level, lysine was severely deficient and chicks performed poorly (diet 2). Lysine fortification to bring total dietary lysine to the corn/soybean-meal control level resulted in a tripling of the rate of weight gain (diet 3). Performance still fell somewhat short of that achieved with the control diet, probably because the corn/sesame-meal diet was deficient in a second-limiting amino acid. However, when the same dietary level of lysine was achieved by adjusting the corn/sesame-meal ratio (diet 4), weight gain—although better than that in chicks fed diet 2—was vastly inferior to that of chicks fed diet 3, the lower-protein, lysine-fortified diet. The increased protein level of diet 4 resulted in an increase in the percentage concentration of lysine required in the diet, as indicated by the finding that even though diet 4 satisfied the so-called lysine requirement, both feed intake and gain responded to supplemental lysine (diet 5).

Therefore, when dealing with amino acid-deficient proteins, it is certainly far better to meet a given requirement by using a crystalline source of the limiting amino acid than to feed excessive levels of the amino acid-deficient protein. The latter case results in an amino acid imbalance, which causes a reduction in voluntary food intake. Moreover, the availability of amino acids in oilseed meals,

Table II. Efficacy of Soybean Meal and Sesame Meal in Furnishing
Available Lysine for Growth of the Chick[a,b]

Diet No.	Content	Dietary lysine (%)	Dietary protein (%)	Gain (g)	Feed intake (g)	Gain/feed
1.	Corn/soybean meal	1.10	20.3	243	368	0.66
2.	Corn/sesame meal	0.56	20.3	68	200	0.34
3.	As 2 + lysine·HCl	1.10	20.3	222	363	0.61
4.	Corn/sesame meal	1.10	38.0	144	327	0.44
5.	As 4 + lysine·HCl	1.46	38.0	181	393	0.46
	Pooled standard error			6.2	11.8	0.01

[a]Data from Baker and Easter (1976).
[b]Results represent the mean of 21 chicks on test from days 8–21 posthatching.

including soy, is probably between 80% and 90% (Netke and Scott, 1970); this, too, would make the *total* requirement higher when intact protein furnishes proportionately more of the amino acid requirement.

3.2. Energy Level

Combs and Romoser (1955) clearly demonstrated that variations in caloric density of a chick diet influence the level of dietary protein necessary for maximal growth. In fact, their work together with subsequent studies suggested a direct relationship between dietary energy level and the protein requirement. The relationship between energy level and the requirement for an indispensable amino acid is somewhat more complicated because dietary protein quantity and quality are confounding factors that influence the results. Therefore, as long as there is an excess of all amino acids in the diet, except the limiting amino acid, the relationship between energy level and the requirement for the limiting amino acid is direct, i.e., as dietary energy level increases, the requirement for the limiting amino acid increases linearly (Baldini and Rosenberg, 1955; Baldini *et al.*, 1957; Schwartz *et al.*, 1958; Leong *et al.*, 1959). Grimminger *et al.* (1957) and Scott and Forbes (1958) demonstrated this nicely when they showed that an amino acid deficiency could be overcome by narrowing the calorie/protein ratio, by adding bulk (cellulose) to the diet.

Cases for which a well-balanced pattern of crystalline amino acids is fed and all amino acids remain in the same ratio one to another demonstrate that energy increases or decreases have very little effect on the dietary requirement for an indispensable amino acid (Boomgaardt and Baker, 1973b). For example, with no dietary excess of amino acids, addition of fat to a diet will decrease voluntary food intake, but intake of all the amino acids including the first-limiting one will be reduced such that response to the first-limiting amino acid will occur only until the second- and then the third-limiting amino acid is provided. Hence, the relationship between energy level and the requirement for an indispensable amino acid is not straightforward. Certainly, caution should be exercised and careful attention given to the source and level of the amino acids that make up a diet before amino acid requirements are related across the board to a dietary concentration of metabolizable energy. The relationship will be linear only if the range of energy levels covered does not result in a level of feed intake that will make a second- or third-limiting amino acid a limiting factor in the response to the first-limiting amino acid.

3.3. Environmental Temperature

There is a similarity between environmental temperature and caloric density in that both factors manifest their effects by influencing voluntary food intake. Thus, with the same qualifying comments with regard to having a sufficient excess of amino acids other than the first-limiting one, as was the case for dietary energy concentration, both protein and individual amino acid requirements have tended to vary directly with environmental temperature. This, too, however, is not clear cut because, unlike the effect of energy, when food intake is decreased in hot weather or increased in cold weather, intake of not only the limiting amino acid, but also of calories, is decreased or increased concomitantly. Therefore, some workers have questioned whether an increased intake of a limiting amino acid can be fully efficacious, as intake of all other nutrients, including energy, has been reduced as well. Early work from Purdue, for example, indicated the negligible effect of environmental temperature on the crude protein requirement (Adams and Rogler, 1968; Adams *et al.*, 1962). In contrast, Reid and Weber (1973) estimated that the SAA requirement of laying hens was increased to 514 mg/day at 32.2°C from 448 mg/day at 21.1°C. For further information on the effect of environmental temperature on egg production, the reader is referred to the report of Mowbray and Sykes (1971) and the review by Payne (1966).

3.4. Age

There is ample evidence in the literature to document that the need for an indispensable amino acid decreases, expressed as a percent of the diet, as growth progresses from birth to maturity. Bird (1953) established that decreased requirement for lysine was directly proportional to decrease in the required level of crude protein. Thus, the lysine requirement remained a constant percent of the crude protein between hatching and 8 weeks posthatching. It would seem logical that amino acid requirements remain a constant percent of the protein with advancing age and weight in light of the finding (Williams *et al.*, 1954) that the amino acid concentration of avian carcass protein remains remarkably constant as a function of age and weight. Recent work by Boomgaardt and Baker (1973a) has confirmed this conclusion with regard to lysine. Sulfur amino acid requirements, however, were found to decrease somewhat as a percent of the protein with increasing age and weight.

The effect of age on the requirement for sulfur amino acids is a subject of controversy. Certainly, if one amino acid should differ with regard to the concept of similar amino acid requirements expressed as a percent of protein with increasing age, it would be sulfur amino acids, because feathers, like all keratoid tissues, are extremely rich in cystine. Therefore, the broiler breed or strain, and even the sex, might influence the time span and extent of rapid feather growth. Bornstein and Lipstein (1964, 1966) for example, reported a constant sulfur amino acid requirement of 3.6% of the protein with increasing age from hatching to 10 weeks posthatching. Graber and co-workers (1971a) obtained similar results. Chung *et al.* (1973) reported a requirement of 2.4% of the protein for chicks during the period 1-3 weeks of age and 3.0% of the protein for chicks at 5-7 weeks of age. Finally, the work of Boomgaardt and Baker (1973a) indicated a decrease in the sulfur amino acid requirement from 3.05% of the protein to 2.56%. It is therefore impossible to generalize as to the exact relationship between age and sulfur amino acid requirements of the chick.

3.5. Criterion of Response

It is generally recognized that requirements for the maximal gain/feed ratio or maximal carcass leanness are somewhat greater than those for maximal rate of weight gain. This has been clearly shown for lysine (Combs, 1968; Bornstein, 1970; Boomgaardt and Baker, 1973a). Results for this aspect of sulfur amino acid requirements have been less clear cut, although data of Sekiz et al. (1975) showed that chicks on a diet slightly deficient in methionine will overconsume the diet. Surprisingly, the increased intake is reflected in increased fat as a percent of the carcass, and not increased weight gain. Hence, a slight deficiency of sulfur amino acids decreased gain/feed, but not gain; therefore, the requirement for maximal gain/feed was greater than for maximal gain.

3.6. Translation into Practice

All requirements must be extrapolated insofar as possible to the practical feeding situation wherein fortified corn/soybean meal diets predominate. The pattern of amino acids in a proper corn/soybean meal mixture is exceptional and represents a good example of the complementary effect of proteins (Table III). The only marginal amino acid in a typical broiler diet (23% protein reduced to

Table III. Performance of Broiler Chicks Fed Varying Ratios of Protein from Dehulled Soybean Meal and Corn[a,b]

Percent of dietary protein		Dietary protein level					
		8.5		11.0		13.5	
SBM[c]	Corn[d]	Gain (g)	Gain/feed	Gain (g)	Gain/feed	Gain (g)	Gain/feed
90	10	128	0.27	206	0.35	269	0.42
80	20	143	0.29	226	0.36	288	0.42
70	30	181	0.32	245	0.38	302	0.44
60	40	168	0.31	228	0.38	304	0.44
50	50	145	0.29	224	0.37	232	0.42

[a] Data from Bray (1962).
[b] Each value represents the mean of 30 chicks fed the experimental diets from days 7–28 posthatching; pooled standard error was 5.1 for gain and 0.005 for gain/feed.
[c] High in lysine and tryptophan; low in sulfur amino acids.
[d] High in sulfur amino acids; low in lysine and tryptophan.

20% at 4 weeks posthatching) is sulfur amino acids. The same situation exists with young turkeys. As little as 0.05% DL-methionine generally corrects the deficiency in broilers, but as much as 0.20% may be needed for poults. In broilers, if protein level is reduced by replacing soybean meal with grain, lysine becomes limiting when it approaches 18% protein for early chick growth, or 15% protein for later growth (4–9 weeks posthatching).

Needless to say, with the multitude of factors that affect requirements it is almost impossible to make general recommendations for universal applications. Therefore, the requirements set forth by the National Research Council (1971) in America and the Agriculture Research Council (1963) in Great Britain are only guidelines, subject to change, should any of the factors discussed above change from a given norm.

4.   Sulfur Amino Acids

Avian species require relatively large intakes of sulfur in various forms to maximize productive functions. The sulfur-bearing amino acid (SAA) cysteine (50% cystine) is the first-limiting amino acid in most purified as well as practical diets for birds (Baker and Bray, 1972). But because of economic considerations, either DL-methionine or its hydroxy analogue is generally used to furnish the limiting nutrient, cysteine. In recent years it has been alleged that provision of sulfur in inorganic form will spare the requirement for SAA. Consequently, interest ranging from the most basic to the level of practical application continues in the preparation of sulfur-fortified diets for avian species.

4.1.   Methionine–Cysteine Interrelationship

The molecular weights of methionine and cysteine are, respectively, 149 and 121. Methionine can be converted to cysteine but the reverse reaction does not take place, presumably because the cystathionine synthase-catalyzed reaction cannot be reversed (Fig. 2). Indeed, Rose and Rice (1939) observed methionine activity from homocysteine, but not cysteine, in young rats almost 40 years ago. Methionine can therefore serve as a source of cysteine, but because of the molecular weight difference it would be expected that the efficiency of methionine in meeting a biological need for cysteine on a weight or concentration basis would be on the order of 80%. Of course, this would assume a molar efficiency of 100%, an assumption that seems to be borne out in practice (Graber and Baker, 1971a). The oxidation product, cystine, would be expected to function similarly to cysteine.*

---

*The reaction between cysteine and cystine is freely reversible. There is neither a dietary nor a physiological requirement for cystine. Cystine is not directly incorporated into proteins but instead is formed when two cysteine molecules are joined together in disulfide linkage within the same peptide chain or between two different amino acid chains.

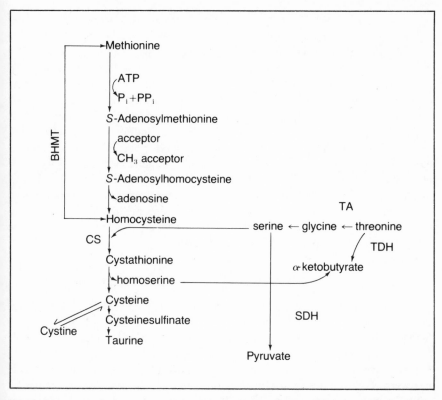

Fig. 2. Scheme of methionine metabolism. Abbreviations: BHMT, betaine-homocysteine methyl-transferase; TA, threonine aldolase; TDH, threonine dehydratase; CS, cystathionine synthase; SDH, serine dehydratase.

Work in our laboratory (Graber and Baker, 1971a) has confirmed the theoretical calculations (Creek, 1968) to the effect that about 1.25 mg methionine is equivalent to 1.0 mg cysteine or cystine in meeting the dietary need for cysteine, when cysteine is the first-limiting dietary factor. Subsequent work (Sowers et al., 1972) with the rat has confirmed this result. Thus, 2 mol methionine yield 2 mol cysteine, the latter of which can condense to give 1 mol cystine.

From these considerations, it is clear that a dietary requirement for SAA expressed in terms of either milligrams or concentration (i.e., percent of diet or percent of calories) would be lower when a proper combination of methionine and cystine is fed than when methionine alone is used to meet the requirement. Again, this has been demonstrated conclusively in practice (Graber and Baker, 1971a). Cystine can safely provide approximately 50% of the SAA requirement for growth, not only of chicks (Graber et al., 1971b; Sasse and Baker, 1974b; Soares, 1974), but of rats (Rose and Rice, 1939) and pigs (Baker et al., 1969) as well.* It would indeed be a rare occasion if cystine were to occupy a greater percentage of the dietary SAA than this in practice. Only in situations where keratoid proteins contributed significantly to the total protein intake would intake of cystine exceed that of methionine. Although not of practical significance, but of great importance from a procedural standpoint in other assays involving SAA, cystine can apparently provide no more than 40% of the requirement for SAA when growth rate is restricted to about 50% of maximum (Sasse and Baker, 1974b). This result, plus the seemingly obvious finding that a methionine-cystine combination is more efficacious weight for weight than methionine alone, helps explain several disparities in the literature with regard to sulfur utilization. To cite just one example of recent interest (Sasse and Baker, 1973b), availabilities of SAA in various intact proteins approaching, or sometimes even exceeding, 100% have generally resulted from using methionine alone to construct a standard curve, even though the SAA provided in the unknown (i.e., the intact protein) consists of both methionine and cystine.

4.2. Methionine Toxicity

Consumption of excess methionine has been studied extensively. The underlying mechanism by which methionine exerts its pathogenicity and causes a variety of biochemical lesions, however, is not well understood. Hardwick *et al.* (1970) proposed depletion of hepatic ATP in adenylating methionine to S-adenosylmethionine as an explanation for its toxicity. These workers were able to reverse both the depression in liver ATP and an attendant hypoglycemia by administering supplemental adenine. It should be noted, however, that guinea pigs (very sensitive to methionine toxicity) were used with no growth data reported. Work in our laboratory (Katz and Baker, 1975b) has likewise shown hypoglycemia and reduced hepatic ATP in chicks fed excess methionine, but supplemental adenine had no efficacy whatever in ameliorating the growth depression, although plasma glucose levels were brought back to normal.

Depletion of methyl acceptors in the conversion of S-adenosyl-methionine to S-adenosylhomocysteine (see Fig. 2) has also been investigated as a causative factor in methionine toxicity. This explanation has proved untenable, however, as little benefit has been observed from supplementing methionine-imbalanced diets with methyl-accepting compounds, such as guanidoacetic acid (Cohen *et al.*, 1958; Sauberlich, 1961) or niacinamide (DeBey *et al.*, 1952; Klain *et al.*, 1963). More recently the metabolism of the labile methyl group of methionine has been proposed as a means by which methionine toxicity is exerted (Benevenga, 1974). An alternate pathway competitively inhibited by S-methyl-L-cysteine may account for a sizable portion of the methionine

---

*Inorganic sulfate (cf. Section 4.3), taurine (Anderson *et al.*, 1975), and glutathione (Dyer and DuVigneaud, 1936) have all been shown capable of exerting a partial sparing effect on the physiological requirement for cysteine.

catabolized when high levels of methionine are fed. One can postulate that an intermediate in this pathway may cause the toxicity.

Work in our laboratory (Katz and Baker, 1975a,b) points to a buildup in plasma and tissue of methionine or homocysteine, or both, as the most likely means by which methionine toxicity is exerted. Although Benevenga and Harper (1967) found homocysteine less toxic than an equimolar concentration of methionine for rats, earlier work with rats (Cohen *et al.*, 1958) and our work with chicks (Table IV) clearly shows homocysteine to be as toxic as methionine. Unpublished data on L-homocysteine from our laboratory indicates only minimal conversion back to methionine (Table V). Therefore, homocysteine toxicity probably is not attributable to its being methylated back to methionine, but rather to its accumulation *per se* or to accumulation of cystathionine, or perhaps even homoserine. Cysteine, in contrast to homocysteine, is considerably less toxic than methionine. The hydroxy analogue of methionine is likewise considerably less toxic than methionine. That glycine becomes first-limiting in a methionine-induced toxicity would seem to implicate the cystathionine synthase-catalyzed reaction, which leads to cystathionine formation from homocysteine as the rate-limiting reaction in methionine catabolism (Baker, 1976a). There is recent biochemical evidence to support this hypothesis (Finkelstein *et al.*, 1974). Glycine is more effective than serine in reversing methionine toxicity (Benevenga and Harper, 1970; Katz and Baker, 1975a), even though serine *per se* is required in the reaction. This probably results from the fact that serine, when

Table IV. Relative Toxicity of Various Organic
Sulfur Compounds for Chick Growth[a]

Addition[b] to basal diet[c]	%	Gain (g)	Gain/feed
None	—	73.9	0.69
DL-methionine	1.25	38.6	0.55
L-methionine	1.25	41.8	0.57
D-methionine	1.25	44.6	0.59
OH-methionine (Ca)	1.42	66.1	0.66
OH-methionine (Ca)	2.84	35.9	0.57
L-cystine	1.01	69.1	0.69
L-cysteine · HCl · H$_2$O	1.47	68.2	0.70
DL-homocysteine	1.13	42.0	0.59
DL-homocystine	1.12	38.6	0.58
Pooled standard error		3.9	0.02

[a] Average of duplicate groups of 8 female chicks for the period 8–14 days posthatching; average initial weight was 79.1 g (Katz and Baker, 1975a).
[b] All sulfur-containing compounds were provided at isosulfurous concentrations; the 2.84% OH-methionine (Ca) addition was provided at twice that level.
[c] Complete crystalline amino acid diet (Sasse and Baker, 1973a) with both glycine and threonine set at recently established requirements for maximal gain of 0.60 and 0.55%, respectively.

Table V. Efficacy of D- and L-Homocysteine as a Source of Either Methionine or Cysteine for Young Chicks Fed a Crystalline Amino Acid Diet[a]

Dietary source and level of sulfur amino acids				Results[e]	
L-methionine	L-cystine[b]	L-homocysteine	D-homocysteine	Gain	Gain/feed
0.30	—	—	—	28.5	0.33
0.30	0.25	—	—	100.8	0.66
—	0.25	0.27[c]	—	23.6	0.36
—	0.25	—	0.27[c]	5.3	0.12
0.30	—	0.28[d]	—	101.5	0.69
0.30	—	—	0.28[d]	75.1	0.60
Pooled standard error				3.9	0.02

[a]Homocysteine activity provided as homocysteine thiolactone hydrochloride, a compound shown in our laboratory to be as active biologically as homocysteine.
[b]A methionine-free diet containing cystine alone as a source of sulfur amino acids was deemed an unnecessary control, as this treatment has been shown on numerous occasions in our laboratory to result in negative growth.
[c]Isosulfurous to 0.30% methionine.
[d]Isosulfurous to 0.25% cystine.
[e]Mean of triplicate groups of 7 male chicks during the period 8–16 days posthatching; average initial weight was 62 g.

fed in excess of its requirement for protein synthesis, stimulates serine-threonine dehydratase activity (Girard-Globa *et al.*, 1972) to such an extent that a secondary deficiency of threonine might be caused. Methionine, too, induces serine-threonine dehydratase (Girard-Globa *et al.*, 1972; Sanchez and Swendseid, 1969), and this likely explains the growth response to glycine as well as threonine when they are added to a methionine-imbalanced diet.* Threonine oxidation is enhanced markedly when excess methionine is fed to chicks (Katz and Baker, 1975a). Because a significant portion of the threonine catabolized in the chick is converted to glycine *via* the threonine aldolase pathway (Baker *et al.*, 1972), it is conceivable that some of the response to threonine may be attributable to the fact that it furnishes glycine.

The growth depression that results from ingestion of excess methionine is largely caused by a reduction in voluntary food intake. However, even when food intake is equalized, weight gain is less in birds fed 1.25% excess methionine than in those fed a methionine-adequate diet. Additional glycine + threonine restores metabolic efficiency (gain/feed ratio), but not food intake.

An alternate ATP-independent pathway of methionine catabolism appears to become operative in the chick when excesses approximating three times the requirement are fed (Harter and Baker, 1977). A metabolite in this alternate pathway (possibly 3-methylthiopropionate) appears to be responsible for a marked hemosiderosis of the spleen. Thus, excess methionine or 3-methylthiopropionate results in decreased blood hemoglobin and increased iron deposition in the spleen.

*Homocysteine would probably also induce serine–threonine dehydratase activity, but to our knowledge this has not been tried.

Excesses of metabolites in the normal ATP-dependent pathway of methionine metabolism (e.g., homocysteine, homoserine, cysteine) have no effect whatever on the spleen lesion.

4.3. Inorganic Sulfate

Recent work at the University of Florida (Ross and Harms, 1970; Hinton and Harms, 1972; Ross *et al.*, 1972; Sloan and Harms, 1972) and elsewhere (Martin, 1972; Sasse and Baker, 1974a) has confirmed the earlier work of Gordon and Sizer (1955) and Machlin and Pearson (1956), which showed that inorganic sulfate can, under certain conditions, stimulate growth of broiler chicks. Radiochemical studies have clearly established $^{35}SO_4$ incorporation into body cystine or taurine, but not into methionine (Huovinen and Gustafsson, 1967; Mason *et al.*, 1965; Martin, 1972). Work in our laboratory (Sasse and Baker, 1974a,b) as well as at Maryland (Soares, 1974) has suggested that up to 15% of the cystine requirement of chicks can be spared by inorganic sulfur provided either as K_2SO_4 or Na_2SO_4. This is probably more a function of sulfate sulfur being incorporated into sulfur-containing compounds for which cysteine may serve as a precursor than of sulfate sulfur incorporation into cysteine itself. It is well established, for example, that sulfate–sulfur can be used in synthesizing taurine as well as the sulfate esters, chondroitin and keratin. Conceivably other sulfur compounds, such as heparin, fibrinogen, sulfolipids, sterol sulfates, and phenol sulfates, could be involved as well.

There is little question from data collected in our laboratory that SO_4 spares cystine rather than methionine and that SO_4 *per se* is not an essential dietary ingredient (Table VI). Chicks were fed purified crystalline amino acid diets devoid of inorganic sulfur and containing SAA at either an adequate level (diet 1) or a deficient level (diets 2, 3, and 4). Deionized water was fed throughout the assays. Chicks fed diets adequate in SAA, deficient in methionine (diet 2), or equally deficient in both methionine and cystine (diet 4) did not respond to K_2SO_4 supplementation. Those fed a cystine-deficient diet, however, responded to K_2SO_4 with more than a 40% response in weight gain. That the response was caused by the SO_4 portion of the molecule *per se* is evidenced by the fact that similar responses have been shown to Na_2SO_4 and other sulfur compounds by several investigators. Most of the increment in gain resulting from SO_4 supplementation appears to result from an increase in voluntary food intake (Sasse and Baker, 1974b).

The model diet selected to give a consistent growth response to SO_4 supplementation contained 0.25% DL-methionine and 0.15% L-cystine. Thus, SAAs were limiting and cystine was more limiting than methionine, both of these conditions being necessary in order to demonstrate a response to inorganic SO_4 as shown in Table VI. The minimal level of K_2SO_4 necessary to give a maximal growth response in chicks fed the model diet was found to be 0.10% (0.055%

Table VI. Effect of Source and Level of Dietary Sulfur Amino Acids on Weight Gain of Chicks Fed Inorganic Sulfate[a]

Dietary DL-methionine		Dietary L-cystine	Gain (g)	
No.	%		0	0.50% K_2SO_4
1.	0.30	0.30	170	170
2.	0.10	0.30	8	9
3.	0.30	0.10	94	135
4.	0.20	0.20	114	101
Pooled standard error			3.15	

[a] Average of triplicate groups of 7 male chicks for the period 8–20 days posthatching; average initial weight was 74 g; data from Sasse and Baker (1974a). The basal diet was a crystalline amino acid diet devoid of inorganic sulfate; only the levels of methionine, cystine, and K_2SO_4 were varied.

SO_4), as indicated in Table VII. Diets adequate in methionine and marginal in cystine (a condition similar to that in practical poultry rations) demonstrate as little as 150–200 ppm SO_4 to maximize growth (Sasse and Baker, 1974a,b; Anderson *et al.*, 1975). In practice, poultry generally consume considerably more than this, particularly in areas in which drinking water is high in SO_4 and where $MnSO_4$ is used to provide supplemental manganese (dicalcium phosphate and fish solubles are also rich in SO_4). Hence, a response to supplemental SO_4, in practice, would seem unlikely (Sasse and Baker, 1974a; Anderson *et al.*, 1975).

Because inorganic sulfate indirectly spares cystine, it would seem logical that presence or absence of dietary SO_4 would affect the degree to which cystine

Table VII. Response of Chicks Fed Graded Levels of K_2SO_4[a]

K_2SO_4 %	Gain (g)	Gain/feed
0.00	86	0.43
0.02	107	0.47
0.04	116	0.47
0.06	128	0.46
0.10	141	0.50
0.20	141	0.49
Pooled standard error	5.4	0.01

[a] Average of triplicate groups of 7 male chicks for the period 8–21 days posthatching; average initial weight was 69 g; data from Sasse and Baker (1974a). The basal crystalline amino acid diet was devoid of SO_4 and contained 0.25% DL-methionine and 0.15% L-cystine.

can spare dietary methionine. Work in our laboratory has shown this to be the case, cystine being capable of furnishing about 50% of the SAA requirement in the presence of SO_4 and about 55% in its absence. It would appear that much of the confusion in the literature with regard to SO_4 efficacy for poultry stems from not knowing (a) the level of contaminating SO_4 in water supplies and basal diets, and (b) the SAA level, and more particularly the methionine and cystine levels in basal diets employed. Hence, there would seem little question that simple-stomached animals can utilize inorganic sulfur for growth and development, but only when certain dietary prerequisites are met. As pointed out previously, these prerequisites are seldom met in practice.

4.4. Utilization of Methionine Hydroxy Analogue

The biological efficacy of the hydroxy analogue of DL-methionine (OH-M) as a source of SAA (see Fig. 3) in practical diets for poultry is well established (Bird, 1952; Gordon et al., 1954). Its metabolism, however, is less well understood. When purified or semipurified diets have been fed, most workers have reported biological activities lower than that for DL-methionine (Tipton et al., 1965, 1966; Smith, 1966; Baker, 1974; Featherston and Horn, 1974; Katz and Baker, 1975c), although some have reported equal efficacy on an isosulfurous basis (Romoser et al., 1976). Early work suggested that the branched-chain amino acids were the principal source of nitrogen in transaminating OH-M to methionine, but recent work from Purdue (Featherston and Horn, 1974) and Illinois (Katz and Baker, 1975c) casts doubt on this. In fact, work with the rat points to the conclusion that glutamine is the most active amino donor in the conversion (Langer, 1965).

DL-OH-M → Keto methionine $\xrightarrow{\text{glutamine}}$ L-methionine → L-cysteine

It would appear from this scheme that OH-M could serve equally well as a source of either methionine or cysteine. However, OH-M must be converted to methionine before it can serve as a source of cysteine. Because of the molecular weight difference between methionine and cysteine that was discussed pre-

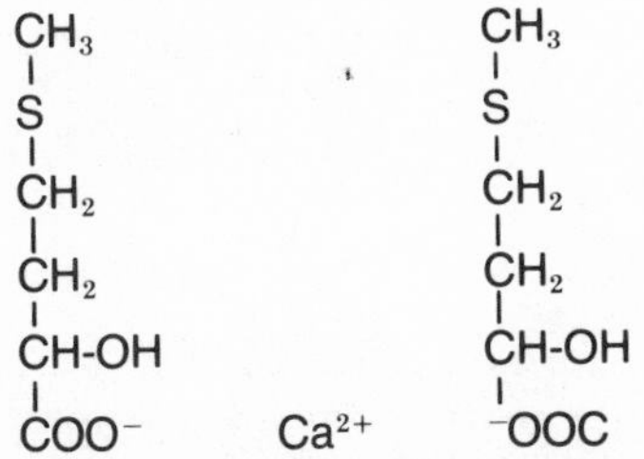

```
CH₃                      CH₃
 |                        |
 S                        S
 |                        |
CH₂                      CH₂
 |                        |
CH₂                      CH₂
 |                        |
CH-OH                    CH-OH
 |                        |
COO⁻        Ca²⁺        ⁻OOC
```

Fig. 3. Methionine hydroxy analogue, calcium. Pure OH-M (Ca) would contain 88.15% OH-M.

viously, one would expect OH-M to be more efficient in furnishing methionine than in furnishing cysteine. Data collected in our laboratory using the Modified Illinois Reference Standard amino acid diet, however, would seemingly make this supposition questionable (Table VIII).

Strikingly different results were obtained concerning apparent OH-M efficacy, depending on the ratio of methionine to cystine in the diet. Addition of 0.30% cystine to the SAA-deficient basal diet (diet 1) resulted in performance superior to that which occurred when an equimolar quantity of OH-M was added to the same diet. Moreover, the methionine–OH-M combination promoted growth far superior to the cystine–OH-M combination or to OH-M alone. Thus, without precisely understanding the methionine–cystine interrelationship, one would be tempted to conclude that OH-M is more efficacious as a source of cystine than of methionine.

One possible explanation might be that the sulfur in OH-M is used to furnish inorganic sulfate, in addition to being converted to one of the SAA. That this is highly unlikely is indicated in our study by the fact that the diets used for the assay in question contained 1740 ppm sulfate from inorganic salts, a level far in excess of the levels suggested by Sasse and Baker (1974a) as being necessary for maximal response. Moreover, other data collected in our laboratory have shown that neither supplemental K_2SO_4, branched-chain amino acids, glutamine, nor glutamic acid enhances OH-M efficacy of any of the diets shown in Table VIII.

A more tenable explanation can be offered to explain the results shown in Table VIII. Work in our laboratory (Baker, 1974; Katz and Baker, 1975c), as well as from Mississippi State (Tipton et al., 1965, 1966), would suggest a biopotency of OH-M (Ca) of close to 70% relative to DL-methionine. Thus, when 0.34% OH-M (Ca) replaces 0.30% cystine, the diet contains 0.30% DL-methionine and 0.24% methionine activity from OH-M (Ca), i.e., 70% of 0.34, or a total of 0.54% usable SAA. In contrast, if methionine is replaced by OH-M

Table VIII. Performance of Chicks Fed Crystalline Amino Acid Diets Containing Various Sources of Sulfur-Bearing Amino Acids[a]

	Dietary level of SAA				
Diet No.	DL-methionine	L-cystine	OH-M (Ca)[b]	Gain (g)	Gain/feed
1.	0.30	—	—	47	0.32
2.	0.30	0.30	—	151	0.62
3.	0.30	—	0.34	131	0.57
4.	—	0.30	0.34	93	0.51
5.	—	—	0.68	89	0.46
Pooled standard error				8.8	0.02

[a] Average of triplicate groups of 7 male chicks for the experimental period 8–21 days posthatching; average initial weight was 71 g. Data from Katz and Baker (1974c).
[b] 0.34% OH-M (Ca) is isosulfurous to 0.30% methionine.

(Ca), the diet then contains 0.24% usable methionine activity from OH-M (Ca) and 0.30% cystine. The 0.30% cystine, however, can only partially be used, because the sparing value of cystine is reduced from 50% to 40% when deficient levels of SAA are fed (Sasse and Baker, 1974b). Thus, only 0.16% of the cystine may be utilized in fulfilling the requirement for SAA, and this results in a total of only 0.40% usable SAA in the diet. Hence, a thorough understanding of the quantitative aspects of the methionine–cystine interrelationship is necessary to arrive at a proper interpretation of data such as those presented in Table VIII.

Perhaps the principal reason why so much confusion exists in the literature with regard to OH-M efficacy for growth is that exact levels of both methionine and cystine in the semipurified and practical diets that have been employed cannot be accurately determined. Certainly, the well-known problems in accurately determining methionine and cystine in intact proteins raise serious interpretive questions. Use of a crystalline amino acid diet avoids these problems. The good agreement concerning OH-M efficacy that exists among workers who have used crystalline amino acid diets (Katz and Baker, 1955c; Smith, 1965; Featherston and Horn, 1974) is reflected in data reported by Smith as shown in Table IX.

The lower molar efficacy of OH-M relative to methionine could result either from incomplete absorption from the gut or from inefficient conversion of OH-M to methionine (Baker, 1976b). Little information is available to support or reject either of these possible causes, although the fact that OH-M is considerably less toxic than an equimolar quantity of methionine (Table IV) would suggest that one or both of them could be involved.

Table IX. Relative Efficacy of DL-Methionine and Its Hydroxy Analogue When Added to a Crystalline Amino Acid Diet[a]

	Source and level of sulfur AA[b]	Gain	
Diet No.	Content	(g)	Gain/feed
1.	Basal A[c] + 0.05% DL-methionine	70	0.62
2.	Basal A + 0.06% OH-M (Ca)	60	0.61
3.	Basal A + 0.10% DL-methionine	95	0.75
4.	Basal A + 0.12% OH-M (Ca)	90	0.71
5.	Basal B[d] + 0.30% OH-M (Ca)	35	0.45
6.	Basal B + 0.36% OH-M (Ca)	59	0.58

[a] Average of triplicate groups of 3 male chicks for the period 8–14 days posthatching; data from Smith (1965).
[b] Isosulfurous quantities of each compound were used within a pair; the OH-M (Ca) used in this assay contained 83% OH-M.
[c] Contained 0.20% L-methionine and 0.35% L-cystine.
[d] Contained 0.35% L-cystine and no methionine.

Methionine appears to be absorbed from the gut over twice as rapidly as OH-M in the chicken (Lerner *et al.,* 1969), and this could influence OH-M utilization when purified diets are fed. Hence, with crystalline amino acid diets where the amino acids are rapidly absorbed into the bloodstream, methionine, as is obvious from Tables VIII and IX, is considerably more efficacious than is OH-M. With diets based on intact protein, however, the advantage of DL-methionine over an isosulfurous quantity of the DL-isomer of OH-M is only slight (Baker, 1975). Our recent work with chicks fed a crystalline amino acid diet shows that the calcium salt of the keto analogue of methionine is less efficacious than DL-methionine, but more efficacious than an isosulfurous quantity of the calcium salt of DL-OH-M. This implies that the conversion of DL-OH-M to keto methionine is not 100% efficient, perhaps because the D-hydroxy analogue is not efficiently converted to the keto compound. Indeed, recent work with the chick (Katz and Baker, 1975d) supports the view that D-methionine, which must also go through a keto derivative to be converted to L-methionine, is less efficacious than L-methionine. Because OH-M is an important commercial product used extensively in foods for both animals and humans, more work is needed to delineate both the mechanism and extent of absorption and any possible alternate routes of metabolism.

4.5. Methionine as a Methyl Donor and Its Relationship to Choline

Growing birds have a high dietary requirement for choline that cannot be spared by high levels of dietary methionine or other methyl donors (Jukes, 1940, 1941; Latshaw and Jensen, 1972; Kushwaha and Jensen, 1974, 1975). The high requirement for choline in the chick decreases with age until at 8 weeks post-hatching a dietary choline requirement is difficult to show (Nesheim *et al.,* 1971). High-producing laying hens, too, apparently synthesize choline in sufficient quantity to meet their requirement for egg production (Lucas *et al.,* 1946; Nesheim *et al.,* 1971; Ringrose and Davis, 1946). In young mammalian species, such as the rat (Treadwell, 1948) and pig (Nesheim and Johnson, 1950), methionine is capable of completely replacing choline. Thus, young avian species, which lack the capacity to actively synthesize choline *in vivo,** differ considerably from young mammalian species, which actively synthesize choline from such precursors as serine and aminoethanol.

Recent work from Cornell (Ketola and Nesheim, 1974) has suggested not only that methionine does not spare choline, but that in excess it may actually increase the dietary requirement for choline. Excess protein was observed to have the same effect. Critical examination of this finding in our laboratory with a

---

*The rate-limiting step in choline biosynthesis in the chick appears to be the first methylation step to involve methylation of aminoethanol.

crystalline amino acid diet indicated a choline requirement for maximal weight gain* in 2-week-old chicks of 601 ppm for the complete crystalline amino acid diet (100% of amino acid requirements), 874 ppm for a 2× level of this diet (200% of amino acid requirements), 541 ppm for a 2× level of methionine (twice the minimal requirement for maximal growth), and 639 ppm for a 3× level of methionine (Molitoris and Baker, 1976a). Hence, our work with excess methionine does not support the Cornell work, as our studies show that excess dietary methionine had no effect whatever on the dietary need for choline. In contrast, and in agreement with their work, excess protein (i.e., amino acids) increased the choline requirement markedly. That proper understanding of the choline–methionine interrelationship is important should be obvious when dealing with poultry. Both nutrients are routinely added to practical diets. Moreover, the recent finding that choline availability in soybean meal may be as low as 60% places additional stress on this nutrient (Molitoris and Baker, 1976b).

Chicks in the 7th week of life also require dietary choline, and neither methionine nor betaine can spare this need (Nesheim *et al.*, 1971; Molitoris and Baker, 1976c). A dietary requirement of 466 ppm for birds of this age was recently established in our laboratory. Excess methionine neither increased nor decreased the requirement. Beyond 8 weeks of age, chickens no longer require a dietary source of choline, but Japanese quail continue to exhibit a dietary choline need throughout adult life (Kushwaha and Jensen, 1974, 1975).

Apart from providing labile methyl groups for choline biosynthesis, methionine is intimately involved in transmethylation reactions. It is well established, for example, that methionine provides labile methyl groups for biosynthesis of ergosterol, carnitine, epinephrine, creatine, and vitamin B_{12} in addition to choline. This does not mean, however, that S-adenosylmethionine is necessarily *the* methylating agent in the synthesis of these compounds. It may mean instead only that methionine can serve as the origin of the one-carbon unit. Nonetheless, methionine is an active contributor to the labile methyl pool, as is choline (*via* betaine); this makes it surprising that methionine cannot at least partially spare choline in avian species.

5. Lysine

In practice, lysine is the second-limiting amino acid in broiler diets. Thus, in a typical 23% protein chick starter diet, cystine is first-limiting and lysine second-limiting. In circumstances wherein crystalline lysine is less expensive than the lysine of soybean meal or of a soybean meal–fishmeal combination, it

*Although choline requirements for poultry tend to be somewhat higher for prevention of perosis than for achievement of maximal weight gain, the former is very subjective and is unsuitable for quantitative measurement.

may be economically feasible to lower the protein level of the diet from 23% to 18 or 20% and supplement with crystalline lysine. As is indicated in Table III, lysine becomes relatively more deficient and sulfur amino acids relatively less deficient as the ratio of protein from corn and soybean meal is widened.

Lysine is the first-limiting amino acid in feed grains and in most oilseed meals, as compared with soybean meal, fishmeal, and single-cell protein (i.e., yeast), which are very rich in lysine. In recent years, intensive efforts have been directed toward selection of high-lysine varieties of feed grains. A negative correlation exists between protein quantity and protein quality (i.e., practically speaking—lysine and tryptophan concentration as a percent of protein) in most feed grains. This has been clearly shown for corn (Sauberlich *et al.*, 1953; Hogan *et al.*, 1955), sorghum grain (Vavich *et al.*, 1959), barley (Munck *et al.*, 1970), and wheat (Lawrence *et al.*, 1958). Oats, in contrast, appear different in this respect in that protein quality apparently remains relatively constant as protein content of the grain increases (Robbins *et al.*, 1971; Maruyama *et al.*, 1975).

Opaque-2 corn was developed in the United States as a new variety of corn, containing more lysine and tryptophan, but somewhat less methionine (Mertz *et al.*, 1964). Because of the lower content of available sulfur amino acids, its use in poultry rations does not appear promising (Cromwell *et al.*, 1968; Chi and Speers, 1973). Dal Oats and other high-protein oat varieties, however, have performed very well in poultry rations (Maruyama *et al.*, 1975). Likewise, high-lysine barley (Munck *et al.*, 1970) and high-lysine sorghum grain (Featherston *et al.*, 1975), both of Ethiopian origin, appear to have much promise for poultry feeding. Sulfur amino acids, as well as lysine, are increased in these three newly developed cereal grains. These developments, although important, are not as yet practical, mainly because of agronomic problems. But for countries in which climate and soil conditions are unsuitable for soybean production, high-protein cereal grains also high in both lysine and methionine offer great future potential for feeding meat-producing animals as well as humans.

6.   Arginine

Unlike most mammalian species who actively synthesize arginine *in vivo*,* avian species do not possess a mitochondrial source of carbamyl phosphate synthetase and thus do not synthesize arginine. Consequently, the dietary need for arginine in birds is about twice that of mammals. In addition to the factors affecting amino acid requirements discussed in Section 3, lysine in excess of its requirement antagonizes arginine such that the arginine requirement is increased markedly when excess lysine is fed. In fact, D'Mello and Lewis (1970a) observed a linear increase in the arginine requirement of the chick as the level of

*In contrast to other mammals, rabbits and feline species apparently synthesize only limited quantities of arginine.

lysine in the diet was increased from 1.10 to 1.85%. Allen and Baker (1972a) likewise reported an increase in the arginine requirement of almost 50% when the dietary lysine level was increased to twice its required level. The lysine–arginine antagonism goes a long way toward explaining why casein-containing diets (high lysine in relation to arginine) for poultry necessitate such high levels of arginine supplementation to achieve optimal performance.

Excess lysine exerts its effects through at least three metabolic mechanisms: induction of renal arginase activity, thus increasing arginine catabolism (Jones *et al.*, 1967; Austic and Nesheim, 1970), depression of hepatic glycine transamidinase activity (Jones *et al.*, 1967; Austic and Nesheim, 1971) and competition of lysine and arginine for reabsorption at the kidney tubule (Jones *et al.*, 1967; Nesheim, 1968; Boorman, 1971). As a result of any one or a combination of these biochemical lesions, food intake is reduced. However, reduced arginine utilization *per se* independent of depressed appetite accounts for some of the growth depression (Allen and Baker, 1972a; Allen *et al.*, 1972; Austic and Scott, 1975).

It is probable that with levels up to 0.75% excess lysine, increased kidney arginase activity and decreased liver transamidinase activity may account for virtually all the manifestations of excess lysine (Austic and Scott, 1975). Both supplemental arginine and creatine have been found efficacious in overcoming the growth depression in this range of excess lysine (Fisher *et al.*, 1956; Jones *et al.*, 1967; Austic and Nesheim, 1971; Allen *et al.*, 1972). With excessive levels of lysine, urinary losses of arginine would become important. Also, food intake depression *per se* (independent of metabolic aberrations involving arginine) caused by excessive intake of lysine would come into play at lysine intakes greatly in excess of the requirement.

Arginine has been shown capable of partially reversing growth depressions caused by excess amino acids other than lysine. Smith (1968), for example, observed responses to supplemental arginine in diets imbalanced with excess lysine, tyrosine, histidine, glycine, cystine, valine, threonine, isoleucine, or phenylalanine, but not with excess leucine, tryptophan or methionine. The mechanism by which arginine functions (uniquely) in the detoxification of amino acid excesses is not known, although stimulation of voluntary food intake is undoubtedly involved. Also, several amino acids in addition to lysine are known to induce kidney arginase activity (Austic and Nesheim, 1970).

7. Tryptophan

Of all the indispensable amino acids required for chick growth, tryptophan is required in smallest quantities. In contrast to rats and swine, but paralleling results with poults, mice, and humans, the D-isomer of this amino acid is very poorly utilized by chicks (Morrison *et al.*, 1956). The requirement for

L-tryptophan appears to be in the range of 0.78–1.0% of the crude protein at all ages between hatching and 9 weeks posthatching (Boomgaardt and Baker, 1971 NRC, 1971; Hunchar and Thomas, 1975; Woodham and Deans, 1975).

Tryptophan in excess of its requirement for protein synthesis is converted to a variety of metabolic end products including niacin. Work from our laboratory has indicated an efficiency of niacin biosynthesis of 45 : 1 (i.e., 2.2%) on a weight basis (Fig. 4) or 27 : 1 on a molar basis (Baker *et al.*, 1973). Recent suggestions that excess leucine impairs the conversion of tryptophan to niacin in mammals (Gopalan, 1970; Gopalan and Srikantis, 1960) have not been reproducable with the chick. Allen *et al.* (1971) fed 3% excess leucine to chicks and observed no effect on either tryptophan or niacin utilization, nor on the conversion efficiency of tryptophan to niacin.

8. Leucine, Isoleucine, and Valine

Practical diets for chicks are rarely, if ever, deficient in leucine, isoleucine, or valine. However, much has been written about the branched-chain amino acids for poultry because of the so-called antagonism that exists among them (Mathieu and Scott, 1968; D'Mello and Lewis, 1970b; Allen and Baker, 1972b;

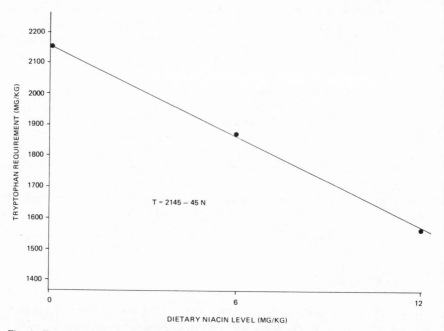

Fig. 4. Tryptophan–niacin interrelationship in young chicks. The regression of the dietary tryptophan requirement (T) on dietary niacin concentration (N) was established by determining the tryptophan requirement for maximal growth at each of three deficient levels of dietary niacin (cf. Baker *et al.*, 1973). $T = 2145\text{-}45\,N$.

Baker, 1974; D'Mello, 1974). With purified diets it has been clearly shown, for example, that excess leucine impairs isoleucine and valine utilization, and excess isoleucine impairs leucine and valine utilization. Valine in excess of its requirement does not impair utilization of either isoleucine or leucine (Allen and Baker, 1972b).

In terms of effects on dietary requirements, caution must be exercised in extrapolation, as the food intake depression resulting from each excess amino acid is variably reversible upon supplementing the imbalanced diet with the particular amino acid antagonized by the excess amino acid in question. Thus, the food intake depression that results from excess leucine can more nearly be overcome than can the depression resulting from excess isoleucine or valine. Hence, 3% excess leucine has little effect on the isoleucine requirement but increases the valine requirement, whether expressed in terms of dietary concentration or absolute intake. A similar excess of isoleucine increases the leucine requirement expressed as a percent of the diet but decreases it expressed as milligrams per day. The valine requirement under conditions of excess isoleucine is unaltered expressed in terms of dietary concentration, but decreases substantially (owing to reduced food intake) when evaluated in terms of absolute intake. Excess valine has very little effect on the dietary requirement (percent of diet) for either leucine or isoleucine, but lesser absolute quantities of both are required. In studies with practical diets, moderate excesses of any one of these amino acids has had very little effect on chick performance (Woodham and Deans, 1975).

In studies of branched-chain amino acid deficiency, it has been observed that the pattern of excess amino acids over and above an isoleucine or valine deficiency depresses voluntary food intake to a greater extent than the pattern of amino acids in excess of an equal deficiency (relative to the requirement) of leucine (Netke et al., 1969; Sugahara et al., 1969). Tryptophan and phenylalanine deficiency produced results similar to isoleucine and valine in that growth depressions from deficiencies of these amino acids caused greater food intake depressions than that occurring when all amino acids were reduced to the same level of deficiency relative to chick requirement (Sugahara et al., 1969).

The unnatural D-isomer of the branched-chain amino acids is utilized to different extents by the chick. Thus, D-leucine is well utilized, D-isoleucine is poorly utilized, and D-valine is intermediate in efficacy (Sugahara et al., 1967).

9. Glycine–Serine

Glycine and serine must be considered together because either amino acid can meet the dietary requirement (Akrabawi and Kratzer, 1968; Baker et al., 1968). Although chicks can synthesize both glycine* and serine, as much as 40%

*The four principal sources of endogenously derived glycine are (a) serine ↔ glycine, (b) choline → betaine → dimethylglycine → sarcosine → glycine, (c) phosphotidyl serine → phosphtidyl amino ethanol → aminoethanol → glycolaldehyde → glycolate → glyoxylate → glycine and (d) threonine → glycine (Meister, 1960; Weisbach and Sprinson, 1953a,b).

of the total physiological requirement of young chicks must be provided orally if maximal growth is to be achieved (Graber and Baker, 1973). Moreover, broiler chicks require a dietary source of one or the other of these amino acids at all ages between hatching and 9 weeks posthatching (Graber and Baker, 1973). Laying hens, however, appear capable of meeting their entire requirement through biosynthesis (Johnson and Fisher, 1956).

Avian species require large quantities of glycine for nitrogen elimination as uric acid, the principal end product of protein catabolism in birds. Glycine is also needed for synthesis of protein, purines other than uric acid, and the important body compounds hemoglobin, creatine, glutathione, cholic acid, and hippuric acid. Serine is synthesized *in vivo* from three-carbon fragments originating mainly from glucose. Synthesis of protein, phospholipids, and glycine (*via* the enzyme hydroxymethyltransferase) accounts for the majority of the serine thus synthesized. Serine, however, is also intimately involved in the biosynthesis of both cysteine and choline.

The dietary requirement of chicks for glycine–serine can be spared, but not eliminated,by several precursor compounds. Thus, choline, betaine, sarcosine, and glycolic acid (Baker and Sugahara, 1970), aminoethanol (Wixom *et al.*, 1958) or threonine (Baker *et al.*, 1972; D'Mello, 1973) all produce responses when added to a diet deficient in either glycine or serine. It is likely that compounds such as creatine and glutathione, which contain glycine as part of their chemical structure, would also possess activity.

Several investigators have observed large variations in the dietary need for glycine–serine, especially with regard to the amino acid pattern of the diet. The lowest recorded dietary level which has allowed maximal growth was 0.30% glycine or 0.42% serine (Baker *et al.*, 1968). Coon *et al.* (1974) confirmed this result. It is significant that a well-balanced, minimal nitrogen, crystalline amino acid diet was used in both of the above-mentioned studies. Diets that contain intact protein require at least four times this quantity (Snetsinger and Scott, 1961; Waterhouse and Scott, 1961a,b; Waterhouse and Scott, 1962; Akrabawi and Kratzer, 1969; Featherston, 1976). Glycine, like arginine, has also been reported to partially alleviate certain amino acid imbalances, especially those that result from excesses of lysine, histidine and phenylalanine (Snetsinger and Scott, 1961) and methionine (Baker, 1976a). Glycine is also very efficacious as a source of nonspecific amino nitrogen (Allen and Baker, 1973).

10. Proline

Early studies with proline indicated that it was dispensable for chick growth (Almquist and Grau, 1944). Later studies from Wisconsin appeared to confirm this result (Adkins *et al.*, 1962). The first convincing evidence for classifying proline as an indispensable amino acid was the work of Greene *et al.* (1962).

Using diets devoid of proline, these workers and subsequent investigators showed that lack of dietary proline causes depressed rate and efficiency of weight gain as well as impaired nitrogen utilization. The proline requirement is small. Reductions in weight gain no greater than 10–20% have been observed as a result of deleting proline from an otherwise complete diet (Greene *et al.*, 1962; Sugahara and Ariyoshi, 1967a; Graber *et al.*, 1970; Bhargava *et al.*, 1971; Graber and Baker, 1973; Austic, 1976). Reduced efficiency of food utilization is a consistent feature of the deficiency.

Graber and co-workers (1970) showed proline to be a dietary essential at all rates of chick growth and in the presence of excess glutamic acid. Bhargava *et al.* (1971) also observed no response upon adding glutamic acid to a proline-free chick diet. Attempts to eliminate the requirement for proline by feeding excess arginine or ornithine have also met with failure (Graber and Baker, 1971b; Austic, 1976). Both glutamic acid (Shen *et al.*, 1972; Austic, 1976) and arginine or ornithine (Klain and Johnson, 1962; Graber and Baker, 1971b; Shen *et al.*, 1973a; Austic, 1976) have been established as metabolic precursors of proline in the young chick. These data coupled with results that suggest that avian liver and kidney tissue contain the necessary enzymes for carrying out proline biosynthesis from arginine or glutamic acid (Austic, 1976) make it surprising, indeed, that proline is indispensable for avian species. Because glutamic acid is undoubtedly the most important precursor, more work is needed on the conversion of glutamic acid to proline to ascertain if the chick differs from the rat in the rate of proline synthesis from this precursor.

Hydroxyproline is generally listed as one of the dispensable amino acids for growth of chicks as well as other species. Paralleling the relationship between cysteine and cystine, hydroxyproline is neither physiologically nor dietarily essential, but instead is formed on elongating peptide chains when proline is hydroxylated. Therefore, although hydroxyproline is present in tissue proteins (keratoid proteins, such as feathers, are particularly rich in both hydroxyproline and proline), it is not *per se* incorporated into these proteins.

11. Other Indispensable Amino Acids

11.1. Phenylalanine–Tyrosine

The aromatic amino acids phenylalanine and tyrosine are rich in virtually all feed ingredients; therefore, a deficiency would be extremely unlikely in practice. Because it can be converted to tyrosine, phenylaianine is capable of meeting the requirement for both phenylalanine and tyrosine. The reverse reaction does not occur. The molar efficiency of tyrosine formation from phenylalanine approaches 100% in the chick (Sasse and Baker, 1972). Also, tyrosine can furnish up to 45% of the total requirement for aromatic amino amino acids (Sasse and

Baker, 1972). The D-isomers of both phenylalanine and tyrosine are well utilized by poultry (Sugahara *et al.*, 1967).

11.2. Histidine

In addition to its function in protein synthesis, histidine is found in muscle as part of the free dipeptide carnosine (β-alanylhistidine). Chicks generally live longer when placed on a histidine-free diet than when fed a diet devoid of any other essential amino acid not synthesized by the chick (Ousterhout, 1960).* It is assumed that endogenous carnosine, as well as hemoglobin (a histidine-rich protein), provide histidine under these circumstances. Ousterhout (1960), for example, observed depletion of both free histidine and carnosine, but not anserine (β-alanylmethylhistidine), in muscle of chicks fed a histidine-free diet. Moreover, hematocrit values were markedly depressed. Carnosine brings about a dramatic growth response when either injected or fed to histidine-deprived rats (du Vigneaud *et al.*, 1937), and when fed to histidine-deficient chicks (Robbins and Baker, 1977).

11.3. Threonine

Threonine is relatively low in cereal grains and soybean meal. At protein levels currently employed for practical chick rations, however, threonine deficiency is seldom encountered. As pointed out in Section 4.2, excess methionine appears to antagonize threonine. Also, like lysine and arginine, the D-isomer of threonine has no nutritional value for chicks (Sugahara *et al.*, 1967).

12. Dispensable Nitrogen

L-Glutamic acid has been clearly established in chicks as an effective source of nitrogen for biosynthesis of dispensable amino acids. In fact, studies with crystalline amino acid diets have shown it to be as efficient as any mixture of dispensable amino acids (Scott *et al.*, 1963). Although both glycine and proline are efficacious sources of dispensable nitrogen for chick growth, inferior performance would likely result if either of these amino acids were relied on solely to furnish nitrogen for synthesis of dispensable amino acids (Allen and Baker, 1973). Glycine, especially, depresses voluntary food intake when fed at high levels. Diammonium citrate is effective in supplying dispensable nitrogen, being close to glutamic acid in overall efficacy (Blair *et al.*, 1972; Lee and Blair, 1972; Allen and Baker, 1973; Featherston and Horn, 1974). Aspartic acid is also very efficacious in furnishing nitrogen to the chick; alanine, however, seems to be

---

*It is likely that chicks could survive longer on a proline- or glycine-free diet than on a histidine-free diet.

utilized less efficiently (Sugahara and Ariyoshi, 1967b). Serine at high levels is toxic (Sugahara and Ariyoshi, 1967b; Shen *et al.*, 1973b). Inorganic ammonium salts appear to be utilized poorly as well (Allen and Baker, 1974).

Indispensable amino acids when in excess of their dietary requirement can also furnish nitrogen for synthesis of dispensable amino acids, but the efficiency with which they do so is rather low (Stucki and Harper, 1961; Allen and Baker, 1973). Indispensable amino acids that enter into transamination reactions are superior in this regard to those that do not. However, whether a given amino acid is glucogenic or ketogenic seems to have no bearing on its utilization as a source of nitrogen (Allen and Baker, 1973).

Urea has been studied extensively as a source of dispensable nitrogen. In order to be utilized, urea must undergo hydrolysis in the gut to free ammonia *via* bacterial urease.* The ammonia must then be absorbed and joined with a carbon skeleton such that it can enter into transamination reactions. Hence, of the roughly 30% of urea used by the chick for dispensable amino acid synthesis (under conditions of a dispensable nitrogen deficit), presumably most go for synthesis of glutamic acid *via* the enzyme glutamate dehydrogenase (Baker and Molitoris, 1974).

Certain purines and pyrimidines can contribute usable nitrogen for chick diets (Baker and Molitoris, 1974). Both adenine and uracil are metabolizable to an extent that yields available nitrogen. Both oral sources (nucleic acids, free purines, and pyrimidines) and endogenous sources (i.e., purines and pyrimidines from nucleic acid turnover) contribute to the available nitrogen pool, although only to a very minor extent relative to total pool size. Neither uric acid nor xanthine contributes usable nitrogen to this pool.

References

Adams, R. L., and Rogler, J. C., 1968, The effects of environmental temperature on the protein requirements and response to energy in slow and fast growing chicks, *Poultry Sci.* **47**:579.

Adams, R. L., Andrews, F. N., Rogler, J. C., and Carrick, C. W., 1962, The protein requirement of 4-week-old chicks as affected by temperature, *J. Nutr.* **77**:121.

Adkins, J. S., Sunde, M. L., and Harper, A. E., 1962, The development of a free amino acid diet for the growing chick, *Poultry Sci.* **41**:1382.

Agricultural Research Council, 1963, The nutrient requirements of farm livestock, No. 1, Poultry, London, Her Majesty's Stationery Office.

Akrabawi, S. S., and F. H. Kratzer, 1968, Effects of arginine or serine on the requirement for glycine by the chick, *J. Nutr.* **95**:41.

Akrawawi, S. S., and Kratzer, F. H., 1969, Dietary effects on the need for glycine by the chick, *Proc. Soc. Exp. Biol. Med.* **130**:1270.

Allen, N. K., and Baker, D. H., 1972a, Effects of excess lysine on the utilization of and requirement for arginine by the chick, *Poultry Sci.* **51**:902.

Allen, N. K., and Baker, D. H., 1972b, Quantitative efficacy of dietary isoleucine and valine for

---

*Intestinal urease (99%) is found in the ceca of chicks (Stutz and Metrokotsas, 1972).

chick growth as influenced by variable quantities of excess dietary leucine, *Poultry Sci.* **51**:1292.

Allen, N. K., and Baker, D. H., 1973, Quantitative evaluation of non-specific nitrogen sources for the growing chick, *Poultry Sci.* **53**:258.

Allen, N. K., Baker, D. H., and Graber, G., 1971, Interrelationships among leucine, tryptophan and niacin in the chick, *Poultry Sci.* **50**:1544 (Abst.).

Allen, N. K., Baker, D. H., Scott, H. M., and Norton, H. W., 1972, Quantitative effect of excess lysine on the ability of arginine to promote chick weight gains, *J. Nutr.* **102**:171.

Almquist, H. J., and Grau, C. R., 1944, The amino acid requirements of the chick, *J. Nutr.* **28**:325.

Anderson, J. O., Warnick, R. E., and Dalai, R. K., 1975, Replacing dietary methionine and cystine in chick diets with sulfate or other sulfur compounds, *Poultry Sci.* **54**:1122.

Austic, R. E., 1976, Nutritional and metabolic interrelationships between arginine, proline, and glutamic acid in the chick, *Fed. Proc.* **35**:1914.

Austic, R. E., and Nesheim, M. C., 1970, Role of kidney arginase in variations of the arginine requirement of chicks, *J. Nutr.* **100**:855.

Austic, R. E., and Nesheim, M. C., 1971, Arginine and creatine interrelationships in the chick, *Poultry Sci.* **51**:1098.

Austic, R. E., and Scott, R. L., 1975, Involvement of food intake in the lysine-arginine antagonism in chicks, *J. Nutr.* **105**:1122.

Baker, D. H., 1974, Amino acids in swine and poultry feeds, *Proc. Cornell Nutr. Conf.,* 5–11.

Baker, D. H., 1975, A comparison of the biological availability of DL-methionine and its hydroxy analogue, *Proc. Md. Nutr. Conf.,* 28–32.

Baker, D. H., 1976a, Nutritional and metabolic interrelationships among sulfur compounds in avian nutrition, *Fed. Proc.* **35**:1917.

Baker, D. H., 1976b, Utilization of methionine analogues, *J. Nutr.* **106**:1376.

Baker, D. H., and Bray, D. J., 1972, Comparison of supplemental sulfur amino acid sources for growing chicks fed a corn–soybean meal diet, *Poultry Sci.* **51**:1782 (Abst.).

Baker, D. H., and Easter, R. A., 1976, Soy protein as a source of amino acids for nonruminant animals, *Proc. World Soybean Conf.* 969–977.

Baker, D. H., and Molitoris, B. A., 1974, Utilization of nitrogen from selected purines and pyrimidines and from urea by the young chick, *J. Nutr.* **104**:553.

Baker, D. H., and B. A. Molitoris, 1975, Lack of response to supplemental tin, vanadium, chromium and nickel when added to a purified crystalline amino acid diet for chicks, *Poultry Sci.* **54**:925.

Baker, D. H., and Sugahara, M., 1970, Nutritional investigation of the metabolism of glycine and its precursors by chicks fed a crystalline amino acid diet, *Poultry Sci.* **49**:756.

Baker, D. H., Sugahara, M., and Scott, H. M., 1968, The glycine–serine interrelationship in chick nutrition, *Poultry Sci.* **47**:1376.

Baker, D. H., Clausing, W. W., Harmon, B. G., Jensen, A. H., and Becker, D. E., 1969, Replacement value of cystine for methionine for the young pig, *J. Anim. Sci.* **29**:581.

Baker, D. H., Hill, T. M., and Kleiss, A. J., 1972, Nutritional evidence concerning formation of glycine from threonine in the chick. *J. Anim. Sci.* **34**:582.

Baker, D. H., Allen, N. K., and Kleiss, A. J., 1973, Efficiency of tryptophan as a niacin precursor in the young chick, *J. Anim. Sci.* **36**:299.

Baker, D. H., Katz, R. S., and Easter, R. A., 1975, The lysine requirement for growing pigs at two levels of dietary protein, *J. Anim. Sci.* **40**:851.

Baldini, J. T., and Rosenberg, H. R., 1955, The effect of productive energy level of the diet on the methionine requirement of the chick, *Poultry Sci.* **34**:1301.

Baldini, J. T., Marvel, J. P., and Rosenberg, H. R., 1957, The effect of productive energy level of the diet on the methionine requirement of the poult, *Poultry Sci.* **36**:1031.

Benevenga, N. J., 1974, Toxicities of methionine and other amino acids, *J. Agr. Food Chem.* **22**:2.

Benevenga, N. J., and Harper, A. E., 1967, Alleviation of methionine and homocystine toxicity in the rat, *J. Nutr.* **93**:44.

Benevenga, N. J., and Harper, A. E., 1970, Effect of glycine and serine on methionine metabolism in rats fed diets high in methionine, *J. Nutr.* **100**:1205.

Bhargava, K. K., Shen, T. F., Bird, H. R., and Sunde, M. L., 1971, The effect of glutamic acid on chick's proline requirement, *Poultry Sci.* **50**:726.

Bird, F. H., 1952, A comparison of methionine and two of its analogues in the nutrition of the chick, *Poultry Sci.* **31**:1095.

Bird, F. H., 1953, The lysine requirement of eight-week old chickens, *Poultry Sci.* **32**:10.

Blair, R., Shannon, D. W. F., McNab, J. M., and Lee, D. J. W., 1972, Effects on chick growth of adding glycine, proline, glutamic acid or diammonium citrate to diets containing crystalline essential amino acids, *Br. Poultry Sci.* **13**:215.

Boomgaardt, J., and Baker, D. H., 1971, Tryptophan requirement of growing chicks as affected by dietary protein level, *J. Anim. Sci.* **33**:595.

Boomgaardt, J., and Baker, D. H., 1973a, Effect of age on the lysine and sulfur amino acid requirement of growing chickens, *Poultry Sci.* **52**:592.

Boomgaardt, J., and Baker, D. H., 1973b, Effect of dietary energy concentration on sulfur amino acid requirements and body composition of young chicks, *J. Anim. Sci.* **36**:307.

Boomgaardt, J., and Baker, D. H., 1973c, The lysine requirement of young chicks fed sesame meal–gelatin diets at three protein levels, *Poultry Sci.* **52**:586.

Boorman, K. N., 1971, The renal reabsorption of arginine, lysine and ornithine in the young cockerel (*Gallus domesticus*), *Comp. Biochem. Physiol.* **39A**:29.

Bornstein, S., 1970, The lysine requirement of broilers during their finishing period, *Br. Poultry Sci.* **11**:197.

Bornstein, S., and Lipstein, B., 1964, Methionine supplementation of practical broiler rations. I. The value of added methionine in diets of varying fishmeal levels, *Br. Poultry Sci.* **5**:167.

Bornstein, S., and Lipstein, B., 1966, Methionine supplementation of practical broiler rations. III. The value of added methionine in broiler finisher rations, *Br. Poultry Sci.* **7**:273.

Bray, D. J., 1962, The optimum combination of corn and soybean protein for chick growth and egg production determined at various levels of dietary protein, *Poultry Sci.* **41**:1630 (Abst.).

Brookes, I. M., Owens, F. N., and Garrigus, U. S., 1972, Influence of amino acid level in the diet upon amino acid oxidation by the rat, *J. Nutr.* **102**:27.

Chi, M. S., and Speers, G. M., 1973, Nutritional value of high-lysine corn for the broiler chick, *Poultry Sci.* **52**:1148.

Chung, E., Griminger, P., and Fisher, H., 1973, The lysine and sulfur amino acid requirements at two stages of growth in chicks, *J. Nutr.* **103**:117.

Cohen, H. P., Choitz, H. C., and Berg, C. P., 1958, Response of rats to diets high in methionine and related compounds, *J. Nutr.* **64**:555.

Combs, G. F., 1968, Amino acid requirements of broilers and laying hens, *Proc. Md. Nutr. Conf.*, 86–96.

Combs, G. F., and Romoser, G. L., 1955, A new approach to poultry feed formulation, *Md. Agr. Exp. Sta. Misc.*, Publ. No. 226.

Coon, C. N., Grossie, V. B., and Couch, J. R., 1974, Glycine-serine requirements of chicks, *Poultry Sci.* **53**:1709.

Creek, R. D., 1968, Nonequivalence in mass in the conversion of phenylalanine to tyrosine and methionine to cystine, *Poultry Sci.* **47**:1385.

Cromwell, G. L., Rogler, J. C., Featherston, W. R., and Cline, T. R., 1968, A comparison of the nutritive value of *opaque-2*, *floury-2* and normal corn for the chick, *Poultry Sci.* **47**:840.

Dean, W. F., and Scott, H. M., 1965, The development of an amino acid reference diet for the early growth of chicks, *Poultry Sci.* **44**:803.

DeBey, H. J., Snell, E. E., and Bauman,C. A., 1952, Studies on the interrelationship between methionine and vitamin B-6, *J. Nutr.* **46**:203.

D'Mello, J. P. F., 1973, Threonine and glycine metabolism in the chick *(Gallus domesticus), Nutr. Metab.* **15**:357.

D'Mello, J. P. F., 1974, Plasma concentrations and dietary requirements of leucine, isoleucine and valine: studies with the young chick, *J. Sci. Food Agric.* **25**:187.

D'Mello, J. P. F., and Lewis, D., 1970a, Amino acid interactions in chick nutrition. 3. Interdependence in amino acid requirements, *Br. Poultry Sci.* **11**:367.

D'Mello, J. P. F., and Lewis, D., 1970b, Amino acid interactions in chick nutrition. 2. Interrelationships between leucine, isoleucine and valine, *Br. Poultry Sci.* **11**:313.

Dobson, D. C., Anderson, J. O., and R. E. Warnick, 1964, A determination of the essential amino acid proportions needed to allow rapid growth in chicks, *J. Nutr.* **82**:67.

DuVigneaud, V., Sifferd, R. H., and Irving, G. W., 1937, The utilization of L-carnosine by animals on a histidine-deficient diet, *J. Biol. Chem.* **117**:589.

Dyer, H. M., and DuVigneaud, V., 1936, Utilization of glutathione in connection with a cysteine-deficient diet, *J. Biol. Chem.* **115**:543.

Featherston, W. R., 1976, The glycine–serine interrelationship in chick nutrition, *Fed. Proc.* **35**:1910.

Featherston, W. R., and Horn, G. W., 1974, Studies on the utilization of the α-hydroxy acid of methionine by chicks fed crystalline amino acid diets, *Poultry Sci.* **53**:680.

Featherston, W. R., Roger, J. C., Axtell, J. D., and Oswalt, D. L., 1975, Nutritional value of high-lysine sorghum grain for the chick, *Poultry Sci.* **54**:1220.

Finkelstein, J. D., Cello, J. P., and Kyle, W. E., 1974, Ethanol-induced changes in methionine metabolism in rat liver, *Biochem. Biophys. Res. Commun.* **61**:475.

Fisher, H., Salander, R. C., and Taylor, M. W., 1956, The influence of creatine biosynthesis on the arginine requirement of the chick, *J. Nutr.* **59**:491.

Girard-Globa, A., Robin, P., and Forestier, M., 1972, Long-term adaptation of weanling rats to high dietary levels of methionine and serine, *J. Nutr.* **102**:209.

Glista, W. A., Mitchell, H. H., and Scott, H. M., 1951, The amino acid requirements of the chick. *Poultry Sci.* **30**:915 (Abstr).

Gopalan, C., 1970, Some recent studies in the nutrition research laboratories, Hyderabad, *Amer. J. Clin. Nutr.* **23**:35.

Gopalan, C., and Srikantis, S. G., 1960, Leucine and pellagra, *Lancet* **1**:954.

Gordon, R. S., and Sizer, I. W., 1955, Ability of sodium sulfate to stimulate growth of the chicken, *Science* **122**:1270.

Gordon, R. S., Maddy, K. H., and Knight, S., 1954, Value of methionine hydroxy analogue supplementation of broiler rations, *Poultry Sci.* **33**:424.

Graber, G., and Baker, D. H., 1971a, Sulfur amino acid nutrition of the growing chick: Quantitative aspects concerning the efficacy of dietary methionine, cysteine and cystine, *J. Anim. Sci.* **33**:1005.

Graber, G., and Baker, D. H., 1971b, Ornithine utilization by the chick, *Proc. Soc. Exp. Biol. Med.* **138**:585.

Graber, G., and Baker, D. H., 1973, The essential nature of glycine and proline for growing chickens, *Poultry Sci.* **52**:892.

Graber, G., Allen, N. K., and Scott, H. M., 1970, Proline essentiality and weight gain, *Poultry Sci.* **49**:692.

Graber, G., Scott, H. M., and Baker, D. H., 1971a, Sulfur amino acid nutrition of the growing chick: The methionine requirement as influenced by age, *Poultry Sci.* **50**:854.

Graber, G., Scott, H. M., and Baker, D. H., 1971b, Sulfur amino acid nutrition of the growing chick: effect of age on the capacity of cystine to spare dietary methionine, *Poultry Sci.* **50**:1450.

Grau, C. R., 1948, Effect of protein level on the lysine requirement of the chick, *J. Nutr.* **36**:99.

Grau, C. R., and Almquist, H. J., 1944, Requirement of tryptophan by the chick, *J. Nutr.* **28**:263.

Greene, D. E., Scott, H. M., and Johnson, B. C., 1962, The role of proline and certain nonessential amino acids in chick nutrition, *Poultry Sci.* **41**:116.

Griminger, P., and Scott, H. M., 1959, Growth rate and the lysine requirement of the chick, *J. Nutr.* **68**:429.

Griminger, P., Scott, H. M., and Forbes, R. M., 1957, Dietary bulk and amino acid requirements, *J. Nutr.* **62**:61.

Hardwick, D. F., Applegarth, D. A., Cockroft, P. M., and Calder, R. J., 1970, Pathogenesis of methionine-induced toxicity, *Metabolism* **19**:381.

Harter, J. M., and Baker, D. H., 1977, Factors affecting the pathogenicity of methionine toxicity, *J. Anim. Sci.* (In press.)

Hegsted, D. M., 1944, Growth in chicks fed amino acids, *J. Biol. Chem.* **156**:247.

Hewitt, D., and Lewis, D., 1972a, The amino acid requirements of the growing chick. 1. Determination of amino acid requirements, *Br. Poultry Sci.* **13**:449.

Hewitt, D., and Lewis, D., 1972b, The amino acid requirements of the growing chick. 2. Growth and body composition of chicks fed on diets in which the proportions of amino acids are well balanced, *Br. Poultry Sci.* **13**:465.

Hinton, C. F., and Harms, R. H., 1972, Evidence for sulfate as an unidentified growth factor in fish solubles, *Poultry Sci.* **51**:701.

Hogan, A. G., Gillespie, G. T., Kocturk, O., O'Dell, B. L., and Flynn, L. M., 1955, The percentage of protein in corn and its nutritional properties, *J. Nutr.* **57**:275.

Hunchar, J. G., and Thomas, O. P., 1975, The tryptophan requirement of 4–7 week old broilers, *Proc. Md. Nutr. Conf.,* 114.

Huovinen, J. A., and Gustafsson, B. F., 1967, Inorganic sulfate, sulfite and sulfide as sulfur donors in the biosynthesis of sulfur amino acids in germ-free and conventional rats, *Biochim. Biophys. Acta* **136**:441.

Huston, R. L., and Scott, H. M., 1968, Effect of varying the composition of a crystalline amino acid mixture on weight gain and pattern of free amino acids in chick tissue, *Fed. Proc.* **27**:1204.

Ishibashi, T., 1973, Phenylalanine requirement of the growing chick, *Nippon Nogeikagaku Kaishi* **47**:153.

Johnson, D., and Fisher, H., 1956, The amino acid requirement of the laying hen. II. Classification of the essential amino acids required for egg production, *J. Nutr.* **60**:275.

Jones, J. D., Petersburg, S. J., and Burnett, P. C., 1967, The mechanism of the lysine-arginine antagonism in the chick: Effect of lysine on digestion, kidney arginase and liver transaminase, *J. Nutr.* **93**:103.

Jukes, T. H., 1940, Effect of choline and other supplements on perosis, *J. Nutr.* **20**:445.

Jukes, T. H., 1941, The effect of certain organic compounds and other dietary supplements on perosis, *J. Nutr.* **22**:315.

Katz, R. S., and Baker, D. H., 1975a, Methionine toxicity in the chick: Nutritional and metabolic implications, *J. Nutr.* **105**:1168.

Katz, R. S., and Baker, D. H., 1975b, Toxicity of various organic sulfur compounds for chicks fed crystalline amino acid diets containing threonine and glycine at their minimal dietary requirements for maximal growth, *J. Anim. Sci.* **41**:1355.

Katz, R. S., and Baker, D. H., 1975c, Factors associated with utilization of the calcium salt of methionine hydroxy analogue by the young chick, *Poultry Sci.* **54**:584.

Katz, R. S., and Baker, D. H., 1975d, Efficacy of D-, L- and DL-methionine for growth of chicks fed crystalline amino acid diets, *Poultry Sci.* **54**:1667.

Ketola, H. G., and Nesheim, M. C., 1974, Influence of dietary protein and methionine levels on the requirement for choline by chickens, *J. Nutr.* **104**:1484.

Klain, G., and Johnson, B. C., 1962, Metabolism of labeled aminoethanol glycine and arginine in the chick, *J. Biol. Chem.* **237**:123.

Klain, G. J., Scott, H. M., and Johnson, B. C., 1959, Utilization of nutrients in a crystalline amino acid diet as influenced by certain nonessential amino acids, *Poultry Sci.* **38**:489.

Klain, G. J., Scott, H. M., and Johnson, B. C., 1960, The amino acid requirements of the growing chick fed a crystalline amino acid diet, *Poultry Sci.* **39**:39.

Klain, G. J., Vaughan, D. A., and Vaughan, L. N., 1963, Some metabolic effects of methionine toxicity in the rat, *J. Nutr.* **80**:337.

Kushwaha, R. P. S., and Jensen, L. S., 1974, *In vivo* and *in vitro* utilization of labeled precursors for choline synthesis in mature Japanese quail as compared to rats, *J. Nutr.* **104**:901.

Kushwaha, R. P. S., and Jensen, L. S., 1975, Effect of choline deficiency on utilization of labeled precursors of choline and turnover of choline in the liver of mature Japanese quail, *J. Nutr.* **105**:226.

Langer, B. W., 1965, The biochemical conversion of 2-hydroxy-4-methylthiobutyric acid into methionine by the rat *in vitro*, *Biochem. J.* **95**:683.

Latshaw, J. D., and Jensen, L. S., 1972, Choline deficiency and synthesis of choline from precursors in mature Japanese quail, *J. Nutr.* **102**:749.

Lawrence, J. M., Day, K. M., Huey, E., and Lee, B., 1958, Lysine content of wheat varieties, species and related genera, *Cereal Chem.* **35**:169.

Lee, D. J. W., and Blair, R., 1972, Effects of chick growth of adding various non-protein nitrogen sources or dried autoclaved poultry manure to diets containing crystalline essential amino acids, *Br. Poultry Sci.* **13**:243.

Leong, K. C., Sunde, M. L., Bird, H. R., and Elvehjem, C. A., 1959, Interrelationships among dietary energy, protein and amino acids for chickens, *Poultry Sci.* **38**:1267.

Lerner, J., Yankelowitz, S., and Taylor, M. W., 1969, The intestinal absorption of methionine in chickens provided with permanent thiryvella fistulas, *Experientia* **25**:689.

Lucas, H. L., Norris, L. C., and Heuser, G. F., 1946, Observations on the choline requirement of hens, *Poultry Sci.* **25**:373.

Machlin, L. J., and Pearson, P. B., 1956, Studies on utilization of sulfate sulfur for growth of the chicken, *Proc. Soc. Exp. Biol. Med.* **93**:204.

Martin, W. G., 1972, Sulfate metabolism and taurine synthesis in the chick, *Poultry Sci.* **51**:608.

Maruyama, K., Shands, H. L., Harper, A. E., and Sunde, M. L., 1975, An evaluation of the nutritive value of new high protein oat varieties (cultivars), *J. Nutr.* **105**:1048.

Mason, V. C., Hanson, J. G., and Weidner, K., 1965, Studies on the quantitative incorporation of sulfate sulfur into methionine, cystine and taurine in the hen, *Acta Agr. Scand.* **15**:3.

Mathieu, D., and Scott, H. M., 1968, Growth depressing effect of excess leucine in relation to amino acid composition of the diet, *Poultry Sci.* **47**:1694 (Abst.).

Meister, A., 1965, *Biochemistry of the Amino Acids,* Vol. II, pp. 636–672, Academic Press, New York.

Mertz, E. T., Bates, L. S., and Nelson, O. E., 1964, Mutant gene that changes protein composition and increases lysine content of maize endosperm, *Science* **145**:279.

Mitchell, J. R., Becker, D. E., Jensen, A. H., Harmon, B. G., and Norton, H. W., 1968, Determination of amino acid needs of the young pig by nitrogen balance and plasma free amino acids, *J. Anim. Sci.* **27**:1327.

Molitoris, B. A., and Baker, D. H., 1976a, Choline utilization in the chick as influenced by levels of dietary protein and methionine, *J. Nutr.* **106**:412.

Molitoris, B. A., and Baker, D. H., 1976b, Assessment of the quantity of biologically available choline in soybean meal, *J. Anim. Sci.* **42**:481.

Molitoris, B. A., and Baker, D. H., 1976c, The choline requirement of broiler chicks during the seventh week of life, *Poultry Sci.* **55**:220.

Morrison, W. D., Hamilton, T. S., and Scott, H. M., 1956, Utilization of D-tryptophan by the chick, *J. Nutr.* **60**:47.

Morrison, A. B., Middleton, E. J., and McLaughlan, J. M., 1961, Blood amino acid studies. II.

Effects of dietary lysine, concentration, sex and growth rate on plasma free lysine and threonine levels in the rat, *Can. J. Biochem. Physiol.* **39**:1675.

Mowbray, R. M., and Sykes, A. H., 1971, Egg production in warm environmental temperatures, *Br. Poultry Sci.* **12**:25.

Munck, L., Karlsson, K. E., Hagberg, A., annd Eggun, B. O., 1970, Gene for improved nutritional value in barley seed protein, *Science* **168**:985.

National Research Council, 1971. Nutrient Requirements of Domestic Animals. 1. *Nutrient Requirements of Poultry,* 6th ed., National Academy of Sciences, Washington, D.C.

Neale, R. J., and Waterlow, J. C., 1974, Critical evaluation of a method for estimating amino acid requirements for maintenance in the rat by measurement of the rate of ^{14}C-labelled amino acid oxidation in vivo, *Br. J. Nutr.* **32**:257.

Nesheim, M. C., 1968, Genetic variation in arginine and lysine utilization, *Fed. Proc.* **27**:1210.

Nesheim, R. O., and Johnson, B. C., 1950, Effect of a high level of methionine on the dietary choline requirement of the baby pig, *J. Nutr.* **41**:149.

Nesheim, M. C., Norvell, M. J., Ceballos, E., and Leach, R. M., 1971, The effect of choline supplementation of diets for growing pullets and laying hens, *Poultry Sic.* **50**:820.

Netke, S. P., and Scott, H. M., 1970, Estimates on the availability of amino acids in soybean oil meal as determined by chick growth assay: Methodology as applied to lysine, *J. Nutr.* **100**:281.

Netke, S. P., Scott, H. M., and Allee, G. L., 1969. Effect of excess amino acids on the utilization of the first-limiting amino acid in chick diets, *J. Nutr.* **99**:75.

Ousterhout, L. E., 1960, Survival time and biochemical changes in chicks fed diets lacking different essential amino acids, *J. Nutr.* **70**:226.

Payne, C. J., 1966. Environmental temperature and egg production, In *Physiology of the Domestic Fowl,* C. Horton-Smith and E. C. Amoroso, eds., p. 235. Oliver & Boyd, Edinburgh.

Potter, L. M., 1976, A look at the amino acid requirements of turkeys. *Proc. Carolina Poultry Nutr. Conf.,* 22.

Price, W. A., Taylor, M. W., and Russell, W. C., 1953, The retention of essential amino acids by the growing chick, *J. Nutr.* **51**:413.

Reid, B. L., and Weber, C. W., 1973, Dietary protein and sulfur amino acid levels for laying hens during heat stress, *Poultry Sci.* **52**:1335.

Ringrose, R. C., and Davis, H. A., 1946, Choline in the nutrition of laying hens, *Poultry Sci.* **50**:820.

Robbins, K. R., and Baker, D. H., 1977, Growth rate, plasma histidine and muscle carnosine as indices of histidine status in the chick, *Fed. Proc.* **36**:1153 (Abst.).

Robbins, G. S., Pomeranz, Y., and Briggle, L. W., 1971, Amino acid composition of oat groats, *J. Agr. Food Chem.* **19**:536.

Romoser, G. L., Wright, P. L., and Grainger, R. B., 1976, An evaluation of the L-methionine activity of the hydroxy analogue of methionine, *Poultry Sci.* **55**:1099.

Rose, W. C., and Rice, E. E., 1939, The utilization of certain sulfur-containing compounds for growth purposes, *J. Biol. Chem.* **130**:305.

Ross, E., and Harms, R. H., 1970, The response of chicks to sodium sulfate supplementation of a corn-soy diet, *Poultry Sci.* **49**:1605.

Ross, E., Damron, B. L., and Harms, R. H., 1972, The requirement for inorganic sulfate in the diet of chicks for optimum growth and feed efficiency, *Poultry Sci.* **51**:1606.

Sanchez, A., and Swenseid, M. E., 1969, Adaptation of the weanling rat to diets containing excess methionine, *J. Nutr.* **99**:299.

Sasse, C. E., and Baker, D. H., 1972, The phenylalanine–tyrosine interrelationship in chick nutrition, *Poultry Sci.* **51**:1531.

Sasse, C. E., and Baker, D. H., 1973a, Modification of the Illinois reference standard amino acid mixture, *Poultry Sci.* **52**:1970.

Sasse, C. E., and Baker, D. H., 1973b, Availability of sulfur amino acids in corn and corn gluten meal for growing chicks, *J. Anim. Sci.* **37**:1351.

Sasse, C. E., and Baker, D. H., 1974a, Factors affecting sulfate–sulfur utilization by the young chick, *Poultry Sci.* **53**:652.

Sasse, C. E., and Baker, D. H., 1974b, Sulfur utilization by the chick with emphasis on the effect of inorganic sulfate on the cystine–methionine interrelationship, *J. Nutr.* **104**:244.

Sauberlich, H. E., 1961, Studies on the toxicity and antagonism of amino acids for weanling rats, *J. Nutr.* **75**:61.

Sauberlich, H. E., Chang, W., and Salmon, W. D., 1953, The comparative nutritive value of corn of high and low protein content for growth in the rat and chick, *J. Nutr.* **51**:623.

Schwartz, H. G., Taylor, H. W., and Fisher, H., 1958, The effect of dietary energy concentration and age on the lysine requirement of growing chicks, *J. Nutr.* **65**:25.

Scott, H. M., 1972, Development and application of amino acid diets, *Poultry Sci.* **51**:9.

Scott, H. M., and Forbes, R. M., 1958, The arginine requirement in relation to diet composition, *Poultry Sci.* **37**:1347.

Scott, H. M., Dean, W. F., and Smith, R. E., 1963, Studies on the non-specific nitrogen requirement of chicks fed a crystalline amino acid diet, *Poultry Sci.* **42**:1305 (Abst.).

Sekiz, S. S., Scott, M. L., and Nesheim, M. C., 1975, The effect of methionine deficiency on body weight, food and energy utilization in the chick, *Poultry Sci.* **54**:1184.

Shen, T. F., Bird, H. R., and Sunde, M. L., 1972, Conversion of glutamic acid to proline in the chick, *Poultry Sci.* **51**:1864 (Abst.).

Shen, T. F., Bird, H. R., and Sunde, M. L., 1973a, Relationship between ornithine and proline in chick nutrition, *Poultry Sci.* **52**:1161.

Shen, T. F., Bird, H. R., and Sunde, M. L., 1973b, Effect of excess dietary L-, DL- and D-serine on the chick, *Poultry Sci.* **52**:1168.

Sloan, D. R., and Harms, R. H., 1972, Utilization of inorganic sulfate by turkey poults, *Poultry Sci.* **51**:1673.

Smith, R. E., 1966, The utilization of L-methionine, DL-methionine and methionine hydroxy analogue by the growing chick, *Poultry Sci.* **45**:571.

Smith, R. E., 1968, Effect of arginine upon the toxicity of excesses of single amino acids in chicks, *J. Nutr.* **95**:547.

Snetsinger, D. C., and Scott, H. M., 1961, Efficacy of glycine and arginine in alleviating the stress induced by dietary excesses of single amino acids, *Poultry Sci.* **40**:1681.

Soares, J. H., 1974, Experiments on the requirement of inorganic sulfate by the chick, *Poultry Sci.* **53**:246.

Sowers, J. E., Stockland, W. L., and Meade, R. J., 1972, L-methionine and L-cystine requirements of the growing rat, *J. Anim. Sci.* **35**:782.

Stockland, W. L., Meade, R. J., and Melliere, A. L., 1970, Lysine requirement of the growing rat: Plasma free lysine as a response criterion, *J. Nutr.* **100**:925.

Stucki, W. P., and Harper, A. E., 1961, Importance of dispensable amino acids for normal growth of chicks, *J. Nutr.* **74**:377.

Stutz, M. W., and Metrokotsas, M. J., 1972, Urease activity in the digestive tract of the chick and metabolism of urea, *Poultry Sci.* **51**:1876 (Abst.).

Sugahara, M., and Ariyoshi, S., 1967a, The nonessentiality of glycine and the essentiality of L-proline in the chick nutrition, *Agr. Biol. Chem.* **31**:106.

Sugahara, M., and Ariyoshi, S., 1967b, The nutritional value of the individual nonessential amino acid as the nitrogen source in the chick nutrition, *Agr. Biol. Chem.* **31**:1270.

Sugahara, M., Morimoto, T., Kobayashi, T., and Ariyoshi, S., 1967, The nutritional value of D-amino acid in the chick nutrition, *Agr. Biol. Chem.* **31**:77.

Sugahara, M., Baker, D. H., and Scott, H. M., 1969, Effect of different patterns of excess amino acids on performance of chicks fed amino acid-deficient diets, *J. Nutr.* **97**:29.

Tipton, H. C., Dilworth, B. C., and Day, E. J., 1965, The relative biological value of DL-methionine and methionine hydroxy analogue in chick diets, *Poultry Sci.* **44**:987.

Tipton, H. C., Dilworth, B. C., and Day, E. J., 1966, A comparison of D-, L-, DL-methionine and methionine hydroxy analogue calcium in chick diets, *Poultry Sci.* **45**:381.

Treadwell, C. R., 1948, Growth and lipotropism. II. The effects of dietary methionine, cystine, and choline in the young rat, *J. Biol. Chem.* **176**:1141.

Vavich, M. G., Kemmerer, A. R., Nimbkar, B., and Stith, L. S., 1959, Nutritive value of low and high protein sorghum grain for feeding chickens, *Poultry Sci.* **38**:36.

Velu, J. G., and Baker, D. H., 1974, Body composition and protein utilization of chicks fed graded levels of fat, *Poultry Sci.* **53**:1831.

Velu, J. G., Baker, D. H., and Scott, H. M., 1971, Protein and energy utilization by chicks fed graded levels of a balanced mixture of crystalline amino acids, *J. Nutr.* **101**:1249.

Waterhouse, H. N., and Scott, H. M., 1961a, Glycine need of the chick fed casein diets and the glycine, arginine, methionine and creatine interrelationships, *J. Nutr.* **73**:266.

Waterhouse, H. N., and Scott, H. M., 1961b, Effect of different protein levels on the glycine need of the chick fed purified diets, *Poultry Sci.* **40**:1160.

Waterhouse, H. N., and Scott, H. M., 1962, Effect of sex, feathering, rate of growth and acetates on the chick's need for glycine, *Poultry Sci.* **41**:1957.

Weisbach, A., and Sprinson, D. B., 1953a, The metabolism of 2-carbon compounds related to glycine. I. Glyoxylic acid, *J. Biol. Chem.* **203**:1023.

Weisbach, A., and Sprinson, D. B., 1953b, The metabolism of 2-carbon compounds related to glycine. II. Ethanolamine, *J. Biol. Chem.* **203**:1031.

Wethli, E., Morris, T. R., and Shresta, T. P., 1975, The effect of feeding high levels of low-quality proteins to growing chickens, *Br. J. Nutr.* **34**:363.

Williams, H. H., Curtin, L. V., Abraham, J., Loosli, J. K., and Maynard, L. A., 1954, Estimation of growth requirements for amino acids by assay of the carcass, *J. Biol. Chem.* **208**:277.

Woodham, A. A., and Deans, P. S., 1975, Amino acid requirements of growing chickens, *Br. Poultry Sci.* **16**:269.

Zimmerman, R. A., and Scott, H. M., 1965, Interrelationship of plasma amino acid levels and weight gain in the chick as influenced by suboptimal and superoptimal dietary concentrations of single amino acids, *J. Nutr.* **87**:13.

Index